Metacognitive Artificial Intelligence

This groundbreaking volume is designed to meet the burgeoning needs of the research community and industry. This book delves into the critical aspects of AI's self-assessment and decision-making processes, addressing the imperative for safe and reliable AI systems in high-stakes domains such as autonomous driving, aerospace, manufacturing, and military applications.

Featuring contributions from leading experts, the book provides comprehensive insights into the integration of metacognition within AI architectures, bridging symbolic reasoning with neural networks, and evaluating learning agents' competency. Key chapters explore assured machine learning, handling AI failures through metacognitive strategies, and practical applications across various sectors. Covering theoretical foundations and numerous practical examples, this volume serves as an invaluable resource for researchers, educators, and industry professionals interested in fostering transparency and enhancing the reliability of AI systems.

PAULO SHAKARIAN is the KG Tan Endowed Professor of Artificial Intelligence at Syracuse University. He has made notable contributions in the areas of logic programming, neurosymbolic AI, security, and data mining. His academic accomplishments include four best-paper awards, over 100 peer-reviewed articles, over 10 issued patents, and 8 published books. He has been featured on CNN and in *The Economist*.

HUA WEI is Assistant Professor at the School of Computing and Augmented Intelligence at Arizona State University. He specializes in data mining, artificial intelligence, and machine learning. His work has been awarded multiple best paper awards and his research has been funded by agencies including the National Science Foundation and the US Department of Energy.

"This book offers a fascinating exploration of the astounding relationship between metacognition and AI. It provides readers with a comprehensive understanding of how AI systems can be designed not only to make accurate predictions but also to learn from their mistakes and improve over time. The authors explore various methods for enhancing trust in AI models by incorporating aspects of human cognitive processes, providing practical insights for building more reliable and transparent AI technologies."

— *Todd C. Hughes, Scientific Systems Chief Innovation Officer*

"This book on metacognitive AI addresses a timely and critical question in the general field of artificial intelligence: how to make AI systems more reliable and self-aware. The book strikes a good balance between theories, methods, and applications. It is an invaluable resource for researchers and practitioners."

— *Hanghang Tong, University of Illinois Urbana-Champaign*

Metacognitive Artificial Intelligence

Edited by

PAULO SHAKARIAN
Syracuse University

HUA WEI
Arizona State University

CAMBRIDGE
UNIVERSITY PRESS

Shaftesbury Road, Cambridge CB2 8EA, United Kingdom

One Liberty Plaza, 20th Floor, New York, NY 10006, USA

477 Williamstown Road, Port Melbourne, VIC 3207, Australia

314–321, 3rd Floor, Plot 3, Splendor Forum, Jasola District Centre,
New Delhi - 110025, India

103 Penang Road, #05–06/07, Visioncrest Commercial, Singapore 238467

Cambridge University Press is part of Cambridge University Press & Assessment,
a department of the University of Cambridge.

We share the University's mission to contribute to society through the pursuit of
education, learning and research at the highest international levels of excellence.

www.cambridge.org
Information on this title: www.cambridge.org/9781009522458

DOI: 10.1017/9781009522472

© Cambridge University Press & Assessment 2025

This publication is in copyright. Subject to statutory exception and to the provisions
of relevant collective licensing agreements, no reproduction of any part may take
place without the written permission of Cambridge University Press & Assessment.

When citing this work, please include a reference to the DOI 10.1017/9781009522472

First published 2025

Cover image: KTSDESIGN/SCIENCE PHOTO LIBRARY/Getty Images

A catalogue record for this publication is available from the British Library

Library of Congress Cataloging-in-Publication Data
Names: Workshop on Metacognitive Prediction of AI Behavior (1st : 2023 : Scottsdale, Ariz.), author. |
Shakarian, Paulo, author. | Wei, Hua (Professor of computer science), author.
Title: Metacognitive artificial intelligence / edited by Paulo Shakarian, Hua Wei.
Description: Cambridge ; New York, NY : Cambridge University Press, 2025. | "This book serves as
proceedings for the Workshop on Metacognitive Prediction of AI Behavior held November 13-15, 2023
in Scottsdale, Arizona." | Includes bibliographical references and index.
Identifiers: LCCN 2024059184 (print) | LCCN 2024059185 (ebook) | ISBN 9781009522458 (hardback) |
ISBN 9781009522465 (paperback) | ISBN 9781009522472 (epub)
Subjects: LCSH: Artificial intelligence–Congresses. | Metacognition–Congresses.
Classification: LCC Q335 .W67 2025 (print) | LCC Q335 (ebook) | DDC 006.3–dc23/eng/20250305
LC record available at https://lccn.loc.gov/2024059184
LC ebook record available at https://lccn.loc.gov/2024059185

ISBN 978-1-009-52245-8 Hardback

Cambridge University Press & Assessment has no responsibility for the persistence
or accuracy of URLs for external or third-party internet websites referred to in this
publication and does not guarantee that any content on such websites is, or will
remain, accurate or appropriate.

For EU product safety concerns, contact us at Calle de José Abascal, 56, 1°, 28003 Madrid,
Spain, or email eugpsr@cambridge.org

Contents

List of Contributors	page viii
Acknowledgments	xi

Part I Introduction   1

1 **Metacognitive AI**   3
HUA WEI, PAULO SHAKARIAN, CHRISTIAN LEBIERE, BRUCE DRAPER, NIKHIL KRISHNASWAMY, SARATH SREEDHARAN, SERGEI NIRENBURG

Part II Taxonomy of Metacognitive Approaches   23

2 **An Architectural Approach to Metacognition**   25
CHRISTIAN LEBIERE, ROBERT THOMSON, ANDREA STOCCO, MARK ORR, DONALD MORRISON

3 **Metacognitive AI through Error Detection and Correction Rules**   44
BOWEN XI, PAULO SHAKARIAN

4 **Mutual Trust in Human–AI Teams Relies on Metacognition**   56
SERGEI NIRENBURG, MARJORIE MCSHANE, THOMAS M. FERGUSON

Part III Neuro-Symbolic Models in AI   81

5 **Learning Where and When to Reason in Neurosymbolic Inference**   83
CRISTINA CORNELIO

| 6 | Assessment of Competency of Learning Agents via Inference of Temporal Logic Formulas | 95 |

ZHE XU, NASIM BAHARISANGARI, JEAN-RAPHAËL GAGLIONE, UFUK TOPCU

Part IV Metacognition with LLMS — 109

| 7 | Metacognitive Intervention for Accountable LLMs through Sparsity | 111 |

TIANLONG CHEN

| 8 | Metacognitive Insights into ChatGPT's Arithmetic Reasoning | 121 |

NOEL NGU, PAULO SHAKARIAN, ABHINAV KOYYALAMUDI, LAKSHMIVIHARI MAREEDU

Part V Metacognition in Learning Agents — 133

| 9 | Uncertainty Quantification's Role in Metacognition | 135 |

GAVIN STRUNK

| 10 | The Role of Predictive Uncertainty and Diversity in Embodied AI and Robot Learning | 148 |

RANSALU SENANAYAKE

Part VI Assured Machine Learning in High-Stakes Domains — 183

| 11 | Toward Certifiably Trustworthy Deep Learning at Scale | 185 |

LINYI LI

| 12 | Metacognition with Neural Network Verification and Repair Using Veritex | 212 |

XIAODONG YANG, TOMOYA YAMAGUCHI, BARDH HOXHA, DANIL PROKHOROV, TAYLOR T. JOHNSON

Part VII Metacognition as a Solution to Handle Failure — 229

| 13 | Reasoning about Anomalous Object Interaction Using Plan Failure as a Metacognitive Trigger | 231 |

NIKHIL KRISHNASWAMY

| 14 | Tractable Probabilistic Reasoning for Trustworthy AI | 245 |

YOOJUNG CHOI

Part VIII Applications of Metacognitive AI   257

15 Robust and Compositional Concept Grounding for Image Generative AI   259
YEZHOU YANG

16 mLINK: Machine Learning Integration with Network and Knowledge   264
SERGEI CHUPROV, RAMAN ZATSARENKO, LEON REZNIK

17 Military Applications of Artificial Intelligence Metacognition   276
BONNIE JOHNSON

Contributors

NASIM BAHARISANGARI
Arizona State University

TIANLONG CHEN
University of North Carolina, Chapel Hill

YOOJUNG CHOI
Arizona State University

SERGEI CHUPROV
Rochester Institute of Technology, New York

CRISTINA CORNELIO
Samsung AI

BRUCE DRAPER
Colorado State University

THOMAS M. FERGUSON
Rensselaer Polytechnic Institute, New York

JEAN-RAPHAËL GAGLIONE
Arizona State University

BARDH HOXHA
Toyota Research Institute of North America (TRINA)

List of Contributors

BONNIE JOHNSON
Naval Postgraduate School, Monterey, California

TAYLOR T. JOHNSON
Vanderbilt University, Tennessee

ABHINAV KOYYALAMUDI
Arizona State University

NIKHIL KRISHNASWAMY
Colorado State University

CHRISTIAN LEBIERE
Carnegie Mellon University, Pennsylvania

LINYI LI
University of Illinois, Urbana-Champaign

LAKSHMIVIHARI MAREEDU
Arizona State University

MARJORIE McSHANE
Rensselaer Polytechnic Institute, New York

DONALD MORRISON
Carnegie Mellon University, Pennsylvania

NOEL NGU
Arizona State University

SERGEI NIRENBURG
Rensselaer Polytechnic Institute, New York

MARK ORR
University of Virginia

DANIL PROKHOROV
Toyota Research Institute of North America (TRINA)

LEON REZNIK
Rochester Institute of Technology, New York

RANSALU SENANAYAKE
Arizona State University

PAULO SHAKARIAN
Syracuse University

SARATH SREEDHARAN
Colorado State University

ANDREA STOCCO
University of Washington

GAVIN STRUNK
Scientific Systems

ROBERT THOMSON
United States Military Academy

UFUK TOPCU
Arizona State University

HUA WEI
Syracuse University

BOWEN XI
Arizona State University

ZHE XU
Arizona State University

TOMOYA YAMAGUCHI
Woven Planet

XIAODONG YANG
Visa

YEZHOU YANG
Arizona State University

RAMAN ZATSARENKO
Rochester Institute of Technology, New York

Acknowledgments

This book serves as proceedings for the Workshop on Metacognitive Prediction of AI Behavior held November 13–15, 2023, in Scottsdale, Arizona. This workshop was funded by the Army Research Office (ARO) under grant W911NF2310345. The views, opinions, and/or findings contained in this report are those of the author(s) and should not be construed as an official Department of the Army or US Government position, policy, or decision, unless so designated by other documentation.

PART I

Introduction

1
Metacognitive AI

Hua Wei, Paulo Shakarian, Christian Lebiere, Bruce Draper,
Nikhil Krishnaswamy, Sarath Sreedharan, Sergei Nirenburg

Metacognition is the concept of reasoning about an agent's own internal processes. The concept was originally introduced in the field of developmental psychology [26] as a description of higher-order cognition. This "cognition about cognition" is regarded by some as a self-monitoring process that is integral to the functioning of the human mind [20]. It has been studied extensively in the fields of manufacturing [47], aerospace [35], transportation [1, 11, 30], and military applications [62]. In this section, we argue for the study of metacognitive artificial intelligence (AI), which deals with the reasoning about an artificial agent's own processes. This idea actually has been studied on and off in the history of AI [14, 15], but recent developments indicate that this area deserves a renewed focus. Specifically, recent advances in AI have led to significant engineering investments by large firms in fields such as digital commerce, autonomous driving, web search, and others. Yet despite the scale of these engineering investments, major AI failures still occur – which indicates that pure engineering solutions are unlikely to solve these fundamental failures. Consider the following examples:

- A large language model falsely accuses a college professor of sexual harassment [54].
- An autonomous robot taxi in San Francisco accidentally drags a woman for 20 feet causing major injury [25].
- A reinforcement learning (RL) model has to be retrained to play with slight changes in the environment [37, 68].
- A robot mistakes a man for a box in South Korea and crushes him to death [4].

Each of these case studies exhibits a different modality of AI failure. The first item illustrates a failure of **Transparency** – the system generated information that was false and could not provide a way to check itself on the facts. The second illustrates a failure in **Reasoning** – how the system synthesizes information and ultimately produces a decision. The third illustrates a failure of **Adaptation** – the system could not accommodate itself in a new environment. The fourth

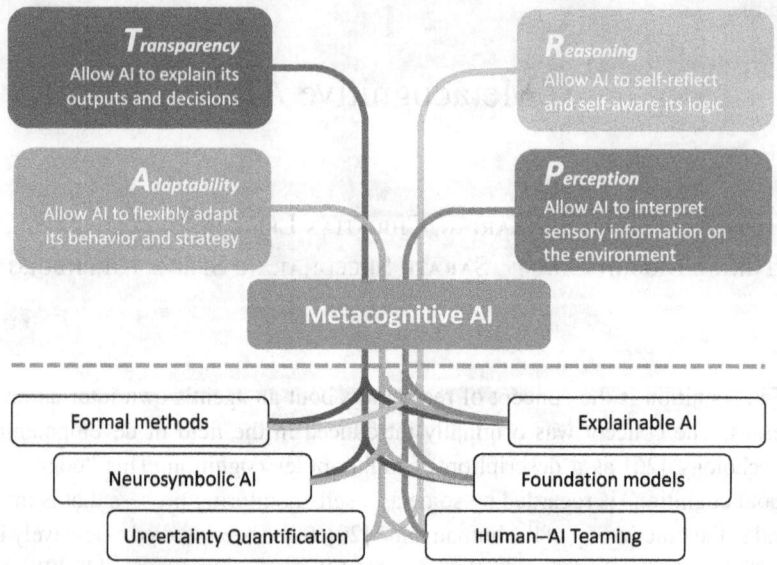

Figure 1.1 Four aspects of metacognitive AI (TRAP) and approaches to achieve metacognition.

illustrates a failure in **Perception** – how the system recognizes entities in its environment. In this introduction, which stemmed from the 2023 ARO-sponsored Workshop on Metacognitive Prediction of AI Behavior, we argue that the study of metacognitive AI should encompass these four areas (**TRAP**), as is shown in Figure 1.1.

1.1 Why Metacognition for AI?

In this section, we identify four areas that metacognitive AI encompasses: Enhanced Decision-Making, Adaptability and Flexibility, Error Detection and Correction, and Trust and Transparency. The interplay of these areas forms the essence of metacognitive AI, driving it toward becoming more than just a tool for automation – it evolves into a collaborative, intelligent partner capable of introspection, learning, and adaptation. In the following subsections, we will explore each area in detail, highlighting key research and applications that exemplify the advancements and potential of metacognitive AI.

A traditional AI system could be simplified as $y = f_\theta(x)$, where x is the input for the AI system; y, depending on applications, could be a description, prediction, or actions to take; and f_θ is most AI systems operational function with parameters θ. A metacognitive AI system could be an additional function g. With different metacognitive areas, g is in different locations concerning f.

1.1.1 Transparency

While traditional AI can sometimes be perceived as a "black box," metacognitive AI enhances trust and transparency by making its decision-making process more understandable to users. This is achieved through the function $g(f(x), \theta)$ or the function $g|f$. The function $g(f(x), \theta)$ represents the process of generating explanations based on both the input x and the parameters θ of the model f, while the function $g|f$ represents the function f with a series of g. This function allows metacognitive AI to explain its decisions in terms of both the input data and its internal parameters, catering to different user expectations and motivations for seeking explanations.

On the one hand, the nature of the explanation can vary significantly depending on whether it's intended for an expert with technical knowledge or a layperson. This distinction influences the complexity and technicality of the explanation provided. Different users could have different expectations of transparency. If they are looking to understand why certain outputs come from a global perspective, then the focus is to have $g(\theta)$ to make the θ transparent; if users are looking to understand certain cases, then the focus is to have $g(f(x))$ to understand a certain prediction $f(x)$ on x.

On the other hand, the motivation behind an explanation could stem from enhancing the performance of the system, reducing bias, and increasing fairness, or simply deriving a clearer understanding of the AI's decision-making process. Each of these purposes necessitates a different approach to how explanations are formulated and presented. Enhancing the performance of the system through transparency could involve using the explanations to correct predictions or induce actions $g|f$ where the understanding outputted from g is then processed by the AI system to better learn f. For example, in a transportation AI application, a metacognitive system could involve several submodular components, with image processing components explaining the traffic condition and detailing the risk assessments and traffic predictions, which could be then used by downstream control components for better adaptive control. This level of transparency not only builds trust among users but also allows for more effective human–AI collaboration, as users can provide informed feedback or override decisions with a deeper understanding of the AI's reasoning process. Mitsopoulos et al. [53] argued building cognitive models of both the AI and the human user that could be introspected upon to adapt explanations according to the discrepancy between the two models, e.g., when the AI decision does not conform to the human model's expectations.

Wang et al. [67] applied XAI framework to a real-world clinical machine learning (ML) use case, that is, an explainable diagnostic tool for intensive care phenotyping. Co-designing with 14 clinicians, they provided five explanation strategies to mitigate decision biases and moderate trust. They implemented an early decision aid system to diagnose patients in an intensive care unit (ICU) and found that users employed a diverse range of XAI facilities, to reason variedly.

1.1.2 Reasoning

Traditional AI systems f often rely on predefined algorithms and datasets for reasoning or decision-making, which can limit their effectiveness in dynamic or unfamiliar scenarios. By contrast, metacognitive AI incorporates self-reflection and self-awareness into its logic, represented by $f(x; g(\theta))$. This indicates how the AI's self-reflection (through g) informs its decision-making process (through f), enhancing its reasoning capabilities. For instance, a metacognitive AI in healthcare could use this approach to evaluate and refine its diagnostic criteria over time, learning to differentiate between complex cases and refining its diagnostic criteria based on outcomes. This leads to decisions that are not only based on data but also enriched by the AI's growing experiential knowledge, resulting in more accurate and reliable outcomes.

Ulam et al. [64] showed that model-based reflection may guide RL with two benefits. The first is a reduction in learning time as compared to an agent that learns the task via pure RL. Second, the reflection-guided RL agent shows benefits over the pure model-based reflection agent, matching the performance of that agent in the metrics measured in addition to converging to a solution in fewer trials. In addition, the augmented agent eliminates the need for an explicit adaptation library such as is used in the pure model-based agent and thus reduces the knowledge engineering burden on the designer significantly.

Andrychowicz et al. [3] presented a novel technique called Hindsight Experience Replay, whose intuition is to reexamine the trajectories with a different goal – while a trajectory may not help us learn how to achieve the desired goal, it tells us something about how to achieve the state in the actual trajectory. They demonstrated this approach on the task of manipulating objects with a robotic arm on three different tasks: pushing, sliding, and pick-and-place, while the vanilla RL algorithm fails to solve these tasks.

1.1.3 Adaptability

Adaptation in metacognitive AI encapsulates the system's ability to detect and correct errors of internal conditions and to flexibly adapt its behavior and strategies. This is represented by $f'(x; g(f(x), \theta))$, where f' is the adapted model based on the metacognitive assessment g of the original model's output $f(x)$ and parameters θ. This notation reflects how metacognitive AI adapts by reassessing its outputs and parameters, allowing for more effective decision-making in the face of uncertainty and changing environments. Additional adaptations could also be implemented with $g(x)?f(x): h(x)$ where the metacognitive process g decides whether to use the main function f or an alternative function h based on its analysis of the input x. This could model AI systems that choose different processing paths based on metacognitive assessment without modifying f. Such systems can adapt to new environments and tasks by understanding their learning process and limitations.

Leibig et al. [45] shows that uncertainty-informed decision referral can improve diagnostic performance. Experiments across different networks, tasks, and datasets show robust generalization. Depending on network capacity and task/dataset difficulty, it surpasses 85% sensitivity and 80% specificity as recommended by the National Health Service when referring 0–20% of the most uncertain decisions for further inspection. By analyzing the causes of uncertainty, they found it is related to intuitions from two-dimensional (2D) visualizations to the high-dimensional image space.

More recently, another metacognitive approach allowing for adaptability known as *error detection and correction rules* (EDCR) was introduced [70]. In this framework, function g results in a set of learned rules that characterize the failure modes of f and how to correct on those failure modes while f' is an inference process conducted using these rules to erase or change the results of the underlying model f. In [70] the authors applied this technique to the classification of geospatial movement trajectories and examined performance improvement on f, where the current state of the art is neural architecture known as LRCN [42]. EDCR was able to both improve over the state of the art and exhibit the ability to improve performance when exposed to out-of-distribution data.

1.1.4 Perception

Perception refers to the ability to interpret sensory information to understand the environment. Perception in metacognitive AI involves the system's ability to interpret and understand sensory information, such as visual and auditory data, in a context-aware manner, represented by $f(g(x), x)$, where context is a metacognitive assessment of the AI's own capacity rather than an external context. Here, f represents the primary perceptual processing function, interpreting sensory data like visuals or audio, while the metacognitive function g specifically evaluates the accuracy and limitations of the AI's own sensory processing. The primary perceptual function f then uses both the original sensory input x and the metacognitive assessment $g(x)$ to refine its interpretation. This dual-input model allows the AI to recognize and compensate for any inherent biases or weaknesses in its perception. This could include AI in autonomous vehicles that must interpret complex visual environments, in medical imaging distinguishing subtle diagnostic details, or in environmental monitoring systems that detect and analyze changes through sensory data.

1.2 Ways toward Metacognitive AI

Following the understanding of why metacognition is crucial for AI, let's explore various methods and techniques that are paving the way toward realizing metacognitive AI. These approaches not only address the challenges identified but also capitalize on the opportunities presented by modern technology.

1.2.1 Formal Methods

Formal methods refer to mathematically based techniques for the specification, development, and verification of software and hardware systems. For applications with dynamic environments, temporal logic can be used to deal with reasoning about temporal information. By applying these rigorous methods to AI, we can ensure the correctness and reliability of AI systems, particularly in their decision-making processes. This is crucial for applications where failure or error can have severe consequences, for example, in verifying the safety of autonomous vehicle algorithms and ensuring the reliability of AI systems in aerospace engineering.

Merits for Metacognitive AI (1) Techniques from formal methods enable reasoning from logical or mathematical specifications of the behaviors of AI systems; they offer rigorous proofs that all system behaviors meet some desirable property [9, 28]. In [9], the authors automatically translate multiagent systems programmed in the logic-based agent-oriented programming language AgentSpeak into Java and then use model checkers to verify the resulting systems. (2) Addressing *transparency* issues related to AI, formal methods use mathematical logic reasoning to produce automatic, quick, and reliable results. Varriano et al. [65] adopts formal methods for the diagnosis of the coronavirus disease to analyze and understand, in a more medical way, the meaning of some radiomic features to connect them with clinical or radiological evidence.

Challenges Even though formal methods have existed for nearly 50 years, they are not in widespread use and are justifiably viewed with skepticism by potential users due to several factors. A key limiting factor in user uptake is the specification problem [58]. Application-level specifications of program logic must be encoded at a relatively low level of abstraction. They require specifications to be encoded using a language outside those in which developers write their programs, often requiring knowledge of specialized constructs and type systems that are not present in mainstream languages.

1.2.2 Neurosymbolic AI (NSAI)

The integration of connections (e.g., neural) with symbolic (e.g., logical) systems is now widely referred to as "neurosymbolic AI" (NSAI). This term was coined in the early 2000s and has gained wider prominence in recent years [19, 32, 40, 59]. The key relationships between NSAI and metacognition relate to the ability to use symbolic knowledge and perceptual models to detect and correct errors in each other (adaptability) and the use of symbolic languages to express information about error modes of a perceptual model (transparency). Similar insights are also shared in Section 2.3.1.

Merits for Metacognitive AI With the introduction of logic tensor networks [6], the canonical paradigm for NSAI has consisted of guiding gradient descent with the addition of soft logical constraints in the loss function – and this was followed by related work such as semantic loss [71], CHMCN [29], and others. In general, these loss-based approaches would not fit the metacognitive paradigm, as in these incarnations of NSAI, the symbolic logic is used as an additional optimization criterion – in much the same way as one would a regularization term. However, more recent views on NSAI do lend themselves to metacognition – in particular with respect to adaptability and transparency. *NSAI for metacognitive adaptability.* The key intuition in the use of NSAI for metacognitive adaptability is to leverage symbolic domain knowledge to explicitly identify errors in a neural model, allowing for some corrective action to be performed. One well-known approach for NSAI metacognitive adaptability is abductive learning (ABL) [17]. Using the paradigm of adaptability introduced in this paper, function f' returns a result based on the combination of a perceptual model, a-priori domain knowledge (i.e., a logic program), and abducted error information (function g in our framework). Here, function g is abducted based on inconsistencies between the perceptual model and domain knowledge and can take the form of additional symbolic structures added to the logic program and/or updates to the perceptual model (f in our framework). ABL has been shown to provide SOTA performance on combined perception-reasoning tasks as well as application to the identification of new concepts as shown in [34]. However, it is noteworthy that more recent work in this area has favored updates to the perceptual model over the addition of symbolic knowledge – hence requiring access to the model f. In [66], the authors use both the addition of symbolic knowledge in the cognitive model's episodic memory and the update of the neural perception model. More recent applications of NSAI to metacognitive adaptability have sought to disentangle perceptual updates from the base perceptual model. Specifically, [13] introduces a framework where an additional transformer model (g in our framework) is used to predict errors in the underlying neural (f) using symbolic knowledge and RL to detect and correct perceptual errors. The work of [70] also addresses perceptual errors but using a rule-learning approach – here rules are learned about the results of the neural model that allow for error detection and correction while providing the byproduct of an explanation of the causes of the errors.

NSAI for Metacognitive Transparency. Complementary to the NSAI work relevant to metacognitive adaptability is NSAI work relating to metacognitive transparency. Here, NSAI is used to reason directly about the inner workings of a perceptual model, often for a downstream task involving an explanation of the perceptual results. One example of such work is [24] where a binarized neural network is used to produce a symbolic theory of perception used in a downstream task of appreciation, which provides an explanation of the perceptual results. Another application of NSAI to metacognitive transparency deals with the use of concept induction [18] to map activations in a neural network to an explanation using description logic, thereby providing transparency.

Challenges Many challenges remain with the application of NSAI to metacognition. With respect to adaptability, a key concern is where the initial symbolic knowledge comes from. Techniques such as inductive logic programming (e.g., [23]) have proven difficult to scale while obtaining knowledge from "common sense" sources has challenges with knowledge integration (e.g., [22]). With respect to transparency, several issues arise in such analysis of neural models, such as the level of labeling required to properly identify concepts based on activations or the loss of perceptual performance when using a binarized neural network. That said, NSAI approaches hold much promise for metacognition due to the expressibility of logical language and the use of complementary perceptual and reasoning systems to allow each to understand errors induced by the other.

1.2.3 Uncertainty Quantification

Uncertainty quantification involves methods to determine and reduce uncertainties in AI models, especially those related to predictions and decision-making. This approach directly addresses the challenge of making reliable decisions under uncertainty, enhancing the AI's ability to identify, assess, and inform operators of potential risks or errors. It is widely used in financial modeling, climate forecasting, and risk assessment in engineering projects. There are two main sources of uncertainty, aleatoric uncertainty and epistemic uncertainty; while the former is caused by noise in data or labels, the latter is caused by data sparsity or out-of-distribution data.

Merits for Metacognitive AI (1) Uncertainty quantification methods provide insight into the system's own uncertainty about its *reasoning* mechanics [36]. For example, [7] discussed the necessary steps to quantify uncertainties in the medical domain, including the collection and selection process of the training data, reasoning with the performance bounds and uncertainties related to the model's performance against the operational data. In [52], the authors utilize the uncertainty to reason on the necessity of adding sensors on certain locations in a road network for better traffic prediction. (2) Adaptation enhances decision accuracy under uncertain conditions and facilitates adaptable AI behavior. Dessai et al. [21] reviews existing frameworks for decision-making under uncertainty for adaptation to climate change and shows how different ways of including uncertainty in decision-making match with uncertainty information provided by the various uncertainty assessment methods. Nozarian et al. [56] incorporate three different uncertainty quantification methods into autonomous driving policies, and their results show that accurate uncertainty quantification can significantly improve driving performance with adaptive control.

Challenges A lot of uncertainty quantification methods are essentially exploiting correlations in the data, without paying attention to any causal link. This may not seem like a limitation since prediction does not need a causal relation. The absence of a causal connection, however, means garnering limited conclusions from AI models; furthermore, it is imperative to understand how the training data must be similar to prediction data. Another precedent challenge for uncertainty quantification methods on metacognitive capabilities is the scale of uncertainty: when should we consider the model is uncertain about its prediction? While an uncertainty threshold could be set by humans, the threshold would be expected to adjust to the complex real world.

1.2.4 Explainable AI (XAI)

Explainable AI aims to make the functioning of AI systems transparent and understandable to humans. XAI is essential for building trust and accountability in AI systems, allowing users to understand and interpret AI decisions, and enabling more effective human–AI collaboration. XAI has been implemented in healthcare for explaining diagnostic decisions [63], in finance for credit scoring systems [10], and in customer service as interpretable chatbots [41]. The two major categories of XAI methods appear to present different polarities within the notion of explainability: perceptive explainability, where saliency maps or signal flows within an AI model are visualized, and explainability by mathematical structures, where the explanations are generated by approximations on the AI model with simpler, understandable functions.

Merits for Metacognitive AI (1) Transparency: One obvious merit XAI techniques bring is making AI processes understandable to users and building trust. Hu et al. [33] tested their XAI approach on a COVID-19 chest X-ray dataset and the ISIC 2017 skin lesion dataset, showing that saliency maps help reveal the image features used by models to determine image similarity. This approach aids clinicians when viewing medical images and addresses an urgent need for interventional tools in response to COVID-19. (2) Adaptability: Another merit XAI brings to metacognition is to allow AI techniques to pay attention to possibly sensitive inputs described by the explanations from XAI. Wu et al. [69] utilizes XAI technique to guide an attacker to exploit the weakness of victim agents, where the weakness is highlighted by the input feature importance through gradient-based interpretation methods [60]. Lebiere et al. [44] provides salience indicators for the various target features for individual attackers (transparency), and uses an indicator of trust to optimize a deceptive signal in real time to a specific attacker (adaptability).

Challenges A series of challenges of XAI still remain insufficiently addressed to date. Specifically, the metrics to evaluate the explainability of AI models are concerning: it requires both subjective, user-centered metrics depending on the users it serves, and objective metrics for reproducible research and evaluations, where some objective metrics like fidelity suffer from distribution shifts in the input samples [12]. Another challenge in XAI technique is its robustness. Researchers [27, 43] have found that small perturbations on the input instance generate large changes in the output interpretations that popular XAI methods generate, and the explanations can be easily manipulable by simple transformations.

1.2.5 Foundation Models

This involves leveraging the vast information processing capabilities of large language models (LLMs), such as GPT-4, or vision language models (VLMs) while embedding metacognitive strategies for self-assessment and improvement. It addresses the need for AI systems to process and understand large volumes of text data, making informed decisions based on that understanding. Successful implementations can be seen in advanced chatbots, automated content creation tools, and sophisticated information retrieval systems.

Merits for Metacognitive AI (1) Adaptability and Flexibility: LLMs are known to be capable of in-context zero-shot or few-shot abilities, which adapt these models to diverse tasks without gradient-based parameter updates [2]. This allows them to rapidly generalize to unseen tasks and even exhibit apparent reasoning abilities with appropriate prompting strategies. Da et al. [16] leverage LLMs to understand and profile the system dynamics for sim-to-real transfer. Accepting the cloze prompt template, and then filling in the answer based on accessible context, the pre-trained LLM's inference ability is exploited and applied to understand how weather conditions, traffic states, and road types influence traffic dynamics, being aware of this, the policies' action is taken and grounded based on realistic dynamics, thus help the agent learn a more realistic policy. (2) Trust and Transparency: foundation models show strong reasoning abilities on textual inputs and enhance understanding of AI's text-based decision-making. Liu et al. [48] leverages the power of LLMs for robot failure explanation, which queries LLM for failure reasoning based on a hierarchical summary of robot past experiences generated from multisensory observations. The failure explanation can further guide a language-based planner to correct the failure and complete the task. Their experiments demonstrate that the LLM-based framework is able to generate informative failure explanations that assist successful correction planning.

Challenges Foundation models have also shown challenges in themselves. Firstly, large models come with large costs and latency. Enterprises are already incurring huge costs of operating or using foundation models for their respective use cases [8, 39]. How to reduce the cost of training, maintaining, and querying foundation models will be a long-lasting question for using foundation models for real-world applications. Another challenge is the hallucination in LLMs. Hallucination has been widely recognized to be a significant drawback for LLMs. While [72] shows that it is impossible to eliminate hallucination in LLMs, there is still ongoing research that attempts to reduce the extent of hallucination empirically [38].

1.2.6 Human–AI Teaming

This approach focuses on the supporting synergistic interaction between humans and AI. It aims to leverage the metacognitive capability of humans to develop AI agents that deserve to be treated as teammates and collaborators. Currently, AI agents in applications involving human–AI collaboration – collaborative robots (cobots) in manufacturing, decision support systems in military operations, and interactive educational tools – are not yet truly metacognitive. Indeed, to take one facet of metacognitive abilities as an example, the crucially important ability of AI agents to explain their decisions. The current approach to this task, introduced and developed through DARPA's XAI program [31], does not directly address this ability. Instead, it concentrated on "post hoc algorithmically generated rationales of black-box predictions, which are not necessarily the actual reasons behind those predictions or related causally to them ... [and which] are unlikely to contribute to our understanding of [a system's] inner workings" [5]. This approach is understandable due to the limitations of the current technology, including the latest generative AI models. A longer view of AI agents for human–AI collaboration envisages the development of a new generation of AI agents [51] that will allow humans to start treating AI agents not as tools, however sophisticated, but as human-level interlocutors, teammates, and negotiation partners.

Developing AI agents with human-level metacognitive capabilities is perhaps the ultimate big scientific problem of AI as a field. It is worth our collective while to devote effort to it alongside the many other tasks that promise more immediate practical returns. In fact, work toward this long-term goal supports the gradual enhancement of AI agent capabilities. Basic research in explainable natural language understanding [50] was demonstrated to support several proof-of-concept AI agent systems.

Merits for Metacognitive AI (1) Reasoning: An alternative view would be an extension of interactive task learning [36] where the human teammate can scaffold the AI reasoning where needed. (2) Adaptability and Flexibility: The role of the human could focus on suggesting new representations and strategies that expand the AI outside of its existing framework.

Challenges The main challenge for this line of work is the acquisition and maintenance of comprehensive repositories of interpretable content [55]. Describing the affordances of objects and causal properties of processes in the real world is an ongoing concern. This includes the task of developing contentful theories and practical models of how humans accumulate and use this knowledge. To better approximate human operation, this work must address not only to the scientific view of the world but also reflect "folk psychology" [61], which seeks to account for the thought processes of a regular person. Knowledge acquisition in AI has a long history (e.g., [46]). It requires a very significant effort that the AI R&D community has not fully supported over the years. A commensurate level of effort was directed at an alternative empirical approach that culminated in the achievements of generative AI. It has been persuasively argued, however, that AI is "unlikely to ever be trustworthy if its makers continue to skip the step of having explicit meaning-based cognitive models of the world" [49]. Acquisition of such models is supported by ergonomic environments and for human knowledge engineers, the use of deep natural language understanding systems that facilitate learning by reading and dialog, and the development of infrastructure for using the latest technological advances, such as large language models to boost the efficiency of knowledge acquisition. While work on each of the above thrusts has been ongoing for decades, though it has been recently out of fashion. The major challenge is to return the task of knowledge acquisition for human-level AI agents to the center of attention in the field.

1.3 Open Questions

In this section, we briefly discuss some directions for future research.

Generalization to Diverse Dynamic Environments Metacognitive AI must be capable of adapting to rapidly changing and unpredictable environments, or at least know when it is incapable. Designing systems that can reassess and adjust their strategies in real-time, in response to environmental changes, poses a unique challenge. Applying metacognitive AI effectively across different domains, each with its specific requirements and constraints, is challenging. It is worth investigating the similarities and differences between domains, requirements, and constraints, and how they affect adaptability. Related works in cognitive psychology indicate there is a "power law of practice," which explains the learning curve and its inherent noise partly as differences in the transfer of knowledge between tasks and how the curve arises out of mechanisms necessary for processing [57]. Tailoring metacognitive capabilities to a wide range of applications requires versatile and adaptable AI designs.

Designing for Continuous Self-Improvement Enabling AI systems not only to identify their weaknesses or errors but also to autonomously modify their behavior and learning strategies for continuous improvement is particularly challenging. This requires advanced algorithms capable of self-modification without human intervention and possibly online adaptation.

Ensuring Ethical and Responsible Metacognition As metacognitive AI systems will have a higher level of autonomy in decision-making, ensuring that these decisions are ethically and morally responsible is a challenge. This includes embedding ethical principles into the AI's metacognitive processes.

Interpreting Metacognitive Processes While explainability in AI is already a challenge, making the metacognitive processes of AI interpretable and understandable to humans adds an extra layer of complexity. This is crucial for ensuring trust and transparency in AI systems.

Benchmarking Datasets and Baselines While benchmarking datasets and environments benefit the development of modern AI and ML, we still see a significant challenge in developing benchmarks and methods for evaluating the effectiveness and accuracy of metacognitive processes in AI. The benchmark includes validating that the AI's self-assessment and adaptive learning capabilities are functioning as intended, which requires humans to justify the correctness of the assessment. Having the data and code available helps researchers in examining the proposed method and makes the presented results reproducible. One concern about benchmarking is that as soon as a benchmark dataset and environment are defined, it stops offering unforeseen circumstances, especially when the core characteristic of metacognition is its ability to function in new unforeseen circumstances. For instance, autonomous driving keeps encountering unforeseen situations and failing to deal with them properly (e.g., the most recent Waymo accident in Arizona where two separate autonomous vehicles hit the same tow truck). Trying to deal with them by finding and patching all edge cases doesn't seem to be working because they do not seem enumerable. The same issue arises when trying to test metacognitive abilities to deal with situations outside the training data. We believe, however, that sharing both the data and code is essential for improving the quality of metacognitive AI.

References

[1] Abduljabbar, Rusul, Dia, Hussein, Liyanage, Sohani, and Bagloee, Saeed Asadi. 2019. Applications of artificial intelligence in transport: An overview. *Sustainability*, **11**(1), 189.

[2] Alayrac, Jean-Baptiste, Donahue, Jeff, Luc, Pauline, et al. 2022. Flamingo: A visual language model for few-shot learning. Pages 23716–23736 of: *Advances in Neural Information Processing Systems*, 35.

[3] Andrychowicz, Marcin, Wolski, Filip, Ray, Alex, et al. 2017. Hindsight experience replay. *Annual Conference on Neural Information Processing Systems 2017, December 4–9, 2017*.

[4] Atkinson, Emily. 2023. Man crushed to death by robot in South Korea. *BBC News*, November 8, 2023. Available at www.bbc.com/news/world-asia-67354709.

[5] Babic, Boris, Gerke, Sara, Evgeniou, Theodoros, and Cohen, I Glenn. 2021. Beware explanations from AI in health care. *Science*, **373**(6552), 284–286.

[6] Badreddine, Samy, d'Avila Garcez, Artur, Serafini, Luciano, and Spranger, Michael. 2022. Logic tensor networks. *Artificial Intelligence*, **303**, 103649.

[7] Begoli, Edmon, Bhattacharya, Tanmoy, and Kusnezov, Dimitri. 2019. The need for uncertainty quantification in machine-assisted medical decision making. *Nature Machine Intelligence*, **1**(1), 20–23.

[8] Bender, Emily M, Gebru, Timnit, McMillan-Major, Angelina, and Shmitchell, Shmargaret. 2021. On the dangers of stochastic parrots: Can language models be too big? Pages 610–623 of *FACCT '21: Proceedings of the 2021 ACM Conference on Fairness, Accountability, and Transparency*. March 3–10, 2021.

[9] Bordini, Rafael H, Fisher, Michael, Visser, Willem, and Wooldridge, Michael. 2006. Verifying multi-agent programs by model checking. *Autonomous Agents and Multi-agent Systems*, **12**, 239–256.

[10] Bussmann, Niklas, Giudici, Paolo, Marinelli, Dimitri, and Papenbrock, Jochen. 2020. Explainable AI in fintech risk management. *Frontiers in Artificial Intelligence*, **3**, 26.

[11] Caesar, Holger, Bankiti, Varun, Lang, Alex H, et al. 2020. Nuscenes: A multimodal dataset for autonomous driving. Pages 11618–11628 of: *Proceedings of the IEEE/CVF Conference on Computer Vision and Pattern Recognition*, Seattle, WA, USA, June 13–19, 2020.

[12] Chen, Zhuomin, Zhang, Jiaxing, Ni, Jingchao, et al. 2024. Interpreting graph neural networks with in-distributed proxies. *arXiv preprint arXiv:2402.02036*.

[13] Cornelio, Cristina, Stuehmer, Jan, Hu, Shell Xu, and Hospedales, Timothy. 2022. Learning where and when to reason in neuro-symbolic inference. In: *The Eleventh International Conference on Learning Representations, Kigali, Rwanda, May 1–5, 2023*.

[14] Cox, Michael T. 2005. Metacognition in computation: A selected history. Pages 1–17 of: *AAAI Spring Symposium: Metacognition in Computation*.

[15] Cox, Michael T, and Raja, Anita. 2011. *Metareasoning: Thinking about Thinking*. MIT Press.

[16] Da, Longchao, Gao, Minchiuan, Mei, Hao, and Wei, Hua. 2023. LLM powered sim-to-real transfer for traffic signal control. *arXiv preprint arXiv:2308.14284*.

[17] Dai, Wang-Zhou, Xu, Qiuling, Yu, Yang, and Zhou, Zhi-Hua. 2019. Bridging machine learning and logical reasoning by abductive learning. Pages 2815–2826 of: *Advances in Neural Information Processing Systems*, 32.

[18] Dalal, Abhilekha, Sarker, Md Kamruzzaman, Barua, Adrita, and Hitzler, Pascal. 2023. Explaining deep learning hidden neuron activations using concept induction. *arXiv preprint arXiv:2301.09611*.

[19] d'Avila Garcez, Artur, and Lamb, Luís C. 2020. Neurosymbolic AI: The 3rd Wave. *CoRR*, abs/2012.05876.

[20] Demetriou, Andreas, Efklides, Anastasia, Platsidou, Maria, and Campbell, Robert L. 1993. The architecture and dynamics of developing mind: Experiential structuralism as a frame for unifying cognitive developmental theories. *Monographs of the Society for Research in Child Development*, 58(5/6), i–202.

[21] Dessai, Suraje, van der Sluijs, Jeroen P, et al. 2007. *Uncertainty and Climate Change Adaptation: A Scoping Study*. Copernicus Institute for Sustainable Development and Innovation, Department of Science Technology and Society.

[22] Du, Li, Ding, Xiao, Liu, Ting, and Qin, Bing. 2021. Learning event graph knowledge for abductive reasoning. Pages 5181–5190 of: Zong, Chengqing, Xia, Fei, Li, Wenjie, and Navigli, Roberto (eds), *Proceedings of the 59th Annual Meeting of the Association for Computational Linguistics and the 11th International Joint Conference on Natural Language Processing (volume 1: Long papers)*. Association for Computational Linguistics.

[23] Evans, Richard, and Grefenstette, Edward. 2018. Learning explanatory rules from noisy data. *Journal of Artificial Intelligence Research*, **61**, 1–64.

[24] Evans, Richard, Bošnjak, Matko, Buesing, Lars, et al. 2021. Making sense of raw input. *Artificial Intelligence*, **299**, 103521.

[25] Farivar, Cyrus. 2023 (Oct). Cruise Robotaxi dragged woman 20 feet in recent accident, local politician says. *Forbes*. Available at www.forbes.com/sites/cyrusfarivar/2023/10/06/cruise-robotaxi-dragged-woman-20-feet-in-recent-accident-local-politician-says/.

[26] Flavell, John H. 1979. Metacognition and cognitive monitoring: A new area of cognitive–developmental inquiry. *American Psychologist*, **34**(10), 906.

[27] Ghorbani, Amirata, Abid, Abubakar, and Zou, James. 2019. Interpretation of neural networks is fragile. *Proceedings of the AAAI Conference on Artificial Intelligence*, **33**, 3681–3688.

[28] Giorgini, Paolo, Mylopoulos, John, Nicchiarelli, Eleonora, and Sebastiani, Roberto. 2003. Formal reasoning techniques for goal models. *Journal on Data Semantics I*, 1–20.

[29] Giunchiglia, Eleonora, and Lukasiewicz, Thomas. 2020. Coherent hierarchical multi-label classification networks. In: *Proceedings of the 34th International Conference on Neural Information Processing Systems*.

[30] Grigorescu, Sorin, Trasnea, Bogdan, Cocias, Tiberiu, and Macesanu, Gigel. 2020. A survey of deep learning techniques for autonomous driving. *Journal of Field Robotics*, **37**(3), 362–386.

[31] Gunning, David. 2017. Explainable artificial intelligence (XAI). *Defense Advanced Research Projects Agency (DARPA), nd Web*, **2**(2), 1.

[32] Hitzler, Pascal, Sarker, Md. Kamruzzaman, and Eberhart, Aaron (eds). 2023. *Compendium of Neurosymbolic Artificial Intelligence*. Frontiers in artificial intelligence and applications, vol. 369. IOS Press.
[33] Hu, Brian, Vasu, Bhavan, and Hoogs, Anthony. 2022. X-MIR: Explainable Medical Image Retrieval. Pages 1544–1554 of: *Proceedings of the IEEE/CVF Winter Conference on Applications of Computer Vision*, Waikoloa, HI, USA.
[34] Huang, Yu-Xuan, Dai, Wang-Zhou, Jiang, Yuan, and Zhou, Zhi-Hua. 2023. Enabling knowledge refinement upon new concepts in abductive learning. *Proceedings of the AAAI Conference on Artificial Intelligence*, vol. 37, no. 7.
[35] Izzo, Dario, Märtens, Marcus, and Pan, Binfeng. 2019. A survey on artificial intelligence trends in spacecraft guidance dynamics and control. *Astrodynamics*, **3**, 287–299.
[36] Jalaian, Brian, Lee, Michael, and Russell, Stephen. 2019. Uncertain context: Uncertainty quantification in machine learning. *AI Magazine*, **40**(4), 40–49.
[37] Jayawardana, Vindula, Tang, Catherine, Li, Sirui, Suo, Dajiang, and Wu, Cathy. 2022. The impact of task underspecification in evaluating deep reinforcement learning. Pages 23881–23893 of: *Advances in Neural Information Processing Systems*, 35.
[38] Ji, Ziwei, Lee, Nayeon, Frieske, Rita, et al. 2023. Survey of hallucination in natural language generation. *ACM Computing Surveys*, **55**(12), 1–38.
[39] Kasneci, Enkelejda, Seßler, Kathrin, Küchemann, Stefan, et al. 2023. ChatGPT for good? On opportunities and challenges of large language models for education. *Learning and Individual Differences*, **103**, 102274.
[40] Kautz, Henry A. 2022. The third AI summer: AAAI Robert S. Engelmore memorial lecture. *AI Magazine*, **43**(1), 105–125.
[41] Khurana, Anjali, Alamzadeh, Parsa, and Chilana, Parmit K. 2021. ChatrEx: Designing explainable chatbot interfaces for enhancing usefulness, transparency, and trust. Pages 1–11 of: *2021 IEEE Symposium on Visual Languages and Human-centric Computing (VL/HCC)*. IEEE.
[42] Kim, Jinsoo, Kim, Jae Hun, and Lee, Gunwoo. 2022. GPS data-based mobility mode inference model using long-term recurrent convolutional networks. *Transportation Research Part C: Emerging Technologies*, **135**, 103523.
[43] Kindermans, Pieter-Jan, Hooker, Sara, Adebayo, Julius, et al. 2019. The (un)reliability of saliency methods. Pages 267–280 of: *Explainable AI: Interpreting, Explaining and Visualizing Deep Learning*. Springer.
[44] Lebiere, Christian, Cranford, Edward, Aggarwal, Palvi, Cooney, Sarah, Tambe, Milind, and Gonzalez, Cleotilde. 2023. Cognitive modeling for personalized, adaptive signaling for cyber deception. Pages 59–82 of: Bao, T, Tambe, M, Wang, C (eds) *Cyber Deception*. Advances in Information Security, vol. 89. Springer.
[45] Leibig, Christian, Allken, Vaneeda, Ayhan, Murat Seçkin, Berens, Philipp, and Wahl, Siegfried. 2017. Leveraging uncertainty information from deep neural networks for disease detection. *Scientific Reports*, **7**(1), 17816.
[46] Lenat, Douglas B. 1995. CYC: A large-scale investment in knowledge infrastructure. *Communications of the ACM*, **38**(11), 33–38.
[47] Li, Bo-hu, Hou, Bao-cun, Yu, Wen-tao, Lu, Xiao-bing, and Yang, Chun-wei. 2017. Applications of artificial intelligence in intelligent manufacturing: A review. *Frontiers of Information Technology & Electronic Engineering*, **18**, 86–96.

[48] Liu, Zeyi, Bahety, Arpit, and Song, Shuran. 2023. Reflect: Summarizing robot experiences for failure explanation and correction. *arXiv preprint arXiv:2306.15724*.
[49] Marcus, Gary, and Davis, Ernest. 2019. *Rebooting AI: Building Artificial Intelligence We Can Trust*. Vintage.
[50] McShane, Marjorie, and Nirenburg, Sergei. 2021. *Linguistics for the Age of AI*. MIT Press.
[51] Mcshane, Marjorie, Nirenburg, Sergei, and English, Jesse. 2024. *Agents in the Long Game of AI: Computational Cognitive Modeling for Trustworthy, Hybrid AI*. MIT Press.
[52] Mei, Hao, Li, Junxian, Liang, Zhiming, Zheng, Guanjie, Shi, Bin, and Wei, Hua. 2023. Uncertainty-aware traffic prediction under missing data. *arXiv preprint arXiv:2309.06800*.
[53] Mitsopoulos, Konstantinos, Somers, Sterling, Schooler, Joel, Lebiere, Christian, Pirolli, Peter, and Thomson, Robert. 2022. Toward a psychology of deep reinforcement learning agents using a cognitive architecture. *Topics in Cognitive Science*, **14**(4), 756–779.
[54] Mok, Aaron. 2023. CHATGPT reportedly made up sexual harassment allegations against a prominent lawyer. *Business Insider*, April 2023.
[55] Nirenburg, Sergei, McShane, Marjorie, and English, Jesse. 2023. Content-centric computational cognitive modeling. *Advances in Cognitive Systems*, **10**, 71–84.
[56] Nozarian, Farzad, Müller, Christian, and Slusallek, Philipp. 2020. Uncertainty quantification and calibration of imitation learning policy in autonomous driving. Pages 146–162 of: *International Workshop on the Foundations of Trustworthy AI Integrating Learning, Optimization and Reasoning*. Springer.
[57] Ritter, Frank E, and Schooler, Lael J. 2001. The learning curve. *International Encyclopedia of the Social and Behavioral Sciences*, **13**, 8602–8605.
[58] Rozier, Kristin Yvonne. 2016. Specification: The biggest bottleneck in formal methods and autonomy. Pages 8–26 of: *Verified Software. Theories, Tools, and Experiments*. Springer.
[59] Shakarian, Paulo, Baral, Chitta, Simari, Gerardo I, Xi, Bowen, and Pokala, Lahari. 2023. *Neuro Symbolic Reasoning and Learning*. Springer Briefs in Computer Science. Springer.
[60] Smilkov, Daniel, Thorat, Nikhil, Kim, Been, Viégas, Fernanda, and Wattenberg, Martin. 2017. Smoothgrad: Removing noise by adding noise. *arXiv preprint arXiv:1706.03825*.
[61] Stich, Stephen P, and Nichols, Shaun. 2003. Folk psychology. Pages 235–255 of: *The Blackwell Guide to Philosophy of Mind*. Blackwell.
[62] Svenmarck, Peter, Luotsinen, Linus, Nilsson, Mattias, and Schubert, Johan. 2018. Possibilities and challenges for artificial intelligence in military applications. Pages 1–16 of: *Proceedings of the NATO Big Data and Artificial Intelligence for Military Decision Making Specialists' Meeting*.
[63] Tjoa, Erico, and Guan, Cuntai. 2020. A survey on explainable artificial intelligence (xai): Toward medical xai. *IEEE Transactions on Neural Networks and Learning Systems*, **32**(11), 4793–4813.
[64] Ulam, Patrick, Goel, Ashok, Jones, Joshua, and Murdock, William. 2005. Using model-based reflection to guide reinforcement learning. *Reasoning, Representation, and Learning in Computer Games*, 107–112.

[65] Varriano, Giulia, Guerriero, Pasquale, Santone, Antonella, Mercaldo, Francesco, and Brunese, Luca. 2022. Explainability of radiomics through formal methods. *Computer Methods and Programs in Biomedicine*, **220**, 106824.
[66] Vinokurov, Yury, Lebiere, Christian, Wyatte, Dean, Herd, Seth, and O'Reilly, Randall. 2012. Unsupervised learning in hybrid cognitive architectures. Pages 36–41 of: *Workshops at the Twenty-Sixth AAAI Conference on Artificial Intelligence*.
[67] Wang, Danding, Yang, Qian, Abdul, Ashraf, and Lim, Brian Y. 2019. Designing theory-driven user-centric explainable AI. Pages 1–15 of: *Proceedings of the 2019 CHI Conference on Human Factors in Computing Systems*.
[68] Wei, Hua, Chen, Jingxiao, Ji, XIyang, et al. 2022. Honor of kings arena: an environment for generalization in competitive reinforcement learning. Pages 11881–11892 of: *Advances in Neural Information Processing Systems*, 35.
[69] Wu, Xian, Guo, Wenbo, Wei, Hua, and Xing, Xinyu. 2021. Adversarial policy training against deep reinforcement learning. Pages 1883–1900 of: *30th USENIX Security Symposium (USENIX Security'21)*.
[70] Xi, Bowen, Scaria, Kevin, and Shakarian, Paulo. 2023. Rule-based error detection and correction to operationalize movement trajectory classification. *arXiv preprint arXiv:2308.14250*.
[71] Xu, Jingyi, Zhang, Zilu, Friedman, Tal, Liang, Yitao, and den Broeck, Guy Van. 2018. A semantic loss function for deep learning with symbolic knowledge. *Proceedings of Machine Learning Research*, **80**, 5502–5511.
[72] Xu, Ziwei, Jain, Sanjay, and Kankanhalli, Mohan. 2024. Hallucination is inevitable: An innate limitation of large language models. *arXiv preprint arXiv:2401.11817*.

PART II

Taxonomy of Metacognitive Approaches

PART II

Taxonomy of Microservices Approaches

2
An Architectural Approach to Metacognition

CHRISTIAN LEBIERE, ROBERT THOMSON, ANDREA STOCCO, MARK ORR,
DONALD MORRISON

2.1 Introduction

2.1.1 Classes of Metacognitive Processes

Confusingly, researchers have used the term "metacognition" to indicate a number of heterogeneous mental phenomena. At the highest possible level, the label encompasses any kind of knowledge and mental processes that have one's own cognition as its object [16].

This definition includes low-level evaluations of one's own memory (feeling or knowing, familiarity, and metamemory judgments) and performance (feeling of confidence in one's own decisions) as well as complex, deliberate reflections about one's own cognition (such as assessing whether you would be able to pass a test), scaling up to complex forms of reasoning (e.g., solving an induction puzzle, such as the "three wise men," which requires considering different scenarios in which the thinker might have different amounts of knowledge).

2.1.2 Automatic vs. Deliberate Forms of Metacognition

To impose some order onto these phenomena, several authors have proposed their own taxonomies of metacognitive processes [16, 44, 65]. For convenience, in this chapter we will draw a line between metacognitive processes that are *deliberate* and metacognitive processes that are *automatic*. In the terms popularized by Daniel Kahneman, deliberate metacognitive processes are examples of "System 2" processing, while automatic metacognitive phenomena are examples of "System 1" processing. Within the metacognition literature, this taxonomy is perhaps closest to Flavell's distinction between "metacognitive experience" and "metacognitive knowledge"; however, as the remainder of the chapter will make it clear, the difference between automatic and deliberate metacognition is more directly related to a computational approach.

Most of the deliberate forms of metacognition can be thought of as a form of reasoning: When agents are engaged in this form of metacognition, they are effectively "thinking about thinking." They might, for example, search their memory to assess whether they possess some particular knowledge or skill, or whether they can explain how they performed a task. In a subset of paradigms that are often singled out as quintessentially metacognitive, the nature of reasoning is *recursive*: When an agent is engaged in a meta-cognitive task, they are internally simulating what they would do in a specific situation. A specific variant of these tasks involves reasoning not only about one's own mind but about other minds as well. This is the case of collaborative or competitive tasks such as the prisoner's dilemma, in which an agent might engage in thinking about what they would do in response to another agent's actions. This particularly elaborate form of reasoning, which involves taking both one's perspective and that of a different agent, is also known as *theory of mind* [34, 59].

The types of metacognitive processes that are automatic, on the other hand, do not involve any form of reasoning. They involve, instead, the perception of specific signals that mark the status of cognitive processes, and which are likely generated spontaneously and automatically.

One such example is the phenomenon known as feeling of knowing [29]. When asked a question, individuals can often respond whether they know the answer faster than they can provide the answer itself. An extreme example of this feeling of knowing is the frustration associated with the "tip of the tongue" effect [8], i.e., the familiar feeling of almost remembering a particular fact (e.g., the name of an actor in a movie) without being actually able to retrieve it. A second example is the feeling of response conflict, or the feeling of mental impasse that occurs when different responses are competing for execution, or when a prepotent one needs to be suppressed. Note that these forms of metacognition are automatic in the sense that they are signals generated without the agent's intention. Their use might still be deliberate, and their perception might spurn higher-level forms of thinking.

One striking distinction between these two classes of metacognitive phenomena is their different speed: deliberate metacognitive processes are slow, while automatic ones are fast. Consider, for example, the case of metacognitive assessment involved with asking someone whether they know the answer to a fact. This typically involves the deliberate scanning of one's own memory and some time to respond. By contrast, the feeling of knowing the effect, which functionally contains the same information (that is, whether a memory is present) is fast and automatic, to the point that individuals can correctly state they know an answer before they have actually retrieved it [64].

Another relevant distinction is that automatic processes have well known, idiosyncratic, and localized neural signatures [49, 75]. By contrast, virtually all the types of reasoning discussed before share similar neural signatures, typically involving the "multi domain system" – a network of interconnected brain regions that is involved in virtually all forms of difficult mental processing, and includes regions involved in attention and working memory [13].

2.1.3 Implications for Cognitive Architectures

From a cognitive architecture perspective, the distinction between automatic and deliberate processes is particularly important. In cognitive architectures, the most automatic mechanisms (for example, those that control the execution of procedural knowledge or the access to declarative memory) can be considered as architectural primitives and fundamental features of the system, while the least automatic and more deliberate processes can usually be simulated using the architecture's built-in primitives [69]. Thus, from an architectural perspective, there is no reason to believe that any deliberate, System 2-like form of metacognition is, in itself, any different from any other form of reasoning. But the level of automaticity that is characteristic of System 1-level processes, together with their specific neuronal signals, suggests that these type of metacognitive signals are themselves part of the fundamental functions of the architecture.

Of course, the distinction we have outlined is not clear-cut. Automaticity *per se* does not require the existence of any primitive; reading, for example, is a highly automated capacity (it is almost impossible *not* to read a word in front of us, which is the reason why the Stroop effect exists: MacLeod [40]) and yet the cognitive system is not born with a "reading" module – it takes years of practice to learn how to read, and the process results in substantial rewiring of large portions of the brain.

Conversely, it might be argued that at least some forms of deliberate, System-2-level metacognitive phenomena do indeed count as architectural primitives. Consider, for example, the set of reasoning tasks that are described as theory of mind. A few paragraphs above, we presented them as some of the higher and most sophisticated forms of reasoning; in fact, younger children and animals consistently fail at them. And yet, many researchers have pointed out that theory of mind tasks might be rooted, at the neural level, in the existence of mirror neurons. Mirror neurons in the primate brain have the unique property of firing both when an animal is executing a specific movement and when the animal sees the same movement executed by a different agent [55]. Because of their unique property, they have been speculated as the foundational mechanisms by which primates, and perhaps other animals, understand other agents' intentions and take other agents' perspective [23]. Mirror neurons are definitely a basic, "architectural" property of the human brain; thus, if they are necessary for TOM tasks, one must conclude even higher-level reasoning tasks might ultimately be rooted in some fundamental metacognitive mechanism in the human brain.

2.1.4 Metacognitive Measures

The author Alfred North Whitehead summarized metacognitive awareness best: "[t]he purpose of thinking is to let the ideas die instead of us."[1] Metacognition provides an imperfect set of internal feedback and calibration signals that help provide robustness to decision-making in the face of resource-limited cognitive capabilities in complex, open-ended environments. Two factors greatly influence this robustness: (1) the relative confidence in our (or others') capability to successfully complete a task and (2) an internal estimate of the effort required to complete a task. Perhaps the most well-studied measure of metacognition is the confidence judgment. For the purpose of this chapter, we will not describe potential differences between confidence and certainty (see [52]). A common feature of confidence judgments is that there is an initial tendency for novice participants to exhibit overconfidence when learning a new task ([60]; also see [62, 70]), followed by a period of underconfidence with practice when the task is challenging [27, 33], and continued overconfidence in relatively easier tasks [58]. Furthermore, when participants know that they must generate a confidence score, their response times tend to slow and their responses tend to get more accurate [60]. This means that explicitly generating a confidence judgment actually alters the decision-making process.

What makes confidence so interesting is that it can be dissociated from accuracy. Confidence leak occurs when the confidence judgment from recent trials intrudes on current confidence judgments, whereas the same is not necessarily the case for accuracy. Similarly, cue salience can differentially impact confidence and accuracy in the same task [66]. In perceptual tasks, a positive evidence bias occurs when confidence over-weights the objective evidence whereas task accuracy is unimpacted (effectively a confirmation bias which occurs for confidence judgments as opposed to accuracy).

A second measure of metacognition is the ability to estimate the effort required to complete a task. Although similar to confidence, this ability is not perfect. This prediction influences how much effort we put toward studying for a test [17, 32] and our relative attention allocation in perception [57]. Similar to

---

[1] Astute readers using their metacognitive skills will recognize that this is a frequent, but apocryphal, attribution to Whitehead, and is probably a paraphrase of a passage in [51].

confidence, novice participants tend to overestimate their own capabilities and underestimate time demands when assessing the effort required to complete a novel task. Recent evidence shows that ready access to the internet may exacerbate this misestimation as the availability of information online provides a false sense of knowing [2, 14].

2.1.5 Functional Benefits

What sets metacognitive capabilities apart is not only that we are able to identify our own (or other's) internal states, but that we are able to evaluate and reason over these states [17, 50]. This provides a robustness that is unique to human decision-making (compared with artificial intelligence [AI]-based methods). One's internal awareness of confidence/certainty works with an estimate of effort to integrate with simulation and experience to form a prediction error feedback signal to drive (often unsupervised) learning [12, 53, 74]. Specifically, metacognition mediates the relationship between executive functioning and self-regulated learning ([18]; for a review see [32]). The major benefits of this is that we are able to allocate resources based on specific goals, and failures of this signal (i.e., mis-calibration) provides motivation to adapt by increasing skill, increasing effort, and/or seeking external resources (e.g., social interactions). It also provides for an estimate of whether the cost of the goal is worthwhile ("is the juice worth the squeeze?").

What specific benefits does this provide? First off, metacognitive signals provide the motivation to change social attitudes, update beliefs, and enhance strategic learning outcomes to match our current goals [7, 50, 67]. Specifically, it provides an estimate to balance effort with reward to maximize a goal, which can also trigger when to learn (e.g., how hard to study; [9]). With resource constraints, it provides the drive to satisfice (see [61] for a discussion). Ackerman [1] showed that our tendency to underestimate the time requirements when studying complex topics is due not only to estimating confidence of how close one was toward achieving a predetermined aptitude (i.e., a judgment of learning), but also a top-down assessment of how much time it should have taken with reduced motivation and premature completion of learning. While possibly seen as a negative, this actually avoids the sunk-cost fallacy when one's judgment of required cognitive effort is miscalibrated.

Another benefit is that we can use metacognitive estimates to distinguish between things that we do not know and things that we have forgotten [24, 26]. If a given event should have been memorable and was not recalled, then it was likely something that one had never known. Conversely, if a given event should not have been memorable and was not recalled, it was more likely to be judged to be forgotten.

While the current section has focused on several functional benefits that metacognitive awareness provides, limits of our ability to use metacognitive signals can sometimes lead us to validate negative perceptions and use bias-prone heuristics [43]. Metacognitive signals provide motivation to change attitudes/beliefs and can enhance learning, but can just as easily cause us to harden our viewpoints when biases get in the way. Blindness to our biases and lack of skills (e.g., the Dunning–Kruger effect; [15]) leads to poor calibration, and poor performance on self-regulation.

2.2 Metacognition in Cognitive Architectures

A major challenge in defining metacognition is to distinguish it from "regular" cognition. Metacognition has been defined as "the ability to monitor and adaptively control one's cognitive processing or thinking about thinking" [48]. However, human cognition has many ways of monitoring and adapting its behavior in various contexts, many of which are viewed as part of everyday behavior. What is needed to properly define metacognition is therefore a reference baseline for the structures and mechanisms that constitute "regular" cognition. Cognitive architectures provide a natural integrative framework for that definition.

Following his insight that a divide-and-conquer approach to modeling human cognition could not provide a road map to its ultimate goal [45], Newell proposed the concept of unified theories of cognition, implemented computationally as cognitive architectures [46]. The last five decades has seen a number of diverse proposals to explore the luxuriant space of cognitive architectures [30]. We will focus here on a particular architecture, ACT-R, specifically aimed at modeling human cognition from a neuro-psychological perspective [5].

The ACT-R cognitive architecture (Figure 2.1) is composed of a set of modules with dedicated functionality, localized in specific brain areas [4]. The central module, procedural memory, controls the flow of information between other modules using condition-action pairs, aka production rules. The interface between modules consists of a set of buffers, collectively known as working memory. Each buffer is attached to a particular module and can hold one piece of information, known as a chunk, at a time. When a production rule is selected, it can change the content of one or more buffers, triggering corresponding processes in the associated module. Example actions include shifting attention or recognizing an object in the visual module, or retrieving a piece of information

An Architectural Approach to Metacognition

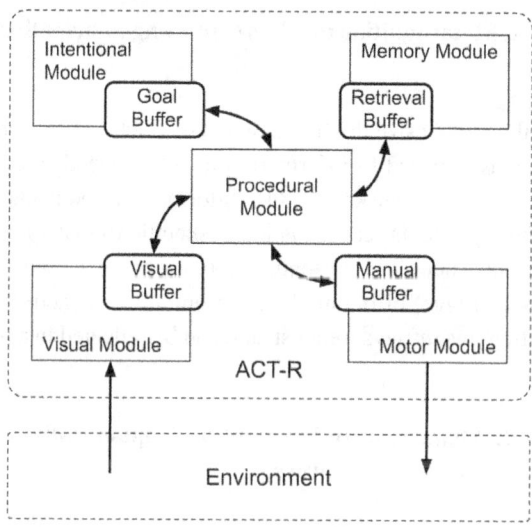

Figure 2.1 ACT-R cognitive architecture.

from the long-term declarative memory module. Those processes typically operate on a combination of symbolic knowledge structures, such as the chunks in memory that are represented as sets of attribute-value pairs, with attached subsymbolic quantities that are learned statistically to reflect the structure of the environment and control access to information. The results of those processes ultimately become available in the corresponding buffer. The procedural module can then detect the asynchronous change and attempt to match the condition part of its production rules against the new state of working memory, and the cognitive cycle repeats.

A common assumption is that awareness is associated with information present in the buffers while the internal contents of the modules are not explicitly accessible. For instance, while we can retrieve information from long-term memory, that process is probabilistic and approximate and we cannot directly know the contents of our memory. In particular, we do not have explicit access to the subsymbolic quantities (e.g., the activation of memory chunks that determine their probability and latency of retrieval) that modulate cognitive processes. However, that absolute encapsulation of the contents of cognitive modules is difficult to reconcile with the limited degree of awareness of our own cognitive processes described in the previous section. Thus, we propose the following conjectures defining the nature of metacognition in the context of cognitive architectures:

Conjecture 1: Metacognition involves extracting information about module processes

It is essential to emphasize the fundamental distinction between information about processes as opposed to information about knowledge, even though of course processes operate on knowledge. Information about knowledge, e.g., the activation of a particular chunk, is local, specific to that specific item, and encapsulated in the module. Information about a process reflects the entirety (or at least a wide range) of the module content, e.g., the activation of all the chunks competing in a retrieval request, and can be reflected in a metacognitive signal.

Conjecture 2: Metacognitive information is quantitative and approximate rather than symbolic in nature

Metacognitive information about processes is not symbolic but rather expresses graded quantities that provide functional insight into those processes such as probability of success, degree of competition, confidence in the answer, and salience of various relevant factors. Despite the quantitative nature of the information, it is not subsymbolic but rather cognitively accessible and actionable.

Conjecture 3: Metacognitive information is available in working memory for cognitive processing

Even though metacognition is a distinct aspect of human cognition, it continues the architectural pattern of making information extracted from cognitive modules available in working memory for further processing. One possibility is to extend the current distinct distinction between content buffers that hold the result of module processes and state buffers that currently hold the state of the module, i.e., whether the module is currently free or busy (to prevent requesting another operation when a module is currently busy). Note that these conjectures are about the monitoring role of metacognition. No additional assumption is made about any additional metacognitive processes that would act on the result of that monitoring:

Conjecture 4: General cognitive processes are sufficient to respond to a situation detected by metacognitive monitoring

Once metacognitive signals are made available into working memory, the information they convey can be processed using the standard mechanisms that an architecture can employ to process any other information. In particular, all the forms of deliberate metacognition (including inference about one's knowledge and theory of mind) can be implemented using the standard tools that allow any cognitive architecture to reason and problem-solve.

2.3 Computational Cognitive Models of Metacognition

We describe here some examples of the kind of metacognitive ability conjectured in the previous section. While those models were developed in the context of the ACT-R cognitive architecture, they were prototypes that were not integrated in the specific way conjectured above. Rather, these examples are intended to illustrate the kind of metacognitive signals that can be extracted within the cognitive architecture framework and the functional benefits that they could confer.

2.3.1 Visual Perception

One of the key functions of visual perception is object categorization. The ACT-R visual module operates in a way reflecting the principles of human vision: it directs attention through a request in the visual-location buffer then requests the object at that location in the visual field to be encoded with the result to be made available in the visual buffer. Object recognition in the ACT-R visual module is fairly primitive and only operates on symbolic representations typical of a computer screen used in cognitive psychology experiments. While in that context object recognition can operate according to a well-defined standard, the same is not true in the real world, where interpreting the content of visual scenes is rife with uncertainty. In that context, obtaining estimates of the quality of the output of the recognition process, such as probability of the accuracy of the judgment, would be highly desirable. However, those estimates would be based on an underlying model whose (in)correctness would likely be correlated with that of the response itself. Instead, indirect measures of correctness such as confidence might be more realistic.

To explore those possibilities, we combined the ACT-R cognitive architecture with a neurally plausible vision model reflecting the structure of the human visual cortex [28]. That model, based on the Leabra neural architecture, is a massively parallel neural network similar to convolutional neural networks but using a neurally plausible algorithm combining top-down error correction with bottom up Hebbian self-organization [47]. The original approach to this integration [72] is to extract global activation quantities from the neural network, including the average activation of the last layer featuring a distributed representation (called the IT layer by analogy to its equivalent in the human visual system) and the maximum activation of the final winner-take-all layer. Those metacognitive quantities are used to build a classifier to detect low-confidence categorizations. When categorization confidence falls below a threshold, a new view of the object is taken and a new cycle of categorization is initiated. The iterative process continues until the confidence threshold is reached and a final categorization is accepted, or a cycle limit is exceeded and the system gives up. The second approach [73] is similar to the first but, instead of extracting scalar metacognitive signals of limited resolution, it introspects into the IT layer to extract the most abstract distributed representation of the object used for categorization. That representation is then associated with the object identification in the declarative memory of the cognitive architecture. The cognitive model can then use memory retrieval processes to determine if the representation is similar enough to other objects of the same type. If that similarity judgment falls below a threshold, then it is determined that the object belongs to a new category. The cognitive model then provides top down input to the visual model to create a new object category and start training it with the current object. This combined system proves capable of accurately learning entirely new object categories without any external supervisory feedback.

Two aspects of this approach to visual metacognition are worth emphasizing. The first is that a hybrid symbolic-neural system provides a natural implementation for the kind of perspective that we are proposing, with the metacognitive monitoring signal extracted from the internal dynamics of the neural subsystem and made available to the symbolic cognitive system. The second noteworthy aspect is that the intervention following the generation of the metacognitive signal detecting an exception condition (a failure to properly recognize an object due an insufficient view or lack of training on that specific category, respectively) does not require additional metacognitive capabilities. Instead, it uses standard cognitive mechanisms such as declarative memory representation and pattern matching to implement strategies designed to remediate the situation (taking another perspective, or learning a new object category, respectively).

2.3.2 Declarative Memory

Declarative memory is the long-term repository of experience and knowledge in the cognitive architecture. ACT-R does not draw a distinction between episodic memory, holding first-person experiences, and semantic memory, holding more abstract forms of knowledge. But, like other modules, it shares the bottleneck of that potentially vast amount of information being accessed through a limited-capacity buffer that can only hold a single chunk at a time. One can think of it as a form of focusing similar to visual attention, where a pattern is placed in the retrieval buffer requesting the most relevant chunk to be retrieved and deposited back in the same buffer. Usually, the pattern requesting the retrieval is underspecified, leaving many potential chunks in declarative memory eligible to be retrieved. As mentioned previously, the eligible chunk with the highest activation is retrieved, reflecting several statistical factors such as frequency and recency of access, associative priming from the current working memory context, and degree of match to the requested pattern. The latency of declarative memory retrieval is typically on the order of hundreds of milliseconds, meaning that only a few chunks of long-term information can be retrieved per second. Providing relevant information given these constraints is an incredibly difficult problem to solve for the memory system. Flawless performance is far from assured, which makes the metacognitive ability to introspect into declarative memory retrieval processes highly desirable in order to inform cognitive strategies designed to improve the performance and robustness of the system. For instance, in the case of the feeling of knowing, a measure of the probability or closeness of success in the event of retrieval failure could trigger future attempts at retrieval, perhaps combined with a strategy such as priming from related information that could improve the probability of a successful retrieval.

We illustrate the kind of metacognitive signals that can be extracted from declarative memory retrieval processes and the use that can be made of them using an example of decision-making applied to cyber security [39]. The cognitive model follows an approach called instance-based learning (IBL; [25]) which makes decisions by generalizing from instances of experiences held in declarative memory using memory retrieval processes, in particular, a mechanism called blending [38] that aggregates multiple chunks to produce a probabilistic expected outcome. The task is a simulated cyber security experiment involving an insider attack on a number of potential targets with distinct characteristics. After a target is chosen, a deceptive signal is given to try to convince the intruder not to attack. The original deceptive signal generated using a Stackelberg game theory paradigm was found to be suboptimal against actual human subjects. Instead a personalized cognitive model aligned against a specific attacker's behavior trace is shown to be a better predictor of individual human behavior. The model is then used to optimize the deceptive signal to a specific attacker by extracting from the model a metacognitive signal of the strength of belief (i.e., trust) in the signal. That level of strength is computed from the activation of the relevant beliefs in memory, specifically past instances of success and failure in attacking in the presence or absence of a signal, which reflects the frequency and recency of past experiences. That metacognitive quantity does not provide direct access to the activation levels of specific memories, which would be cognitively implausible, but rather the relative strength of one set of experiences against another.

A second metacognitive signal called cognitive salience is extracted from the model to quantify the relative reliance of the various target characteristics in the selection process. It is computed as the derivative of the output of the blending process used to generate expected reward for attacking each target with respect to the various features defining those targets. Again, it does not provide access to any specific subsymbolic quantity in memory but rather reflects the overall state of memories relevant to the retrieval process. This concept of cognitive salience was originally defined for the purposes of explainable AI to explain to a human user the underlying basis of decisions made by an intelligent agent [42]. In this case, the cognitive salience values can be used to shift coverage toward targets whose characteristics are more salient to a specific user. As for the perceptual modules described in the previous section, the interventions driven by signals such as trust or salience extracted from memory retrieval processes are not metacognitive in nature but rather can be enacted by the same kind of cognitive decision strategies that are the object of the introspection.

2.4 Discussion

In this chapter, we have argued that metacognitive phenomena can be divided into automatic and deliberate, and that, from a cognitive architecture point of view, the first ones can be conceived as fundamental architectural primitives, while the second ones can be achieved using the architecture's standard processes, once signals of the first type are detected. Although the conjectures and the theory of metacognition exposed in this chapter were formulated in reference to cognitive architectures, they can be extended at lower and higher levels of analysis.

At the lower level, our conjectures are broadly compatible with the principles of predictive coding in neural systems [21, 54], that is, the idea that the brain is organized to maximize homeostasis (or, equivalently, to minimize its "free energy": Friston [20]) by minimizing surprisal and maximizing the predictability of the surrounding environment. Violations of expectations can be used by an agent to modify its behavior – for instance, by changing its own decision policy or moving to a new environment [22].

Within this framework, it is possible to conceive of the specific metacognitive signals that we have reviewed as violations of expectations about how well the cognitive system itself is working. For example, the feeling of knowing would be a violation of expectations about retrieving a memory; the "Aha!" experience in problem solving would correspond to a sudden change in the expectation of solving a problem; and the sense of confidence or uncertainty about a decision would corresponding to a sudden change in the perceived effectiveness of a decision.

One obstacle to reconciling this hypothesis with a cognitive architecture viewpoint is that predictive coding, as its name implies, requires the cognitive system to be continually making predictions, and cognitive architectures are not explicitly built upon this principle. That does not mean, however, that cognitive architectures do not make predictions: in fact, many aspects of a cognitive architecture can be seen as implicit predictions about future states of the world. In ACT-R, for example, each declarative memory has an associated scalar variable called activation, which represents the log odds of this memory being needed at that particular moment in time [3]. The distribution of activations across all memories, therefore, represents an implicit prediction about what should be expected in the environment [6]. Similarly, each procedural memory has an associated scalar quantity, its utility, that is computed through reinforcement learning and computes the expected future reward of the corresponding skill. Again, such a term implicitly defines a prediction, and is temporally adjusted by reducing the mismatch between predicted and actual rewards [63]. Thus, it is possible to relate our architectural view of metacognition to the larger framework of predictive coding and active inference.

At a higher level, a wide range of capabilities have been integrated into cognitive architectures over the last 50 years of exploration of the design space [30]. A number of architectures incorporate metacognitive features such as monitoring of internal resources and extracting confidence values, e.g., CLARION [68], Companions [19], and Soar [35]. Many architectures also include other features that are often associated with metacognition but that we do not consider here to be inherently metacognitive, such as temporal representation of alternative solutions, changing task priorities, storing and using traces of execution, improving analogy or problem solving, and general aspects of theory of mind.

A complementary direction of research is to apply the insights into metacognition achieved in the context of cognitive architectures to AI systems that share many commonalities despite apparent differences. As previously mentioned, cognitive salience is a technique based on blended retrievals which can be used to extract which features contribute to a given decision. When applied to AI algorithms via model tracing (having a cognitive model make the same decisions as an AI, or human, agent), it is possible to investigate the relative contribution of each feature to the given decision, in a manner somewhat analogous to SHAP values [41]. A preliminary investigation of using cognitive salience in the context of models of intrusion detection in a network defense application has shown potential as a metacognitive signal to understand feature importance [71].

A recent convergence in the broad diversity of cognitive architectures has recently prompted an attempt at formalizing an emerging consensus called the Common Model of Cognition (CMC; [36]). A working group organized in the context of that effort summarized various aspects of metacognition [31] but, because of the relative lack of maturity and absence of consensus in approaches to metacognition, it has not been included in current proposals. However, recent efforts to elaborate and integrate a theory of emotions with the CMC have led to a proposal for a treatment of metacognition focused on appraisal theory that is generally compatible with the approach of the current chapter [37, 56] and builds upon some earlier research integrating affective components to cognitive models via modeling the physiological substrate of cognition [10, 11]. Integrating physiological modeling into cognitive models provides a mechanism to model affective features such as emotion in addition to factors such as fatigue and stress. These enterprises further reinforce the belief that metacognition is a highly active area of research that can be integrated within the existing framework of cognitive architectures and is likely to bear much fruit in coming years.

2.5 Acknowledgments

This research was sponsored by the Department of Defense Basic Research Office award HQ00342110002, the Army Research Office MURI grant Number W911NF-17-1-0370, and the Office of Naval Research MURI Award N0001422MP00465. The views expressed in this work are those of the authors and do not necessarily reflect the official policy or position of the United States Military Academy, Department of the Army, Office of Naval Research, Department of Defense, or U.S. Government.

References

[1] Ackerman, Rakefet. 2014. The diminishing criterion model for metacognitive regulation of time investment. *Journal of Experimental Psychology: General*, **143**(3), 1349.

[2] Ackerman, Rakefet, and Goldsmith, Morris. 2011. Metacognitive regulation of text learning: On screen versus on paper. *Journal of Experimental Psychology: Applied*, **17**(1), 18.

[3] Anderson, John R. 1990. *The Adaptive Character of Thought*. Lawrence Erlbaum Associates.
[4] Anderson, John R. 2007. *How Can the Human Mind Occur in the Physical Universe?* Oxford University Press.
[5] Anderson, John R, and Lebiere, Christian. 1998. *The Atomic Components of Thought*. Lawrence Erlbaum Associates.
[6] Anderson, John R, and Schooler, Lael J. 1991. Reflections of the environment in memory. *Psychological Science*, **2**(6), 396–408.
[7] Biggs, John. 1988. The role of metacognition in enhancing learning. *Australian Journal of Education*, **32**(2), 127–138.
[8] Brown, Alan S. 1991. A review of the tip-of-the-tongue experience. *Psychological Bulletin*, **109**(2), 204.
[9] Colombo, Barbara, Iannello, Paola, and Antonietti, Alessandro. 2010. Metacognitive knowledge of decision-making: Pages 445–472 of: Efklides, Anastasia and Misailidi, Plousia (eds.), *Trends and Prospects in Metacognition Research*. Springer. https://doi.org/10.1007/978-1-4419-6546-2_20.
[10] Dancy, Christopher L. 2013. ACT-RΦ: A cognitive architecture with physiology and affect. *Biologically Inspired Cognitive Architectures*, **6**, 40–45.
[11] Dancy, Christopher L, Ritter, Frank E, Berry, Keith A, and Klein, Laura C. 2015. Using a cognitive architecture with a physiological substrate to represent effects of a psychological stressor on cognition. *Computational and Mathematical Organization Theory*, **21**, 90–114.
[12] Desender, Kobe, Boldt, Annika, and Yeung, Nick. 2018. Subjective confidence predicts information seeking in decision making. *Psychological Science*, **29**(5), 761–778.
[13] Duncan, John. 2010. The multiple-demand (MD) system of the primate brain: Mental programs for intelligent behaviour. *Trends in Cognitive Sciences*, **14**(4), 172–179.
[14] Dunn, Timothy L, Gaspar, Connor, McLean, Daev, Koehler, Derek J, and Risko, Evan F. 2021. Distributed metacognition: Increased bias and deficits in metacognitive sensitivity when retrieving information from the internet. *Technology, Mind, and Behavior*, 2(3).
[15] Dunning, David. 2011. The Dunning–Kruger effect: On being ignorant of one's own ignorance. Pages 247–296 of: Olson, James M, and Zanna, Mark P (eds.), *Advances in Experimental Social Psychology*, vol. 44. Academic Press.
[16] Flavell, John H. 1979. Metacognition and cognitive monitoring: A new area of cognitive–developmental inquiry. *American Psychologist*, **34**(10), 906.
[17] Fleming, Stephen M. 2024. Metacognition and confidence: A review and synthesis. *Annual Review of Psychology*, **75**, 241–268.
[18] Follmer, D Jake, and Sperling, Rayne A. 2016. The mediating role of metacognition in the relationship between executive function and self-regulated learning. *British Journal of Educational Psychology*, **86**(4), 559–575.
[19] Forbus, Kenneth D, and Hinrichs, Thomas R. 2006. Companion cognitive systems: A step toward human-level AI. *AI Magazine*, **27**(2), 83–83.
[20] Friston, Karl. 2010. The free-energy principle: A unified brain theory? *Nature Reviews Neuroscience*, **11**(2), 127–138.

[21] Friston, Karl, and Kiebel, Stefan. 2009. Predictive coding under the free-energy principle. *Philosophical Transactions of the Royal Society B: Biological Sciences*, **364**(1521), 1211–1221.

[22] Friston, Karl, FitzGerald, Thomas, Rigoli, Francesco, et al. 2016. Active inference and learning. *Neuroscience & Biobehavioral Reviews*, **68**, 862–879.

[23] Gallese, Vittorio, and Goldman, Alvin. 1998. Mirror neurons and the simulation theory of mind-reading. *Trends in Cognitive Sciences*, **2**(12), 493–501.

[24] Ghetti, Simona. 2003. Memory for nonoccurrences: The role of metacognition. *Journal of Memory and Language*, **48**(4), 722–739.

[25] Gonzalez, Cleotilde, Lerch, Javier F, and Lebiere, Christian. 2003. Instance-based learning in dynamic decision making. *Cognitive Science*, **27**, 591–635.

[26] Grimaldi, Piercesare, Lau, Hakwan, and Basso, Michele A. 2015. There are things that we know that we know, and there are things that we do not know we do not know: Confidence in decision-making. *Neuroscience & Biobehavioral Reviews*, **55**, 88–97.

[27] Hanczakowski, Maciej, Zawadzka, Katarzyna, Pasek, Tomasz, and Higham, Philip A. 2013. Calibration of metacognitive judgments: Insights from the underconfidence-with-practice effect. *Journal of Memory and Language*, **69**(3), 429–444.

[28] Jilk, David J, Lebiere, Christian, O'Reilly, Randall C, and Anderson, John R. 2008. SAL: An explicitly pluralistic cognitive architecture. *Journal of Experimental and Theoretical Artificial Intelligence*, **20**(3), 197–218.

[29] Koriat, Asher. 2000. The feeling of knowing: Some metatheoretical implications for consciousness and control. *Consciousness and Cognition*, **9**(2), 149–171.

[30] Kotseruba, Iuliia, and Tsotsos, John K. 2020. 40 years of cognitive architectures: Core cognitive abilities and practical applications. *Artificial Intelligence Review*, **53**(1), 17–94.

[31] Kralik, Jerald D, Lee, Jee Hang, Rosenbloom, Paul S, et al. 2018. Metacognition for a common model of cognition. *Procedia Computer Science*, **145**, 730–739.

[32] Krieger, Florian, Azevedo, Roger, Graesser, Arthur C, and Greiff, Samuel. 2022. Introduction to the special issue: The role of metacognition in complex skills-spotlights on problem solving, collaboration, and self-regulated learning. *Metacognition and Learning*, **17**(3), 683–690.

[33] Kubik, Veit, Jemstedt, Andreas, Eshratabadi, Hassan Mahjub, Schwartz, Bennett L, and Jönsson, Fredrik U. 2022. The underconfidence-with-practice effect in action memory: The contribution of retrieval practice to metacognitive monitoring. *Metacognition and Learning*, **17**(2), 375–398.

[34] Kuhn, Deanna. 2000. Theory of mind, metacognition, and reasoning: A life-span perspective. Pages 301–326 of: Mitchell, Peter and Riggs, Kevin J (eds), *Children's Reasoning and the Mind*. East Sussex Psychology Press.

[35] Laird, John E. 2019. *The Soar Cognitive Architecture*. MIT Press.

[36] Laird, John E, Lebiere, Christian, and Rosenbloom, Paul S. 2017. A standard model of the mind: Toward a common computational framework across artificial intelligence, cognitive science, neuroscience, and robotics. *AI Magazine*, **38**(4), 13–26.

[37] Larue, Othalia, West, Robert, Rosenbloom, Paul S, et al. 2018. Emotion in the common model of cognition. *Procedia Computer Science*, **145**, 740–746.

[38] Lebiere, Christian. 1999. The dynamics of cognition: An ACT-R model of cognitive arithmetic. *Kognitionswissenschaft*, **8**, 5–19.
[39] Lebiere, Christian, Cranford, Edward, Aggarwal, Palvi, Cooney, Sarah, Tambe, Milind, and Gonzalez, Cleotilde. 2023. Cognitive Modeling for personalized, adaptive signaling for cyber deception. Pages 59–82 of: Bao, Tiffany, Tambe, Milind, Wang, Cliff (eds), *Cyber Deception*. Advances in information security, vol. 89. Springer International Publishing.
[40] MacLeod, Colin M. 1991. Half a century of research on the Stroop effect: An integrative review. *Psychological Bulletin*, **109**(2), 163.
[41] Marcílio, Wilson E, and Eler, Danilo M. 2020. From explanations to feature selection: Assessing SHAP values as feature selection mechanism. Pages 340–347 of: *2020 33rd SIBGRAPI Conference on Graphics, Patterns and Images (SIBGRAPI)*. IEEE.
[42] Mitsopoulos, Konstantinos, Somers, Sterling, Schooler, Joel, Lebiere, Christian, Pirolli, Peter, and Thomson, Robert. 2021. Toward a psychology of deep reinforcement learning agents using a cognitive architecture. *Topics in Cognitive Science*, **14**(4), 756–779.
[43] Moreno, Lorena, Briñol, Pablo, and Petty, Richard E. 2022. Metacognitive confidence can increase but also decrease performance in academic settings. *Metacognition and Learning*, **17**(1), 139–165.
[44] Nelson, Thomas O, and Narens, L. 1990. Metamemory: A theoretical framework and new findings. Pages 125–173 of: *Psychology of Learning and Motivation*, vol. 26. Elsevier.
[45] Newell, Allen. 1973. You can't play 20 questions with nature and win: Projective comments on the papers of this symposium. Page 283–308 of: *Visual Information Processing*. Academic Press.
[46] Newell, Allen. 1990. *Unified Theories of Cognition*. Harvard University Press.
[47] O'Reilly, Randall C, and Munakata, Yuko. 2000. *Computational Explorations in Cognitive Neuroscience*. MIT Press.
[48] Parrish, AE, and Brosnan, SF. 2012. Primate cognition. Pages 174–180 of: Ramachandran, VS (ed.), *Encyclopedia of Human Behavior (Second Edition)*. Academic Press.
[49] Paynter, Christopher A, Reder, Lynne M, and Kieffaber, Paul D. 2009. Knowing we know before we know: ERP correlates of initial feeling-of-knowing. *Neuropsychologia*, **47**(3), 796–803.
[50] Petty, Richard E, Briñol, Pablo, Tormala, Zakary L, and Wegener, Duane T. 2007. The role of metacognition in social judgment. *Social Psychology: Handbook of Basic Principles*, **2**, 254–284.
[51] Popper, Karl R. 1968. Epistemology without a knowing subject. Pages 333–373 of: Van Rootselaar, B, and Staal, J F (eds.), *Studies in Logic and the Foundations of Mathematics*, vol. 52. Elsevier.
[52] Pouget, Alexandre, Drugowitsch, Jan, and Kepecs, Adam. 2016. Confidence and certainty: Distinct probabilistic quantities for different goals. *Nature Neuroscience*, **19**(3), 366–374.
[53] Ptasczynski, Lena Esther, Steinecker, Isa, Sterzer, Philipp, and Guggenmos, Matthias. 2022. The value of confidence: Confidence prediction errors drive

value-based learning in the absence of external feedback. *PLOS Computational Biology*, **18**(10), e1010580.

[54] Rao, Rajesh PN, and Ballard, Dana H. 1999. Predictive coding in the visual cortex: A functional interpretation of some extra-classical receptive-field effects. *Nature Neuroscience*, **2**(1), 79–87.

[55] Rizzolatti, Giacomo, and Craighero, Laila. 2004. The mirror-neuron system. *Annual Review of Neuroscience*, **27**, 169–192.

[56] Rosenbloom, Paul S, Laird, John E, Lebiere, Christian, Stocco, Andrea, Granger, Richard H, and Huyck, Christian. 2024. A proposal for extending the common model of cognition to emotion. In: *Proceedings of the 22nd Annual Meeting of the International Conference on Cognitive Modeling*.

[57] Rummel, Jan, and Meiser, Thorsten. 2013. The role of metacognition in prospective memory: Anticipated task demands influence attention allocation strategies. *Consciousness and Cognition*, **22**(3), 931–943.

[58] Scheck, Petra, and Nelson, Thomas O. 2005. Lack of pervasiveness of the underconfidence-with-practice effect: Boundary conditions and an explanation via anchoring. *Journal of Experimental Psychology: General*, **134**(1), 124.

[59] Schneider, Wolfgang, and Lockl, Kathrin. 2008. Procedural metacognition in children: Evidence for developmental trends. *Handbook of Metamemory and Memory*, **14**, 391–409.

[60] Schoenherr, Jordan R, Leth-Steensen, Craig, and Petrusic, William M. 2010. Selective attention and subjective confidence calibration. *Attention, Perception, & Psychophysics*, **72**(2), 353–368.

[61] Schoenherr, Jordan Richard, and Lacroix, Guy L. 2020. Performance monitoring during categorization with and without prior knowledge: A comparison of confidence calibration indices with the certainty criterion. *Canadian Journal of Experimental Psychology/Revue canadienne de psychologie expérimentale*, **74**(4), 302–315.

[62] Schoenherr, Jordan Richard, and Thomson, Robert. 2021. Persuasive features of scientific explanations: Explanatory schemata of physical and psychosocial phenomena. *Frontiers in Psychology*, **12**, 644809.

[63] Schultz, Wolfram, Dayan, Peter, and Montague, P Read. 1997. A neural substrate of prediction and reward. *Science*, **275**(5306), 1593–1599.

[64] Schunn, Christian D, Reder, Lynne M, Nhouyvanisvong, Adisack, Richards, Daniel R, and Stroffolino, Philip J. 1997. To calculate or not to calculate: A source activation confusion model of problem familiarity's role in strategy selection. *Journal of Experimental Psychology: Learning, Memory, and Cognition*, **23**(1), 3.

[65] Shea, Nicholas, Boldt, Annika, Bang, Dan, Yeung, Nick, Heyes, Cecilia, and Frith, Chris D. 2014. Supra-personal cognitive control and metacognition. *Trends in Cognitive Sciences*, **18**(4), 186–193.

[66] Shekhar, Medha, and Rahnev, Dobromir. 2021. Sources of metacognitive inefficiency. *Trends in Cognitive Sciences*, **25**(1), 12–23.

[67] Spada, Marcantonio M, Nikčević, Ana V, Moneta, Giovanni B, and Wells, Adrian. 2008. Metacognition, perceived stress, and negative emotion. *Personality and Individual Differences*, **44**(5), 1172–1181.

[68] Sun, Ron. 2016. *Anatomy of the Mind: Exploring Psychological Mechanisms and Processes with the Clarion Cognitive Architecture.* Oxford University Press.
[69] Taatgen, Niels A. 2017. Cognitive architectures: Innate or learned? In: *AAAI Technical Report FS-17-05.*
[70] Thomson, Robert, and Frangia, William. 2023. Investigating the use of belief-bias to measure acceptance of false information. Pages 149–158 of: *International Conference on Social Computing, Behavioral-Cultural Modeling and Prediction and Behavior Representation in Modeling and Simulation.* Springer.
[71] Thomson, Robert, Cranford, Edward, Somers, Sterling, and Lebiere, Christian. 2024. A novel approach to intrusion detection using a cognitively-inspired algorithm. In: *Proceedings of the 57th Hawaii International Conference on System Sciences*, Pittsburgh, PA, USA, September 20–22, 2023. Springer-Verlag.
[72] Vinokurov, Yury, Lebiere, Christian, Herd, Seth, and O'Reilly, Randall. 2011. A metacognitive classifier using a hybrid ACT-R/Leabra architecture. In: *Workshops at the Twenty-fifth AAAI Conference on Artificial Intelligence.*
[73] Vinokurov, Yury, Lebiere, Christian, Wyatte, Dean, Herd, Seth, and O'Reilly, Randall. 2012. Unsupervised learning in hybrid cognitive architectures. In: *Workshops at the Twenty-Sixth AAAI Conference on Artificial Intelligence, Workshop, Toronto, ON, Canada, July 23, 2012.* AAAI Press.
[74] Yeung, Nick, and Summerfield, Christopher. 2012. Metacognition in human decision-making: Confidence and error monitoring. *Philosophical Transactions of the Royal Society B: Biological Sciences*, **367**(1594), 1310–1321.
[75] Yeung, Nick, Botvinick, Matthew M, and Cohen, Jonathan D. 2004. The neural basis of error detection: Conflict monitoring and the error-related negativity. *Psychological Review*, **111**(4), 931.

3
Metacognitive AI through Error Detection and Correction Rules

BOWEN XI, PAULO SHAKARIAN

Classification of movement trajectories has many applications in transportation. Supervised neural models represent the current state of the art (SOTA). Recent security applications require this task to be rapidly employed in environments that may differ from the data used to train such models for which there is little training data. We provide a neurosymbolic rule-based framework to conduct error correction and detection of these models to support eventual deployment in security applications.

3.1 Introduction

The identification of a mode of travel for a time-stamped sequence of global position system (GPS) known as "movement trajectories" has important applications in travel demand analysis [5], transport planning [10], and analysis of sea vessel movement [4]. The current SOTA has relied on supervised neural models [9]. More recently this problem has been of interest for security applications such as leading to efforts such as the IARPA HAYSTAC program.[1] In this domain, models may be deployed in environments with different geography, transportation infrastructure, and socio-cultural dynamics than in the training data and expected to adapt to such environments with little or no labeled data specific to those circumstances. Further, such deployments may happen rapidly, precluding extensive data engineering or model retraining.

We extend the current supervised neural methods with a lightweight error detection and correction rule (EDCR) framework providing an overall neurosymbolic system. The key intuition is that training and operation data can be used to learn rules that predict errors and correct errors in the supervised model. Once trained, the rules are employed operationally in two phases: first detection rules identify potentially misclassified movement trajectories. A second type of rule to reclassify the trajectories ("correction rules") is then used

---

[1] www.iarpa.gov/research-programs/haystac

to reassign the sample to a new class. Our key contributions are as follows: (1) We present a strong theoretical framework for EDCR rooted in logic and rule mining and formally prove how quantities related to learned rules (e.g., confidence and support) are related to changes in class-level machine learning metrics such as precision and recall. (2) We conduct experiments where rules trained on the same data as the original model can improve machine learning metrics across a variety of settings and model types, including the SOTA LRCN model. Specifically, the employment of EDCRs leads to 1.7% improvement in accuracy over the original LRCN model when data leakage between training and testing is minimized. (3) By excluding 40% of the classes during the training process, we achieve an enhancement of 5.2% (zero-shot) and a substantial 23.9% improvement (few-shot) compared to the SOTA model. This progress is accomplished without necessitating any retraining of the underlying base model. (4) In addition to offering domain knowledge akin to other papers, we furnish a neural network-incorporated condition, characterized by its overarching generality, thereby enhancing the versatility of EDCR for diverse problem domains.

The rest of the chapter is outlined as follows. In Section 3.2, we describe the movement trajectory classification problem (MTCP) and associated classification approaches. Then, we introduce our error detecting and correcting rule framework (Section 3.3), which formalizes our strategy for EDCR and provides analytical results that support our algorithm development. This is followed by experimental results in Section 3.4 followed by a discussion on related work and future directions.

3.2 Technical Preliminaries

In this section, we introduce MTCP and describe the vector embeddings used for a neural-based classifier [3].

3.2.1 Movement Trajectory Classification Problem

We define the MTCP problem as given a sequence of GPS points, ω, and assign a movement class from C. The number of classes in C is n. In this chapter, as per others (e.g., [3, 9]) we define $C = \{$walk, bike, bus, drive, train$\}$, though we will typically not refer to specific classes outside of the description of the experiments for purposes of generalizability. The current paradigm for the MTCP problem is to create a neural model f_θ that maps sequences to movement classes using a set of weights, θ. In this approach traditional methods (i.e., gradient descent) find a set of parameters such that a loss function is minimized based on some training set $\mathcal{T}$ (where each sample $\omega \in \mathcal{T}$ is associated with a ground truth class $gt(\omega)$). Formally: $\arg\min_\theta \mathbb{E}_{\omega \in \mathcal{T}} Loss(f_\theta(\omega), gt(\omega))$. We also note that with each sample ω, we will associate three predicates for each class i: $pred_i$, $corr_i$, and $error_i$ that we will later use to describe a logic for reasoning about error correction.

- $pred_i$: if the model predicted class i: $pred_i(\omega)$ is true if and only if $f_\theta(\omega) = i$.
- $corr_i$: the correct movement class for ω: $corr_i(\omega)$ is true if and only if $gt(\omega) = i$.
- $error_i$: if the model had an error: $error_i(\omega)$ is true if and only if $f_\theta(\omega) \neq gt(\omega)$. In other words: the model is wrong and predicted class i.

Vector Embedding The current SOTA approaches that we examine for f_θ rely on an embedding of a sequence ω that consists of a stack of vectors describing the velocity, acceleration, jerk (time rate of change of acceleration), and bearing rate. In this chapter, we based these calculations on prior work [3, 9].

3.3 Error Detection and Correction Rules

A key issue with the deployment of model f_θ is that it may encounter sequences whose distribution differs from the data used to train the model. Further, in our target application, there may not be sufficient labeled data or time to properly retrain f_θ. We also note that in some cases, f_θ may be inaccessible for fine-tuning (e.g., behind an API). Additionally, understanding why the results of f_θ change is also important for our envisioned security application. As such, we are employing a rule-based approach to correcting f_θ. The intuition is that using limited data, we will learn a set of rules (denoted Π) that will be able to detect and correct errors of f_θ by logical reasoning [1]. Then, upon deployment for some new sequence ω, we would first compute the class $f_\theta(\omega)$ and then use the rules in set Π to conclude if the result of f_θ should be accepted and if not, provide an alternate class in an attempt to correct the mistake. In this section, we formalize the error correcting framework with a simple first-order logic (FOL) and provide analytical results relating aspects of learned rules that inform our analytical approach to learning such error detecting and correcting rules. We complete the section with a discussion on how various potential "failure conditions" are extracted to create the rules to correct errors.

Throughout this section, we shall assume a set O of operational sequences for which there is ground truth available after model training. The size of set O is N and generally, this is expected to be much smaller than $\mathcal{T}$ (the set of training data). Later, in our experiments, we look at cases where $O = \mathcal{T}$ and

$\mathcal{T} \subseteq O$ – however these are not requirements as our results are based on model performance on O – and we envision use-cases where O is significantly different from $\mathcal{T}$. On these samples, for each class i, the model (f_θ) returns class i for N_i of the samples, and for each class i we have the number of true positives, false positives, true negatives, and false negatives TP_i, FP_i, TN_i, FN_i. We have precision $P_i = TP_i/N_i$, recall $R_i = TP_i/(TP_i + FN_i)$, and prior of predicting class i: $\mathcal{P}_i = N_i/N$.

Language We assume simple first-order language where samples are represented by constant symbols, and we have unary predicates associated with each sample. This language includes (for each movement class) the predicates $pred_i, corr_i, error_i$ described earlier in addition to set C of m "condition" predicates $cond_1, \ldots, cond_m$ associated with each sample that can be either true or false. These can be derived from another model.

3.3.1 Rules

The set of rules Π will consist of two rules for each class: one "error detecting" and one "error correcting." Error detecting rules will determine if a prediction by f_θ is invalid. In essence, we can think of such a rule as changing the movement class assigned by f_θ to some sample ω from i to "unknown." For a given class i, we will have an associated set of detection conditions DC_i that is a subset of conditions, the disjunction of which is used to determine if f_θ gave an incorrect classification.

$$error_i(\omega) \leftarrow pred_i(\omega) \land \bigvee_{j \in DC_i} cond_j(\omega). \quad (3.1)$$

After the application of the error detection rules for each class, we may consider reassigning the samples to another class using a second type of rule called the "corrective rule." Such rules are formed based on a subset of conditions-class pairs $CC_i \subseteq C \times C$.

$$corr_i(\omega) \leftarrow \bigvee_{q,r \in CC_i} \left(cond_q(\omega) \land pred_r(\omega) \right). \quad (3.2)$$

Associated with the rules of both types are the following values – both are defined as zero if there are no conditions.

Support (s): fraction of samples in O where the body is true.

Support with respect to class i (s_i): given the subset of samples where the model predicts class i, the fraction of those samples where the body is true (note the denominator is N_i).

Confidence (c): the number of times the body and head are true together divided by the number of times the body is true.

Now we present some analytical results that inform our learning algorithms. Our strategy for learning involves first learning detection rules (which establish conditions for which a given classification decision by f_θ is deemed incorrect) and then learning correction rules (which then correct the detected errors by assigning a new movement class to the sample). We formalize these two tasks as follows.

Improvement by error detecting rule. For a given class i, find a set of conditions DC_i such that precision is maximized and recall decreases by, at most ϵ.

Improvement by error correcting rule. For a given class i, find a subset CC_i of $C \times C$ such that both precision and recall are maximized.

3.3.2 Properties of Detection Rules

First, we examine the effect on precision and recall when an error detecting rule is used. Our first result shows a bound on precision improvement. If class support (s_i) is less than $1 - P_i$, which we would expect (as the rule would be designed to detect the $1 - P_i$ portion of results that failed), then we can also show that the quantity $c \cdot s_i$ gives us a lower bound on the improvement in precision. In the appendix, we also note that precision will always increase under a reasonable condition (specifically when $c \geq 1 - P_i$).

Theorem 3.1 *Under the condition $s_i \leq 1 - P_i$, the precision of model f_θ for class i, with initial precision P_i, after applying an error detecting rule with support s_i and confidence c increases by a function of s_i and c and is greater than or equal to $c \cdot s_i$.*

The error detecting rules can cause the recall to stay the same or decrease. Our next result tells us precisely how much recall will decrease.

Theorem 3.2 *After applying the rule to detect errors, the recall will decrease by $(1 - c)s_i \frac{R_i}{P_i}$.*

It turns out that both quantities identified in these theorems are submodular and monotonic – a property we can use algorithmically. Specifically, we can see that the selection of a set of rules to maximize $c \cdot s_i$ subject to the constraint that $(1 - c)s_i \frac{R_i}{P_i} \leq \epsilon$ is a special case of the "Submodular Cost Submodular Knapsack" (SCSK) problem and can be approximated with a simple greedy algorithm [6] with approximation guarantee (Theorem 4.7 of [6]).

Properties of Corrective Rules In what follows, we shall examine the results for corrective rules. Here, the error correcting rule with predicate $corr_j$ in the head will have a disjunction of elements of set $CC_i \subseteq C \times C$. Also, note that here the support s is used instead of class support (s_i). Here we find that both precision and recall increase with rule confidence (Theorem 3.3).

Theorem 3.3 *For the application of error correcting rules, both precision and recall increase if and only if rule confidence (c) increases.*

It is clear that confidence is the right quantity to optimize for error correcting rules. However, while this is not a monotonic function over CC_i, it is submodular a formal statement and proof can be found in the appendix. With these results in mind, we can optimize both precision and recall using an error correcting rule (with respect to the class specified in the rule head) but optimizing for confidence. Note that this does not consider the precision and recall for the class specified in the rule body (however, we shall assume that the impact on precision and recall for the class in the body was handled with the application of the initial error detection rules). The property of submodularity allows us to utilize known optimization approaches. However, it is noteworthy that confidence is not monotonic as we add conditions to set CC_i as the precision can decrease. We will consider an initial set of condition-class pairs CC_{all} that is a subset of $C \times C$. For a given class for which we create an error correcting rule, we select CC_i from this larger set. To do so, we adapt the simple "Deterministic USM" algorithm of [2] for nonmonotonic submodular optimization.

3.3.3 Learning Detection and Correction Rules Together

Error correcting rules will provide optimal improvement to precision and recall for the rule in the target class, but in the case of multi-class problems, it will cause recall to drop for some other classes. However, we can combine detecting and correcting rules to overcome this difficulty. The intuition is first to create error detecting rules for each class, which effectively reassigns any sample into an "unknown" class. Then, we create a set CC_{all} (used as input for Error correcting rules) based on the conditions selected by the error detecting rules. In this way, we will not decrease recall beyond what occurs in the application of error detecting rules.

3.3.4 Conditions for Error Detection and Correction

In this section, we describe the methods we used to create conditions (set C) from dataset O. As mentioned in Section 15.1, in addition to offering domain-specific knowledge akin to conventional papers, our contribution extends to

Algorithm 1: DetCorrRuleLearn

Require: Recall reduction threshold ϵ, Condition set C
Ensure: Set of rules Π
    $\Pi := \emptyset$
    $CC_{all} := \emptyset$
    for Each class i **do**
        $DC_i :=$ DetRuleLearn(i, ϵ, C)
        if $DC_i \neq \emptyset$ **then**
            $\Pi := \Pi \cup \{error_i(\omega) \leftarrow pred_i(\omega) \land \bigvee_{j \in DC_i} cond_j(\omega)\}$
            $\Pi := \Pi \cup \{incorr_i(X) \leftarrow pred_i(X) \land \bigvee_{j \in NC_i} cond_j(X)\}$
        end if
        for $cond \in DC_i$ **do**
            $CC_{all} := CC_{all} \cup \{(cond, i)\}$
        end for
    end for
    for Each class i **do**
        $CC_i :=$ CorrRuleLearn(i, CC_{all})
        if $CC_i \neq \emptyset$ **then**
            $\Pi := \Pi \cup \{corr_i(\omega) \leftarrow \bigvee_{q,r \in CC_i} (cond_q(\omega) \land pred_r(\omega))\}$
        end if
    end for
    return Π

the provision of a condition integrated with a neural network, referred to as the model based on our paper. This condition, marked by its comprehensive generality, serves to amplify the adaptability of the EDCR across a spectrum of diverse problem domains.

The field of deep learning witnesses a continuous influx of new and improved models for solving complex problems. The prevailing trend involves the adoption of the latest and supposedly superior models, often leading to the abandonment of previously successful ones. We present a method that challenges this paradigm, proposing a technique to harness the potential of older, proven models to augment the performance of the latest and most advanced models. We employ a collection of diverse preexisting neural models as a set of conditions to enhance the efficacy of the current model. More specifically, a more coarse-grain model can also provide insight into the conditions. As such, we utilized a binary classifier for each class for a given sample. Hence, given class i, we have a binary classifier g_i which returns "true" for sample ω if g_i assigns it as i and "false" otherwise. In this way, for each sample ω we have a $g_i(\omega)$ condition for each of the classes. To illustrate, for the drive class, we divided the training data $\mathcal{T}$ into two distinct datasets: one exclusively containing samples labeled as drive, and the other encompassing samples labeled as walk, bike, bus, train, collectively forming the non_drive class. We employ this binary class classification approach to establish a set of conditions C.

3.4 Experimental Evaluation

GeoLife Dataset The proposed methodology is validated and assessed using GPS trajectories obtained from the GeoLife project [19].

	No Overlap	
LRCN (prev. SOTA)	0.749	0.747
LRCN+EDCR (ours)	**0.761** (+1.6%)	**0.760** (+1.7%)

Table 3.1 *Accuracy when all classes are represented in training and test sets under various data leakage cases. EDCR means "error detecting and correcting rules" were used on the model output and numbers in parentheses show the percent change in accuracy from EDCR over the base model. Bold numbers are the best in each case.*

Training and Test Splits Previous work such as [9] is known to have data leakage based on the split between training and test primarily due to segments of a movement sequence existing in both training and test sets resulting from random assignment to each. To address this data leakage issue, we examine our algorithms under various conditions based on ordering and overlap.

All Classes Observed In our first set of experiments, we examined how EDCR can affect the performance of the underlying model. In Table 3.1, we examine the accuracy of the LRCN model, both with and without EDCR. Of particular importance, in the "no overlap" case – the least likely to exhibit data leakage – EDCR improves the performance of LRCN 1.7%.

Hyperparameter Sensitivity In the "all classes observed" set of experiments, we also examined hyperparameter sensitivity for ϵ. Recall that ϵ is interpreted as the maximum decrease in recall. We observed and validated the theoretical reduction (TR) in recall empirically and the experiments show us that in all cases, recall was no lower than the threshold specified by the hyperparameter ϵ though recall decreases as ϵ increases. In many cases, the experimental evaluation reduced recall significantly less than expected. In Figure 3.1, as the value of ϵ (x-axis) ranges from 0 to 0.10, it is evident that the decline in recall for all classes remains within the confines of 0.10. Likewise, precision only increases with ϵ, which is aligned with our theoretical results. We show precision, recall, and F1 by class for the "no overlap – sequential" of LRCN in Figure 3.1. Though the algorithm DetCorrRuleLearn calls for a single ϵ

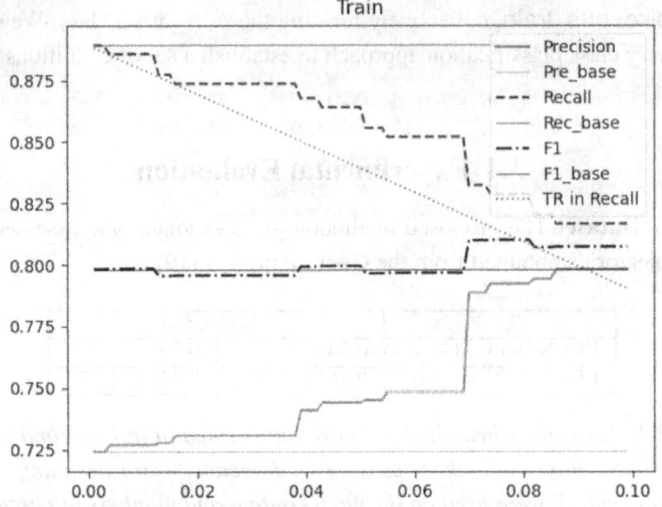

Figure 3.1 LRCN results for application of error detection and correction rules as a function of ϵ on class *train* (no overlaps with sequential selection). TR in recall is the theoretical reduction in recall based on analytic results.

hyperparameter, it is possible to set it differently for each class (e.g., lower values for classes where recall is important, higher values for classes where false positives are expensive). This may be beneficial as F1 for different classes seemed to peak for different values of ϵ. We leave the study of heterogeneous ϵ settings to future work.

Removal of Movement Classes from Training Our experimental focus was on assessing how the introduction of EDCR impacts model performance in scenarios where certain movement classes are excluded from training. In Figure 3.2, we trained the LRCN models without incorporating the walk and drive classes. Remarkably, employing EDCR without any supplementary data yielded a 5.2% (zero-shot) improvement over the base models, and a 23.9% (few-shot) improvement over the SOTA model without resorting to retraining of the base model, with even more pronounced results than in the initial experiment set. Utilizing a mere 30% of data from previously unseen classes, EDCR demonstrates a 21.3% to elevate the performance of the baseline model, all achieved without the need for direct access to the model itself. This outcome implies the potential for conducting few-shot learning, enabling the adaptation of f_θ to novel scenarios with impressive efficacy. This enhancement significantly boosts accuracy using limited data for unseen samples, without extensive

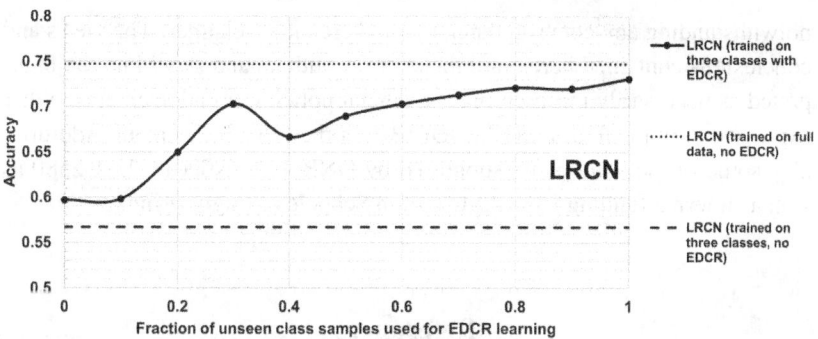

Figure 3.2 Results for experiments with two movement classes removed from training.

model modifications. This is crucial when direct model access is limited, for example through an API.

3.5 Related Work and Conclusion

There are several directions for future work. A key near-term direction for future work is the employment of these methods in government-administered tests of the IARPA HAYSTAC program which will provide an assessment of utility more closely related to real-world use cases. Likewise, an extension related to the aforementioned IARPA program would be to identify a sequence of movement classes in the case where an agent's mode of transit may change. For example, here we would look to apply our error detection and correction framework to recently introduced models such as those described in [18]. Separately, we framed rule learning as a pair of submodular maximization problems, but there are several options for algorithms beyond this chapter. Finally, the use of rules for error detection and correction of machine learning models presented here may be useful in domains such as vision.

3.6 Acknowledgments

This research is supported by the Intelligence Advanced Research Projects Activity (IARPA) via the Department of Interior/Interior Business Center (DOI/IBC) contract number 140D0423C0032. The U.S. Government is authorized to reproduce and distribute reprints for Governmental purposes notwithstanding any copyright annotation thereon. Disclaimer: The views and conclusions contained herein are those of the authors and should not be interpreted as necessarily representing the official policies or endorsements, either expressed or implied, of IARPA, DOI/IBC, or the U.S. Government. Additionally, some of the authors are supported by ONR grant N00014-23-1-2580 as well as internal funding from ASU Fulton Schools of Engineering.

References

[1] Aditya, Dyuman, Mukherji, Kaustuv, Balasubramanian, Srikar, Chaudhary, Abhiraj, and Shakarian, Paulo. 2023. PyReason: Software for open world temporal logic. In: *Proceedings of the AAAI Spring Symposium on Challenges Requiring the Combination of Machine Learning and Knowledge Engineering, Hyatt Regency, San Francisco Airport, California, USA*.

[2] Buchbinder, Niv, Feldman, Moran, Naor, Joseph, and Schwartz, Roy. 2012. A tight linear time (1/2)-approximation for unconstrained submodular maximization. Pages 649–658 of: *2012 IEEE 53rd Annual Symposium on Foundations of Computer Science*.

[3] Dabiri, Sina, and Heaslip, Kevin. 2018. Inferring transportation modes from GPS trajectories using a convolutional neural network. *Transportation Research Part C: Emerging Technologies*, **86**, 360–371.

[4] Fikioris, Giannis, Patroumpas, Kostas, Artikis, Alexander, Pitsikalis, Manolis, and Paliouras, Georgios. 2023. Optimizing vessel trajectory compression for maritime situational awareness. *GeoInformatica*, **27**(3), 565–591.

[5] Huang, Haosheng, Cheng, Yi, and Weibel, Robert. 2019. Transport mode detection based on mobile phone network data: A systematic review. *Transportation Research Part C: Emerging Technologies*, **101**, 297–312.

[6] Iyer, Rishabh K, and Bilmes, Jeff A. 2013. Submodular optimization with submodular cover and submodular knapsack constraints. Pages 2436–2444 of: *Proceedings of the 26th International Conference on Neural Information Processing Systems – Volume 2*.

[7] Janner, Michael, Li, Qiyang, and Levine, Sergey. 2021. Offline reinforcement learning as one big sequence modeling problem. Pages 1273–1286 of: *Advances in Neural Information Processing Systems*, 34.

[8] Jothimurugan, Kishor, Bansal, Suguman, Bastani, Osbert, and Alur, Rajeev. 2021. Compositional reinforcement learning from logical specifications. Pages 10026–10039 of: *Advances in Neural Information Processing Systems*, 34.

[9] Kim, Jinsoo, Kim, Jae Hun, and Lee, Gunwoo. 2022. GPS data-based mobility mode inference model using long-term recurrent convolutional networks. *Transportation Research Part C: Emerging Technologies*, **135**, 103523.

[10] Lin, Miao, and Hsu, Wen-Jing. 2014. Mining GPS data for mobility patterns: A survey. *Pervasive and Mobile Computing*, **12**, 1–16.

[11] Ma, Meiyi, Gao, Ji, Feng, Lu, and Stankovic, John. 2020. STLnet: Signal temporal

logic enforced multivariate recurrent neural networks. Pages 14604–14614 of: *Advances in Neural Information Processing Systems*, 33.

[12] Ramanagopal, Manikandasriram Srinivasan, Anderson, Cyrus, Vasudevan, Ram, and Johnson-Roberson, Matthew. 2018. Failing to learn: Autonomously identifying perception failures for self-driving cars. *IEEE Robotics and Automation*, **3**(4), 3860–3867.

[13] Simoncini, Matteo, Taccari, Leonardo, Sambo, Francesco, Bravi, Luca, Salti, Samuele, and Lori, Alessandro. 2018. Vehicle classification from low-frequency GPS data with recurrent neural networks. *Transportation Research Part C: Emerging Technologies*, **91**, 176–191.

[14] Vanschoren, Joaquin. 2018. Meta-learning: A survey. *arXiv preprint arXiv: 1810.03548*.

[15] Vaswani, Ashish, Shazeer, Noam, Parmar, Niki, et al. 2017. Attention is all you need. Pages 5998–6008 of: *Advances in Neural Information Processing Systems*, 30.

[16] Vincenty, T. 1975. Direct and inverse solutions of geodesics on the ellipsoid with application of nested equations. *Survey Review*, **23**(176), 88–93.

[17] Wang, Hao, Liu, GaoJun, Duan, Jianyong, and Zhang, Lei. 2017. Detecting transportation modes using deep neural network. *IEICE Transactions on Information and Systems*, **100**(5), 1132–1135.

[18] Zeng, Jiaqi, Yu, Yi, Chen, Yong, Yang, Di, Zhang, Lei, and Wang, Dianhai. 2023. Trajectory-as-a-Sequence: A novel travel mode identification framework. *Transportation Research Part C: Emerging Technologies*, **146**, 103957.

[19] Zheng, Yu, Li, Quannan, Chen, Yukun, Xie, Xing, and Ma, Wei-Ying. 2008. Understanding mobility based on GPS data. Pages 312–321 of: *Proceedings of the 10th International Conference on Ubiquitous Computing*.

[20] Zhou, Kaiyang, Liu, Ziwei, Qiao, Yu, Xiang, Tao, and Loy, Chen Change. 2022. Domain generalization: A survey. *IEEE Transactions on Pattern Analysis and Machine Intelligence*, **45**(4), 4396–4415.

4
Mutual Trust in Human–AI Teams Relies on Metacognition

SERGEI NIRENBURG, MARJORIE MCSHANE, THOMAS M. FERGUSON

The long game of artificial intelligence (AI) aims at developing agents that are progressively more human-like in an ever-growing number of facets. Such agents must be able to explain the causes and effects of events and attitudes of agents in their world, including their own attitudes. This state of affairs can only be brought about if the agents are endowed with metacognitive abilities. In this chapter, we highlight the importance of metacognition for modeling the phenomenon of trust. Specifically, we present the case for the interdependence of metacognition and mutual trust between members of human–AI teams. We also argue that metacognition based on causality and contentful explanations requires knowledge support modeling human semantic and episodic memories as well as knowledge of language. We illustrate the above point with examples from systems developed using the OntoAgent cognitive architecture.

4.1 Introduction

AI-based members (agents) in human–AI teams are becoming progressively more sophisticated. The long-term objective of agent development includes making sure that human members of such teams can – justifiably – treat AI agents not just as (possibly, sophisticated) automation tools but as full-blown, human-level teammates. This will involve the need to reliably assess and maintain trust among the members of a human–AI team. In this chapter, we suggest some peculiarities of trust-related issues in this environment, discuss ways of operationalizing trust-related operations by AI agents, and present examples of such operations in several proof-of-concept agent systems developed at the RPI LEIA Lab (e.g., see chapter 8 of [9]). Our main objective in this chapter is to demonstrate the crucial importance of assessments of the trustworthiness of metacognitive reasoning, that is, the agent's ability to reason about its own reasoning as well as other agents' reasoning. This capability is known as "mindreading" [8, 21]. In our earlier work, we extended the scope of mindreading to include "reading" emotions and physical states [11]. Metacognitive reasoning, in turn, relies on the agent possessing knowledge about the world, the situation, itself, and other agents.

Trust is a complex phenomenon, as evidenced by a significant volume of research contributions on the topic, including three recently edited volumes [6, 7, 19]. It comes as no surprise that different authors interpret the phenomenon of trust differently. Indeed, it is a good example of what Marvin Minsky (2006) called suitcase words. As a starting point, we can take a slightly modified formulation of Hardin's 2002 generic definition of trust:

Trustor A trusts Trustee B with respect to action or attitude C.

When an instance of the trust relation exists, it justifies the potential risk involved in the trustor requesting the trustee to carry out a task or assume an attitude (for example, to keep a secret the trustor shares). This task or attitude is related to a particular goal (intention) pursued by the trustor.

The trustor gauges the level of risk by assessing the probability of the trustee not advancing the trustor's goals by not performing a particular action or not holding a particular belief or attitude. The operation of trust assessment is, thus, a necessary component of the trustor's decision-making. It accompanies any goal-directed activity. Within a team, any teammate operates as trustor or trustee in joint tasks with other teammates. But this is also the case even when an agent operates individually. Indeed, an agent must assess whether it trusts itself to successfully perform a particular task. If it does not, then it must replan and either request assistance or choose a plan to attain the current goal that does not involve the action that it does not trust itself to complete. It is clear from this brief discussion that trust maintenance is a metacognitive operation routinely accompanying agents' decision-making. In fact, trust maintenance can be viewed as an extension of the scope of the agent's metacognitive assessment of its *confidence* in the quality and availability of the knowledge resources supporting all its decision-making. Confidence computation in implemented systems developed at the LEIA Lab concentrated so far on assessing the quality of interpretation of language expression containing unexpected input (e.g., unknown words) and the quality of decisions in situations when only a subset of decision function arguments was available at decision time ([14], especially chapter 2).

Malle and Ullman [7] proposes a more contentful definition of trust than Hardin's, emphasizing a distinction between *performance- and morality-based* trust:

Trust is a dyadic relation in which one person accepts vulnerability because they expect that the other person's future action will have/reflect certain characteristics; these characteristics include some mix of performance (ability, reliability) and/or morality (honesty, integrity, and benevolence).

It is clear that this definition applies to human–human trust relations and relations between humans and "high-end" AI agents that are capable of intentional behavior. Such agents (existing or futuristic) are often anthropomorphized by humans. This allows people to impart such agents with qualities (characteristics) that are expected in humans playing the same role in a team as the agent. This saves people some cognitive effort that otherwise would have had to be expended on assessing the agents' trustworthiness without making an analogy with humans. But this economy of effort comes at the price of what can be called "the Eliza effect" – imparting human-level characteristics to AI agents without justification. This widespread phenomenon of human suggestibility elicits what Sullins 2020 calls *thin trust* and Coeckelbergh [1] calls *virtual trust or quasi-trust*.

[Thin trust] is not well justified and possibly manipulated and ... [arises in] situations where systems act as if they deserve trust or appear as trustworthy fellow agents when in fact they are nothing other than semiautonomous machines acting in the interests of their makers (2020, pp. 314–315).

Thin trust in LLMs may be sufficient in noncritical application areas, such as advertisement or entertainment. Unfortunately, thin trust also facilitates success in some nefarious application areas – from cheating on homework and exams to making gullible humans believe many kinds of disinformation. It is clear that assessments of trust by members of human–AI teams must be more reliable and explanatory. In what follows we suggest steps toward this goal.

Trust is not always connected with anthropomorphization. When dealing with automation devices that are not perceived as intentional agents [3], trust is adequately assessed by addressing only the performance characteristics of Malle and Ullman – ability and reliability. People will fully trust a calculator when it always correctly performs arithmetic computations or a thermostat when it correctly and consistently controls the temperature for which it is set. One can envisage arranging all intelligent agents (artificial or otherwise) on a spectrum of the degree of their *anthropomorphism*, that is, the degree to which their cognitive processes are believed to support behavior approximating human behavior. This spectrum extends from devices with virtually no self-reflection or intentionality (e.g., thermostats) to agents simulating ever more human-like capabilities, such as generation of coherent texts in natural languages (e.g., LLMs) to agents modeling themselves and other agents (e.g., LEIAs) to future AI systems that exhibit human-like intelligent behavior at the level of quality similar to – or even, as some futurists project – exceeding that of humans.

The higher an agent's position on the spectrum, the larger the number of characteristics (features) relevant to assessing its trustworthiness.

So, when we say that we trust an agent, we mean quite different things depending on where on the above spectrum this agent belongs. This highlights an interpretation problem: when an agent (human or artificial) states that it trusts some other agent (or self!) with respect to some action, task or belief, the recipient of this message must perform a rather complex interpretation operation: to fully understand the meaning and intention of the speaker, it is necessary to assess the level of anthropomorphism of the trustee in question.

For human–AI teams to be efficient, **mutual trust** must be maintained among all human and agent teammates. When humans entrust an agent with a task, they must trust the agent to perform it. Conversely, when a human teaches an agent to carry out a task, the agent must trust the human to convey correct information (note that there might be more than one correct way: as in when the furniture building robot [5, 17] is taught different operation sequences for the same goal). This underscores the need to include trust maintenance (or trustworthiness computation) in the arsenal of agents worthy of being full-blown members of human–AI teams.

While Malle and Ullman's definition allows for mutual trust, the definition must be extended so that it applies beyond dyadic relations between individuals. In a team environment, trust must be treated as not just a dyadic but also as a **polyadic** relation. An individual may trust a group of individuals (a "juridical person"), a group may trust its members or another group. So, the trustor and the trustee can be individuals or groups.

4.2 Trust Assessment

A large number of different classifications of trust-related features of agents and of the world (environment) can be found in literature (e.g., [20] and the metaanalyses in [2, 4] among many others). The classifications address trust in human–human interactions as well as human interactions with automation devices and robots. Malle and Ullman [7] offers a detailed analysis of the earlier contributions to motivate the choice of the quintet mentioned in their definition of trust. Since their objective was to develop an instrument for a human experiment on assessing trust, they then further reduced this feature set by combining the features of integrity and benevolence into a single feature. Next, they suggested four English language realizations (e.g., SINCERE: sincere, genuine, candid, authentic) for each of the four features of trust and asked subjects to use the features and their sample realizations to rate the level of trust expressed in English sentences describing robot behavior.

Trust assessment by AI agents requires a more detailed conceptual and operational set-up. This task is better served by taking as a starting point Dimock's 2020 definition of trust as a five-place relation:

A trusts B with X in circumstances C for reason(s) R.

To support realistic intelligent agents operating in human–AI teams this conception must be further enhanced. In real life the value of trust is not always binary. Agent A may not completely trust the agent with X^1 – but may still have act as if it has complete trust in B with respect to X because this might still be the best course of action available to A. This means that the value of trust should be expressed as a point or a region on an abstract [0,1] scale. This value will then be used as a parameter in the agent's decision functions for choosing plans of action. In this chapter, we will concentrate on the task of establishing this value of trust. To establish and maintain mutual trust, both A and B must be able to continuously assess and update:

- the competencies (= *ability to successfully complete actions*) X for which A and B can be trusted
- the circumstances (= *the situation, an instance of the state of the world at a particular time*) C in which cooperation occurs
- the reasons (= *previously established levels of trust through experience and observation*) R for expecting that the above two conditions hold

Reasoning in trust-aware AI agents is guided by decision functions that combine evidence from an inventory of heuristic parameters. Time-sensitive values of these parameters are obtained through dedicated computational processes, complete with their own decision functions, that realize a variety of microtheories of the world, language and the agent's personality profile and physical and mental states. Decision functions are expected to operate even when only a subset of these parameters can be efficiently computed or retrieved from memory. To operationalize this capability, the agent establishes an *actionability threshold* – the metacognitive estimate of how little an agent needs to know in order to have sufficient confidence in the corresponding decision. For more on the concept of actionability, see, e.g., Section 1.4.2 of [9] and chapter 2 of [14].

The decision function used for trust maintenance will be referred to as a *trust assessment function (TAF)*. The heuristic parameters used in a particular application of the TAF reflect the *trust-relevant features* in a given situation,

---

[1] X can be an action with A or B as the agent (can the agent be neither A nor B?); X can be a fact or opinion about anything at all.

i.e., those aspects of the situation – along with their relative importance – that are salient to calculating trust. The process of assessing trust, thus, involves:

- selecting an inventory of **trust-relevant features** (TRFs),
- developing procedures for computing timely values of these features,
- computing relative importance ("weight") of TRFs for an agent,
- constructing decision functions computing the level of trust between a trustor and a trustee on the basis of weighted contributions of TRFs,
- defining an inventory of trust assessment triggering situations (TATSes) and developing corresponding procedures to trigger trust assessment,
- implementing an intelligent control scheme for efficient triggering of trust assessment.

TRFs are a subset of states of affairs occurring before and after the agent makes a decision about an action. For example, before making a decision about delegating a step in a plan (a complex event) a trustee, an agent, acting as a trustor assesses whether the potential trustee is indeed trustworthy. If the trustor in fact assesses the trustee as trustworthy and assigns that plan step to the latter, and if the action is not successfully performed, this will be a reason for the trustor to reduce its trust in the trustee, at least with respect to the trustee's ability to perform this action. (Note that the above type of trust assessment pertains also to the situation where one and the same agent is both the trustor and the trustee.)

Trust-relevant situations can also occur as effects of actions by other agents. For example, if a trustee generates a natural language text correctly describing some object or process, then a potential trustor can assess the content of that text and if that content conforms to the trustor's prior knowledge, the latter's trust in the trustee's competence in the particular area of discussion will be enhanced. Conversely, if in a training environment, an agent obtains the description of an object or a process with which it is not familiar from an agent that is deemed trustworthy on authority (of being designated an instructor), then in the simplest case the learner agent will have high confidence in the correctness and appropriateness of the material conveyed by this instructor. Trust in authority is not a bug but a feature. It is pervasive because it licenses cognitive load savings due to less fine-grained and frequent application of trust maintenance with respect to a "high-reputation" trustee.

A more sophisticated version of trust maintenance may take into account the human susceptibility to the halo bias, which would in the above example cause the trustor to mistrust the trustee not only with respect to the action in question but in general. Incorporating this feature would make the agent more human-like, though this capability might well be of limited use in practical agent applications. In this chapter, we discuss trust maintenance in structured cooperative teams in which all agents know the capabilities, responsibilities, and levels of authority of all other team members. This scope narrowing simplifies trust assessment – for example, there is typically no need for an agent to ascertain that another agent is authorized to perform or delegate a particular action.

An Example

MVP ([9], chapter 8) is a system for training medical personnel (MDs) through dialogs with virtual patients (VPs). It is implemented as a particular configuration of the OntoAgent content-centric cognitive architecture ([14], Section 2.1). Consider the following excerpt of the dialog occurring during a VP's visit to an MD:

MD: I suggest an EGD, which is a diagnostic procedure.
VP: How risky is it?
MD: It's not risky at all.
VP: Is it painful?
MD: It's only a little uncomfortable.
VP: Ok, I'll agree to that.

What is going on? Some background is needed. The VP is an instance of an (unembodied) intelligent agent (Figure 4.1) that is endowed with a "body" – a model of human physiology and pathology, and a mind – the capability of understanding language, reasoning, and decision-making. The system sustains the life of the VP over a long (simulated) period of time during which it experiences physical and personality changes due to the progression of the disease in its pathology model and as a result of a variety of medical interventions.

The VP came to the MD with a complaint of difficulty swallowing (which was established in the dialog before the above excerpt). The MD needs to run some tests before establishing the diagnosis. Esophagogastroduodenoscopy (EGD) is a standard diagnostic procedure for VP's complaints. The MD's overall goal is to cure the VP, the proximate goal is to have the VP agree to undergoing EGD. So, MD utters the first dialog turn. The VP then runs its language understanding module on it, generates its meaning representation (TMR) and realizes that the latter contains an unknown entity. Indeed, at the start of the dialog, this particular VP does not know what EGD is. But its ontology already contains the concept diagnostic procedure, so, the VP triggers its learning module (not shown in Figure 4.1; see chapter 7 in [14]) that, on the basis of the incomplete TMR learns the new concept egd, attaches it as a child of diagnostic-procedure in the ontological hierarchy. At this time the VP does not yet know anything about it, so the values of all the properties in this new concept

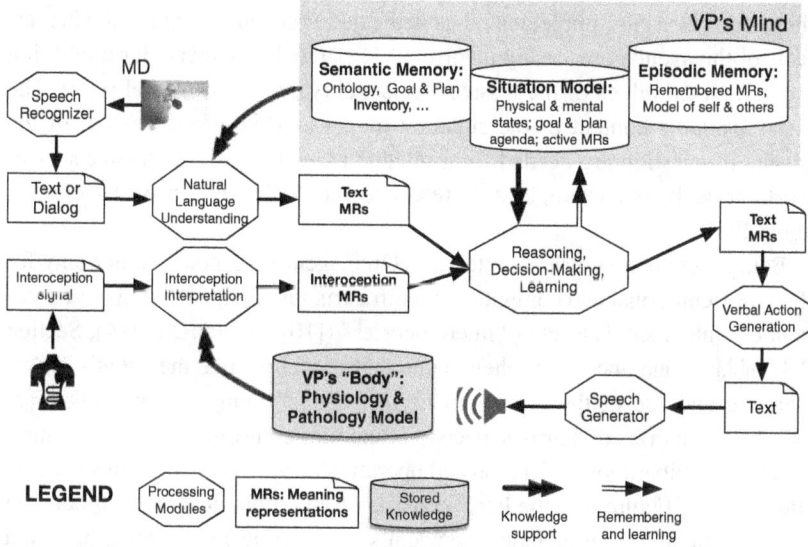

Figure 4.1 The virtual patient (VP) is a particular instantiation of the OntoAgent content-centric cognitive architecture. The model of physiology exists independently of the VP's mind and interacts with the latter through interoception signals about physical states that are interpreted into ontologically grounded meaning representations (MRs), which facilitates reasoning and decision-making. Language inputs are similarly interpreted into text meaning representations (TMRs). The MVP system requires only verbal action to be overtly present. Other VP actions are incorporated only by modifying the VP's physiological model, which occurs due to natural disease progression as well as medical interventions (e.g., surgery) and lifestyle modifications (e.g., diet).

are at this time inherited from its parent. (The values of two of these properties will be subsequently learned by running the learning module on the TMRs generated for the MD's answers to the VP's questions in the remainder of the dialog.)

At the start of the visit, the VP instantiates an initial model of the MD. As this is the first time the VP interacts with the MD, it has not yet constructed a model of this MD's capabilities, personality, and biases. No memories of past interactions with this MD exist at this time either. So, the VP's initial model of the MD is a new instance of the VP's ontological knowledge about MDs. At this time the model holds a default value of the trust parameter (X in Dimock's definition) for any physician in the circumstances (S) of treatment recommendation. The VP will be adjusting parameter values in this model as a result of its interactions with the MD. In our architecture, these operations are implemented as dedicated procedures triggered when effects of the agent's processing involve TATSes (such as triggers for modifying the parameter values of the trustee's personality profile or mental state). The latest available content of the situation model and the agent's model of the trustee in question are needed for a reliable assessment of the trustee's trustworthiness. In our example, this refers to the VP's trust in MD's opinion about EGD.

Being a content-centric architecture, OntoAgent is responsible not only for the representational and computational infrastructure of agents. A core component of OntoAgent is a set of "microtheories" ([10], Section 2.6, [14], Section 2.4) of language and world phenomena that operationalize the agent's understanding and use of these phenomena in its functioning. The microtheories involve inventories of heuristic decision functions controlling perception interpretation, deliberation and action and operating over a knowledge substrate. As illustrated in Figure 4.1, the latter consists of an ontological world model, an episodic memory that includes the agent's model of itself and other agents, a situation model describing entities "in play" at a particular time, and knowledge support for perception interpretation (e.g., lexicons for languages the agent can understand).

Microtheories in OntoAgent are typically operationalized using (a) the inventory of parameters, with their value sets, relevant for interpreting the phenomena addressed by the microtheory; (b) algorithms for determining the value(s) for these parameters at a particular time of the agent's functioning (this task may involve the agent engaging in clarification and negotiation dialogs with the other agents); and (c) decision functions supporting specific choices among alternative actions and determining the agent's confidence in its decisions. These functions can assume a variety of shapes (weighted sums, weighted products, etc.) and a variety of approaches, including probabilistic (e.g., Bayesian) ones. The above apparatus operationalizes the agent's understanding of phenomena in the scope of every microtheory. It also provides the basis for the agent's ability to explain its reasoning and beliefs. The choice of the most appropriate approach to implementing decision functions, including TAFs, is beyond the scope of this chapter. Here, we concentrate on the content and application issues of TAFs.

Different realizations of OntoAgent require different content for their microtheories. The OntoAgent microtheory of trust *for virtual patients* (when circumstances C in Dimock's definition are concretized as interactions of a VP with an MD) currently involves five parameters from the VP's ontology, its episodic memory, personality profile, and physical and mental state profiles. At the start of the above example dialog, the particular VP:

(i) does not have a high level of trust in diagnostic suggestions by MDs (remember that this information is stored in the ontological model);
(ii) is *somewhat* hypochondriac, that is, is *moderately* worried about the risk of the proposed procedure impacting its quality of life (a personality trait);
(iii) has a *low* pain threshold (a personal bias); and
(iv) at the moment is not experiencing pain (a physical state) or
(v) fear (a mental state).

After interpreting the meaning of the initial statement by the MD, the VP performs the following operations:

(i) learns about egd:
  - records *egd* as a child of diagnostic-procedure in its ontology; and
  - records *egd* as a noun in its lexicon and maps its meaning to egd.
(ii) runs the TAF (a weighted combination of the above five parameters) which returns a value below the actionability threshold because:
  - the value of trust in MDs is below a threshold; if the VP had a high value of trust in this particular MD, this would have tipped the function above the threshold);
  - the value of medical-procedure-risk parameter for the newly learned egd concept is currently not precise, as it is inherited from diagnostic-procedure, and some diagnostic procedures are quite risky; so even if the weight of the hypochondria parameter is moderate, its value in this circumstance is low;
  - the same applies to the pain parameter, though its weight is higher due to the VP's low pain threshold; and
  - mental and physical states of VP are neutral, and do not impact the decision-making.

The overall goal of the VP is to be healthy, so at this point in its functioning, it must choose between inquiring about whether EGD is necessary or whether it is the only option and the metacognitive option of learning more about EGD to better motivate its decision (and provide the reasons R in Dimock's definition). The VP chooses the latter option and proceeds to inquire about risk and pain to learn more precise parameter values of the respective properties of the ontological concept of egd. The VP next reruns the above TAF and this time obtains a result above the actionability threshold, which gives it a reason for agreeing to the procedure.

In the continuation of the MVP system run, the EGD procedure is simulated, which generates new content for the VP's physiology model. The interoception interpretation module then generates from this new data its meaning representation of the level of pain caused by the procedure. If the latter was, in fact, painful, the value of the pain parameter in the VP's ontological concept egd will be adjusted downward. This will be recognized as a trust-relevant situation because the earlier value of pain was recorded by the VP on the MD's authority. A TAF will be rerun, and the VP's value of trust in the MD with respect to diagnostics will be lowered. However, had the EGD not been, in fact, accompanied by pain, then the VP can opt to increase its value of trust in the MD as a diagnostician (if the VP models typical human biases, then, implementing the halo bias, it can increase its value of trust in the MD with respect to more actions, facts, and dispositions X in any circumstance C). This latter trust reassessment may lead to the VP agreeing to the MD's suggestions right away, with fewer additional questions.

The situation described in the above example contains at least one additional trust-relevant situation. The VP recognizes that after being informed by the MD about the properties of EGD, it is trusted by the MD to have understood and learned their concept of EGD and the English acronym. (Incidentally, the VP will still not know that EGD means esophagogastroduodenoscopy.) If the MD in the MVP system were represented not by a human but by another agent, then the latter would compute and record this trust assessment in its episodic memory. Of course, if subsequent interaction demonstrates that the concept was not correctly learned, the value of MD's trust would be adjusted down, which might trigger the need for additional interagent interactions to remedy this state of affairs and for the MD to make sure that the VP learns adequate content for the concept of EGD.

4.3 Metacognition for Trust, Actionability, Confidence, and Explainability

Metacognition refers to an agent's awareness and ability to use knowledge and beliefs about its own and other team members' world model, obligations, duties, beliefs, desires, intentions, plans, memories, personality traits (including ethics-related ones), and biases (including interpersonal relations between the agent and others). "Full-blown" treatment of trust can be implemented if and only if the system supports metacognition: the agent's ability to assess – and, if necessary, explain (a) its own state, decisions, knowledge, and intentions and (b) their conception of both the cognitive and observable states and intentions of other agents. This latter functionality is known as mental model ascription or mindreading [21]. The above example illustrates how metacognition is used in the agent's treatment of trust as a phenomenon of human cognitive functioning. Our implementation of a microtheory of trust as a component of the OntoAgent architecture operationalizes the agents' functioning as full-blown trustors and trustees. As a result, their behavior becomes more human-like than the behavior of agents lacking metacognition.

Trust maintenance is one of many facets of using metacognitive abilities to make agents progressively more human-like. OntoAgent currently also includes the metacognitive microtheories of *confidence* and *actionability* that impact the agents' decision-making at all stages of perception interpretation, reasoning, and action specification. The microtheory of confidence (e.g., [9], Section 2.6, [14], Section 8.4) operationalizes the agent's belief in the appropriateness or correctness of its decisions, its reasoning, and its knowledge. Levels of confidence are associated with knowledge elements and algorithms in OntoAgent and are used as parameters in a variety of decision-making heuristic functions.

Conceptually, the microtheory of confidence can, in fact, be reinterpreted as a component of the microtheory of trust (specifically, as a microtheory of *intra-agent* trust as opposed to *interagent* trust, which is the more traditional scope for a microtheory of trust) covering situations when an agent is both the trustor and the trustee with respect to some action, fact, or disposition.

The phenomenon of actionability reflects the human propensity for following the least effort principle by arriving at decisions before considering a full complement of internal or situational reasons for and against. In very general terms, the cost of establishing prerequisites for assessing such reasons may be adjudged as too high in particular circumstances. Considerations of actionability, at least in part, account for why people interrupt others in a conversation: they believe that they understood the meaning of the interlocutor's utterance before it is completed. They also account for why people often make decisions without fully analyzing the situation. Examples of this latter behavior can often be observed during the process of language understanding: people often conclude that they have extracted enough of the meaning of a text or a dialog turn before engaging in complicated situation-dependent reasoning, such as the procedures for determining discourse and pragmatic facets of text meaning. The microtheory of actionability in OntoAgent (e.g., [12], [9], Section 2.5) operationalizes this facet of human behavior.

Metacognition is also an integral part of the capability of OntoAgent agents to communicate with their human teammates. Understanding the intentions of others in dialog turns by interlocutors, making decisions about what to say in response, and about how to phrase these responses all require metacognitive abilities. McShane and Nirenburg [9] details the OntoAgent approach to this topic. One important use of the agents' language understanding and verbal action generation is the support of the agent's capability to *explain* its own and other agents' perception, reasoning, actions, and attitudes. Explanation capabilities are necessary for the agents to be able to engage in such important activities as human-style learning, teaching, negotiating, and persuading. Explanation-related issues are discussed in some detail in Chapter 1.

The key to operationalizing metacognitive reasoning is the availability of a variety of types of enabling static and procedural knowledge. In what follows, we present a brief overview of the knowledge resources in OntoAgent and illustrate their use for metacognitive processing.

4.4 Prerequisites for Metacognition: Knowledge Resources

The very concept of metacognition refers to an agent's ability to reflect on its own behavior, properties, and history. To do so, then, the agent must maintain a record of the above entities as well as its beliefs about behavior, properties, and history of other agents it knows or knows about and a model of the world in which it operates. This content supports all the agent's processing, including the metacognitive reasoning that guides the agent's assessment of its past and future choices, provides the basis for its powers of causal explanation and facilitates its mindreading, communication, and learning capacities. In this section, we present a very brief overview of the knowledge infrastructure of metacognition-capable agents implemented in the OntoAgent architecture (see chapter 3 of [9] for a more detailed treatment). Unlike "traditional" cognitive architectures, OntoAgent is a content-centric architecture in which the knowledge organization model is an integral part of the architecture design. The agent's knowledge resources in OntoAgent (Figure 4.1) include a situation model, a long-term semantic memory and an episodic memory. The types of information stored in each of these knowledge bases are briefly presented below. The situation model ("working memory") of the agent contains:

- the specific elements of the world (ontological concept – object and event – instances) that the agent has perceived and interpreted in the current situation, including their spatiotemporal parameters;
- the MRs of all utterances in any ongoing active dialogs or other interactions with other agents;
- the agent's agenda of *active* goals, their associated subgoals and plans;
- the agent's assessment of the active goal and plan agendas of other agents in the situation, including the trust parameters covering its confidence in

its assessment of its own and other agents' ability to carry out specific plans;
- the current values of the parameters comprising the models of the agent's physical (*energy, pain*, etc.) and mental/emotional (*fear, anger*, etc.) states;
- the agent's assessment of the current values of the parameters comprising the models of the physical and mental/emotional states of other agents in the situation, including the levels of trust in the other agents' shared situation assessment and intentions;
- links between all the elements of the situation model and their underlying background knowledge stored in the long-term semantic and episodic memories that can be activated to support the operation of any of the agent's processing modules.

The long-term semantic memory includes the agent's long-term knowledge about types of elements constituting models of the world and of agents:

- the ontological world model:
  - an inventory of axiomatic properties defined for describing the world and agency;
  - an inventory of specifications of the semantics of (types of) physical, mental, and social objects, events, and states represented as named collections of properties with their value sets;
  - an inventory of agent goals and their associated subgoals and plans;
- knowledge resources supporting the interpretation of sensory inputs:
  - the *lexicon* to support language understanding and generation;
  - the *opticon* to support image and video interpretation;
  - a physiology simulation model to support the interpretation of the results of interoception – the agent's perception of its bodily signals, such as pain;
  - knowledge resources to support the interpretation of nonlanguage audio, olfactory, gustatory, and tactile/haptic perception (not yet implemented in OntoAgent);
- an inventory of heuristic decision functions supporting the operation of each of the processing modules in the agent system; the strong preference for declarative specification of this knowledge facilitates explainability of the OntoAgent agents' operation.

The agent's long-term episodic memory includes:

- instances of past events and states remembered by the agent, with their associated object instances; the above is annotated with spatial and temporal

information about each event, state, and object instance, and a time stamp indicating when the knowledge was added to agent memory;
- an inventory of profiles of agents known to a given agent (including a profile of self); the information in these profiles (including the profile of self!) is not always expected to be correct or reliable, which reflects the state of affairs with humans; agent profiles include knowledge/beliefs about:
  – other agents' ontological world model (typically, only differences from the agent's own ontological model are recorded); this memory component will help the agent to gauge other agents' knowledge about a particular domain or task;
  – other agents' episodic memory content; this memory component will only contain the agent's memory of what past events, states, and objects other agents' are expected to remember;
  – the scope of agents' sensory input interpretation resources; this memory component will contain, for example, the agent's beliefs about other agents' vocabulary coverage in particular languages; and
  – parameters of personality traits of self and other agents,
  – typical value sets for the physical and mental states, behavioral preferences, peculiarities and biases of self and other agents.

4.4.1 Some Details about the OntoAgent Ontology, Lexicon, and Opticon

In what follows we present a brief sketch of the organization of the OntoAgent ontology, lexicon, and opticon (see e.g., Section 2.3 of [9] or chapter 3 of [14] for more detailed descriptions of OntoAgent knowledge resources). The OntoAgent **ontology** contains a collection of concepts that represent types of objects and events in the world. These concepts are described using specific values of a particular subset of basic ontological properties. For example, physical objects are described using, among others, physical properties (size, shape, material, mass, etc.) but also meronymical properties (has-object-as-part) and affordances (e.g., a car can be an instrument of the event drive whose agent must be human). Events are described in the ontology using, among others, such properties as case roles (agent, theme, beneficiary, instrument, etc.) and has-event-as-part which specifies and orders component events of complex events (also known as scripts). The organizational backbone of an ontological model is provided by the subsumption hierarchy, realized by including in each concept an is-a property (see examples in Figures 4.2–4.4). The subsumption hierarchy is realized as a directed acyclic graph that supports multiple inheritances (Figure 4.5).

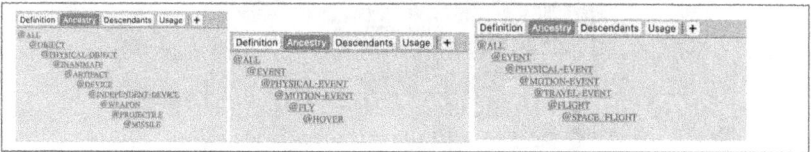

Figure 4.2 The ontological ancestry of the concepts MISSILE, HOVER, and SPACE_FLIGHT.

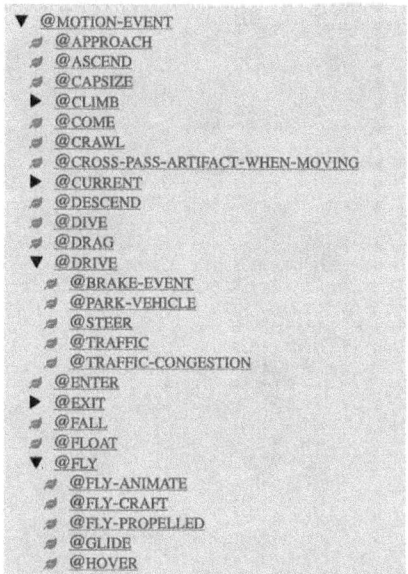

Figure 4.3 Ontological descendants of MOTION-EVENT, including descendants of FLY.

Semantic differences between concepts, including their ontological ancestors and descendants, are specified by the particular set of ontological properties defined for them, coupled with specific value sets defined for those properties. Thus, drive-nails differs from drive, in part, in that values of a subset of its properties constrain the values of the corresponding properties of drive. For example, the default value of the theme property of the event drive-nails is constrained to the concept nail (a descendant of artifact, which is, in turn, a descendant of physical-object) and its default instrument, to the concept hammer (and yes, there is a way to represent the fact that you can drive a nail using, say, a rock). Figures 4.2–4.5 give a partial illustration of the *content* of ontological concepts.

Figure 4.4 Ontological descendants of ARTIFACT, including descendants of AIRPLANE.

The structure of the entries in the ontological-semantic **lexicon** is illustrated in Figure 4.6. Lexicon entries include a variety of types of information that supports semantic and discourse/pragmatic analysis of language input. To give just one example, the entries encode the syntax-semantic context interface that facilitates explainable lexical disambiguation, one of the core tasks in extracting and representing meanings of texts and dialog turns. The SYN-STRUC

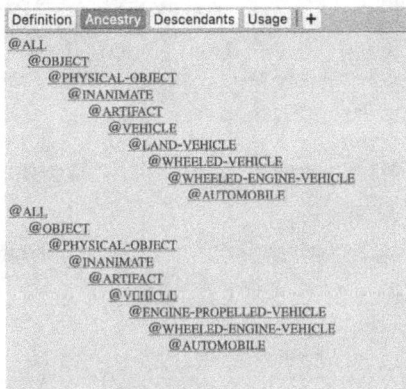

Figure 4.5 An illustration of multiple inheritance in the ontology. The concept of AUTOMOBILE is a descendant of both LAND-VEHICLE and ENGINE-PROPELLED-VEHICLE.

and SEM-STRUC zones of a lexicon entry are cross-referenced and provide a connection between syntactic dependency structures (e.g., Subject-Verb-DirectObject) and the corresponding semantic dependency structures (e.g., Agent-Event-Theme) characterizing every word or phrase sense recorded in the lexicon.

An OntoAgent **opticon** entry for any object, event, or scene includes:

- a head that is a set of one or more visual representations (static images or video clips) that serve as exemplars; this set covers the possible;
- *subtypes* of a particular object (e.g., the cup concept covers both cups with a handle and those without; cups of different geometrical shapes; etc.) as well as different viewpoints of each of the subtypes (e.g., a side view, a top view, a bottom view, etc.);
- a visual representation of the components of the object, event, or scene, along with their spatial relations and links to the components' own opticon entries;
- [optionally] a meaning procedure that helps the agent to recognize the object, event, or scene and its parts;
- [optionally] a meaning procedure that helps the agent to recognize the individual optical features that distinguish the object, event, or scene.

The above knowledge resources support a variety of tasks in implemented systems, including ontology-based perception interpretation, goal and plan selection and prioritization, and action specification and learning. Facilitating acquisition and maintenance of these resources is an ongoing concern. Chapter 9 of [14] contains a detailed description of the various approaches to overcoming

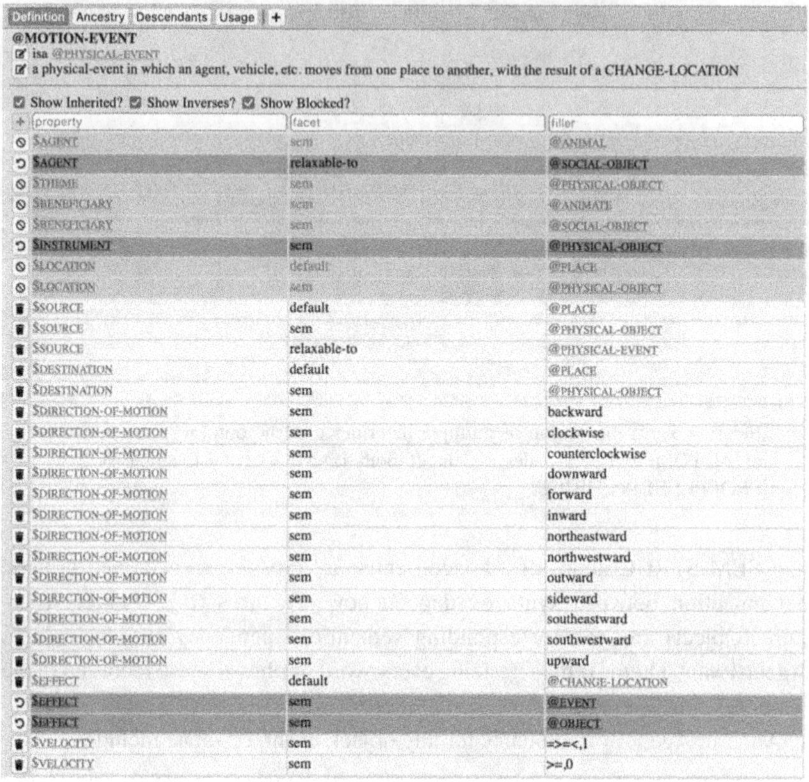

	property	facet	filler
	Definition	Ancestry	Descendants	Usage	+		
	@MOTION-EVENT isa @PHYSICAL-EVENT a physical-event in which an agent, vehicle, etc. moves from one place to another, with the result of a CHANGE-LOCATION		
	☑ Show Inherited? ☑ Show Inverses? ☑ Show Blocked?		
	property	facet	filler
⊘	$AGENT	sem	@ANIMAL
⊃	$AGENT	relaxable-to	@SOCIAL-OBJECT
⊘	$THEME	sem	@PHYSICAL-OBJECT
⊘	$BENEFICIARY	sem	@ANIMATE
⊘	$BENEFICIARY	sem	@SOCIAL-OBJECT
⊃	$INSTRUMENT	sem	@PHYSICAL-OBJECT
⊘	$LOCATION	default	@PLACE
⊘	$LOCATION	sem	@PHYSICAL-OBJECT
🗑	$SOURCE	default	@PLACE
🗑	$SOURCE	sem	@PHYSICAL-OBJECT
🗑	$SOURCE	relaxable-to	@PHYSICAL-EVENT
🗑	$DESTINATION	default	@PLACE
🗑	$DESTINATION	sem	@PHYSICAL-OBJECT
🗑	$DIRECTION-OF-MOTION	sem	backward
🗑	$DIRECTION-OF-MOTION	sem	clockwise
🗑	$DIRECTION-OF-MOTION	sem	counterclockwise
🗑	$DIRECTION-OF-MOTION	sem	downward
🗑	$DIRECTION-OF-MOTION	sem	forward
🗑	$DIRECTION-OF-MOTION	sem	inward
🗑	$DIRECTION-OF-MOTION	sem	northeastward
🗑	$DIRECTION-OF-MOTION	sem	northwestward
🗑	$DIRECTION-OF-MOTION	sem	outward
🗑	$DIRECTION-OF-MOTION	sem	sideward
🗑	$DIRECTION-OF-MOTION	sem	southeastward
🗑	$DIRECTION-OF-MOTION	sem	southwestward
🗑	$DIRECTION-OF-MOTION	sem	upward
🗑	$EFFECT	default	@CHANGE-LOCATION
⊃	$EFFECT	sem	@EVENT
⊃	$EFFECT	sem	@OBJECT
🗑	$VELOCITY	sem	=>=<,1
🗑	$VELOCITY	sem	>=,0

Figure 4.6 A partial view of the definition of the ontological concept MOTION-EVENT. Property-filler lines marked in dark gray are *inherited* and their values are overridden as indicated in corresponding lines marked in gray or white. Duplicate listings of properties with different fillers reflects multiple inheritance.

the "knowledge bottleneck." These approaches cover a spectrum of human–AI collaboration – from "manual" acquisition aided by ergonomic support systems to using text analytics and LLM support [18] to using the existing OntoAgent language analyzer and the available ontology and lexicon to bootstrap learning by understanding and ontologically interpreting text meaning [13, 15, 16].

4.5 Concluding Remarks

The long game of AI aims at developing agents that are progressively more human-like in an ever growing number of facets. Such agents will be social, proactive, and intentional. They will be able to explain the causes and effects

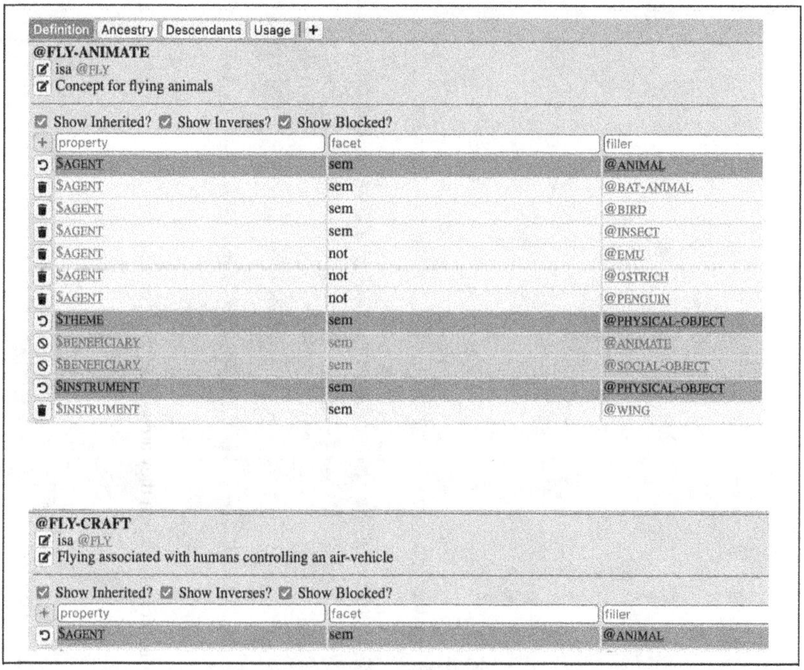

Mutual Trust in Human–AI Teams 75

Definition	Ancestry	Descendants	Usage	+

@FLY-ANIMATE
- isa @FLY
- Concept for flying animals

☑ Show Inherited? ☑ Show Inverses? ☑ Show Blocked?

+ property	facet	filler
↻ $AGENT	sem	@ANIMAL
▮ $AGENT	sem	@BAT-ANIMAL
▮ $AGENT	sem	@BIRD
▮ $AGENT	sem	@INSECT
▮ $AGENT	not	@EMU
▮ $AGENT	not	@OSTRICH
▮ $AGENT	not	@PENGUIN
↻ $THEME	sem	@PHYSICAL-OBJECT
⊘ $BENEFICIARY	sem	@ANIMATE
⊘ $BENEFICIARY	sem	@SOCIAL-OBJECT
↻ $INSTRUMENT	sem	@PHYSICAL-OBJECT
▮ $INSTRUMENT	sem	@WING

@FLY-CRAFT
- isa @FLY
- Flying associated with humans controlling an air-vehicle

☑ Show Inherited? ☑ Show Inverses? ☑ Show Blocked?

+ property	facet	filler
↻ $AGENT	sem	@ANIMAL

Figure 4.7 The agent of the event FLY-ANIMATE is a BAT, an INSECT or a BIRD, but not an EMU, an OSTRICH or a PENGUIN. The theme of FLY-CRAFT is an AIR-VEHICLE, but its AGENT is HUMAN (even if it is sometimes guided by an autopilot).

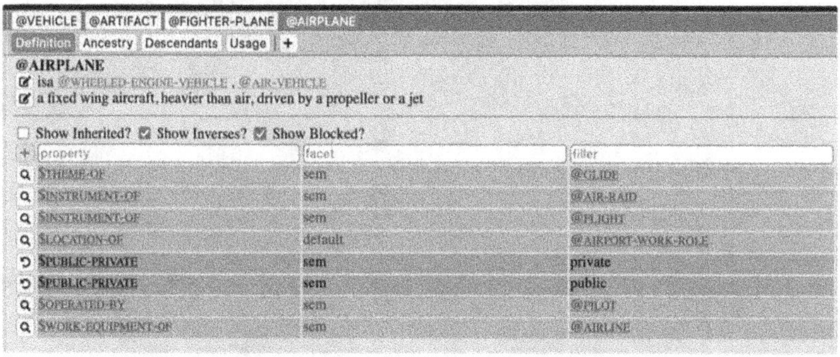

@VEHICLE	@ARTIFACT	@FIGHTER-PLANE	@AIRPLANE

Definition	Ancestry	Descendants	Usage	+

@AIRPLANE
- isa @WHEELED-ENGINE-VEHICLE, @AIR-VEHICLE
- a fixed wing aircraft, heavier than air, driven by a propeller or a jet

☐ Show Inherited? ☑ Show Inverses? ☑ Show Blocked?

+ property	facet	filler
⌕ $THEME-OF	sem	@GLIDE
⌕ $INSTRUMENT-OF	sem	@AIR-RAID
⌕ $INSTRUMENT-OF	sem	@FLIGHT
⌕ $LOCATION-OF	default	@AIRPORT-WORK-ROLE
↻ $PUBLIC-PRIVATE	sem	private
↻ $PUBLIC-PRIVATE	sem	public
⌕ $OPERATED-BY	sem	@PILOT
⌕ $WORK-EQUIPMENT-OF	sem	@AIRLINE

Figure 4.8 Ontological definitions of objects contain specifications of their affordances. For example, an instance of the concept AIRPLANE can fill the THEME case role of GLIDE and the INSTRUMENT case role of AIR-RAID and FLIGHT, etc.

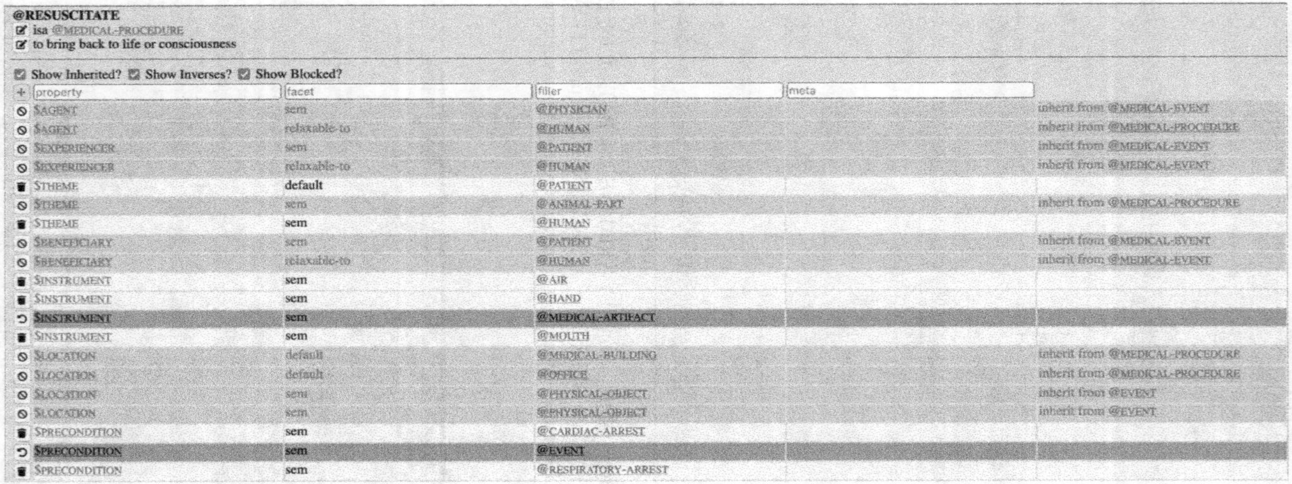

@RESUSCITATE
☑ isa @MEDICAL-PROCEDURE
☑ to bring back to life or consciousness

☑ Show Inherited? ☑ Show Inverses? ☑ Show Blocked?

+	property	facet	filler	meta
⊘	SAGENT	sem	@PHYSICIAN	inherit from @MEDICAL-EVENT
⊘	SAGENT	relaxable-to	@HUMAN	inherit from @MEDICAL-PROCEDURE
⊘	SEXPERIENCER	sem	@PATIENT	inherit from @MEDICAL-EVENT
⊘	SEXPERIENCER	relaxable-to	@HUMAN	inherit from @MEDICAL-EVENT
■	STHEME	default	@PATIENT	
⊘	STHEME	sem	@ANIMAL-PART	
■	STHEME	sem	@HUMAN	
⊘	SPATIENT-CHART	sem	@PATIENT	inherit from @MEDICAL-EVENT
⊘	SBENEFICIARY	relaxable-to	@HUMAN	inherit from @MEDICAL-EVENT
⊘	SINSTRUMENT	sem	@AIR	
■	SINSTRUMENT	sem	@HAND	
C	SINSTRUMENT	sem	@MEDICAL-ARTIFACT	
⊘	SINSTRUMENT	sem	@MOUTH	
⊘	SLOCATION	default	@MEDICAL-BUILDING	inherit from @MEDICAL-PROCEDURE
⊘	SLOCATION	default	@OFFICE	inherit from @MEDICAL-PROCEDURE
⊘	SLOCATION	sem	@PHYSICAL-OBJECT	inherit from @EVENT
⊘	SLOCATION	sem	@PHYSICAL-OBJECT	inherit from @EVENT
⊘	SPRECONDITION	sem	@CARDIAC-ARREST	
C	SPRECONDITION	sem	@EVENT	
■	SPRECONDITION	sem	@RESPIRATORY-ARREST	

Figure 4.9 Concepts describing events list preconditions for when they are typically triggered.

Mutual Trust in Human–AI Teams 77

```
~FLY-V1                                    ~FLY-V4
DEFINITION                                 DEFINITION
  ☑ None.                                    ☑ None.
EXAMPLE                                    EXAMPLE
  ☑ ▶ birds fly                              ☑ ▶ pilots fly airplanes
COMMENTS                                   COMMENTS
  ☑ None.                                    ☑ None.
SYNONYMS                                   SYNONYMS
  + [synonym      ]                          + [synonym      ]
HYPONYMS                                   HYPONYMS
  + [hyponym      ]                          + [hyponym      ]
FLAGS                                      FLAGS
  ☐ Subject to PREP swapping?                ☐ Subject to PREP swapping?
  ☐ Subject to idiomatic creativity?         ☐ Subject to idiomatic creativity?
  ☐ Supports ellipsis?                       ☐ Supports ellipsis?
  ☐ Permits mod with elided head in          ☐ Permits mod with elided head in larger syntactic construct?
    larger syntactic construct?              ☐ Subject to dynamic expansion via transformation?
  ☐ Subject to dynamic expansion via
    transformation?                        SYN-STRUC v-trans
                                             ≡ ☑ $VAR1 ...... SUBJECT  · lemmas
SYN-STRUC v+SUBJ                             ≡ ☐ $VAR0 ...... ROOT     · lemmas
  ≡ ☑ $VAR1 ...... SUBJECT  · lemmas         ≡ ☑ $VAR2 ...... DOBJ     · lemmas
  ≡ ☐ $VAR0 ...... ROOT     · lemmas         + new element
  + new element
                                           SYNTAX PROCEDURES
SYNTAX PROCEDURES                            +
  +
                                           SEM-STRUC concept-with-variables
SEM-STRUC concept-with-variables             ≡ FLY-CRAFT
  ≡ FLY-ANIMATE                                ≡ AGENT
    ≡ AGENT                                      ▮ VALUE         = ^$VAR1
      ▮ VALUE         = ^$VAR1                 ≡ THEME
                                                 ▮ VALUE         = ^$VAR2
```

Figure 4.10 Lexicon entries for two senses of the English word *fly*, semantically anchored in different ontological concepts, FLY-ANIMATE and FLY-CRAFT respectively.

of events in their world and attitudes of agents in that world, including their own attitudes. This will allow them to trust and be trusted by their teammates in human–AI teams. All of the above capabilities can only be brought about if the agents are endowed with metacognition. Trust is a complex phenomenon. In this chapter, we detailed a computational approach to trust maintenance to highlight the importance of metacognitive abilities for advanced AI agents. Specifically, we have laid out a case for the interdependence of metacognition and mutual trust between members of human–AI teams. The assessment and maintenance of trust involves metacognitive capabilities, on the one hand, while many metacognitive processes (e.g., mindreading) require that an agent can assess the trust it places in its target (as well as its own capabilities) (Figures 4.7–4.10).

Our computational model of trust assessment is still in early stages of development. This line of work began when we reframed the existing microtheory of agent confidence to form a component of the microtheory of trust devoted to the agent's assessment of its trust in its own capabilities and attitudes. We extended this model to cover instances of mindreading other agents, concentrating first on the knowledge of the roles of all team members in team tasks (Nirenburg and Lesser 1986). We will continue to work on expanding the scope of trust assessment properties and associated heuristics. This process follows the general methodology of microtheory development (chapters 2 and 5 of [14]). This approach facilitates implementation of microtheories in operational agent systems while these microtheories are still under development.

Acknowledgments This research was supported in part by Grant #N00014-23-1-2060 from the U.S. Office of Naval Research. Any opinions or findings expressed in this material are those of the authors and do not necessarily reflect the views of the Office of Naval Research.

References

[1] Coeckelbergh, Mark. 2012. Can we trust robots? *Ethics and Information Technology*, **14**, 53–60.

[2] Hancock, Peter A, Billings, Deborah R, Schaefer, Kristin E, Chen, Jessie YC, De Visser, Ewart J, and Parasuraman, Raja. 2011. A meta-analysis of factors affecting trust in human-robot interaction. *Human Factors*, **53**(5), 517–527.

[3] Hoff, Kevin Anthony, and Bashir, Masooda. 2015. Trust in automation: Integrating empirical evidence on factors that influence trust. *Human Factors*, **57**(3), 407–434.

[4] Khavas, Zahra Rezaei. 2021. A review on trust in human-robot interaction. *arXiv preprint arXiv:2105.10045*.

[5] Knepper, Ross A, Layton, Todd, Romanishin, John, and Rus, Daniela. 2013. Ikeabot: An autonomous multi-robot coordinated furniture assembly system. Pages 855–862 of: *2013 IEEE International Conference on Robotics and Automation*. IEEE.

[6] Krueger, Frank. 2021. *The Neurobiology of Trust*. Cambridge University Press.

[7] Malle, Bertram F, and Ullman, Daniel. 2021. A multidimensional conception and measure of human-robot trust. *Trust in Human-Robot Interaction*, 3–25.

[8] McShane, Marjorie. 2014. Parameterizing mental model ascription across intelligent agents. *Interaction Studies*, **15**(3), 404–425.

[9] McShane, Marjorie, and Nirenburg, Sergei. 2021. *Linguistics for the Age of AI*. MIT Press.

[10] McShane, Marjorie, Beale, Stephen, Nirenburg, Sergei, Jarrell, Bruce, and Fantry, George. 2012. Inconsistency as a diagnostic tool in a society of intelligent agents. *Artificial Intelligence in Medicine*, **55**(3), 137–148.

[11] McShane, Marjorie, Nirenburg, Sergei, Beale, Stephen, Jarrell, Bruce, Fantry, George, and Mallott, David. 2013. Mind-, body-and emotion-reading. Pages 15–17 of: *Proceedings of IACAP*. Available at www.iacap.org/conferences/iacap2013/iacap_2013_proceedings/.

[12] McShane, Marjorie, Nirenburg, Sergei, and English, Jesse. 2018. Multi-stage language understanding and actionability. *Advances in Cognitive Systems*, **6**, 1–20.

[13] McShane, Marjorie, Beale, Stephen, and Nirenburg, Irene. 2019. Applying deep language understanding to open text: Lessons learned. Pages 796–802 of: *Proceedings of the Annual Meeting of the Cognitive Science Society*, vol. 41.

[14] Mcshane, Marjorie, Nirenburg, Sergei, and English, Jesse. 2024. *Agents in the Long Game of AI: Computational Cognitive Modeling for Trustworthy, Hybrid AI.* MIT Press.
[15] Nirenburg, Sergei, and Wood, Peter. 2017. Toward human-style learning in robots. In: *AAAI Fall Symposium on Natural Communication with Robots.*
[16] Nirenburg, Sergei, Oates, Tim, and English, Jesse. 2007. Learning by reading by learning to read. Pages 694–701 of: *International Conference on Semantic Computing (ICSC 2007).* IEEE.
[17] Nirenburg, Sergei, McShane, Marjorie, Beale, Stephen, et al. 2018. Toward human-like robot learning. Pages 73–82 of: *Natural Language Processing and Information Systems: 23rd International Conference on Applications of Natural Language to Information Systems, NLDB 2018, Paris, France, June 13–15, 2018, Proceedings 23.* Springer.
[18] Oruganti, Sanjay, Nirenburg, Sergei, English, Jesse, and McShane, Marjorie. 2023. Automating knowledge acquisition for content-centric cognitive agents using LLMs. *Proceedings of the AAAI Symposium Series*, **2**(1), 379–385.
[19] Simon, Judith. 2020. *The Routledge Handbook of Trust and Philosophy.* Routledge.
[20] Simpson, Jeffry A, and Vieth, Grace. 2021. Trust and psychology: Psychological theories and principles underlying interpersonal trust. Pages 15–35 of: *The Neurobiology of Trust.* Cambridge University Press.
[21] Spaulding, Shannon. 2020. What is mindreading? *Wiley Interdisciplinary Reviews: Cognitive Science*, **11**(3), e1523.

PART III

Neuro-Symbolic Models in AI

5
Learning Where and When to Reason in Neurosymbolic Inference

CRISTINA CORNELIO

In the rapidly evolving field of metacognitive artificial intelligence (AI), ensuring the reliability and trustworthiness of neural network predictions stands as a fundamental imperative. By integrating hard constraints into neural network outputs, we not only improve the reliability of AI systems but also pave the way for meta-cognitive capabilities that ensure the alignment of predictions with domain-specific knowledge. This topic has received a lot of attention, however, existing methods either impose the constraints in a "weak" form at training time, with no guarantees at inference, or fail to provide a general framework that supports different tasks and constraint types. We tackle this open problem from a neuro-symbolic perspective, developing a pipeline that enhances a conventional neural predictor with (1) a symbolic reasoning module capable of correcting structured prediction errors and (2) a neural attention module that learns to direct the reasoning effort to focus on potential prediction errors, while keeping other outputs unchanged. This framework provides an appealing trade-off between the efficiency of constraint-free neural inference and the prohibitive cost of exhaustive reasoning at inference time that satisfies the rigorous demands of metacognitive assurance.

5.1 Introduction

Despite the rapid advancement of machine learning (ML), deep learning architectures face challenges in addressing certain classes of problems especially when requiring nontrivial symbolic reasoning (e.g., automated theorem proving or scientific discovery). This limitation extends to typical deep learning applications like image processing, where imposing hard symbolic constraints on model outputs proves challenging (see Figure 5.1). The failure of ML systems to adhere to domain knowledge constraints is not only a performance setback but also erodes public trust in AI.

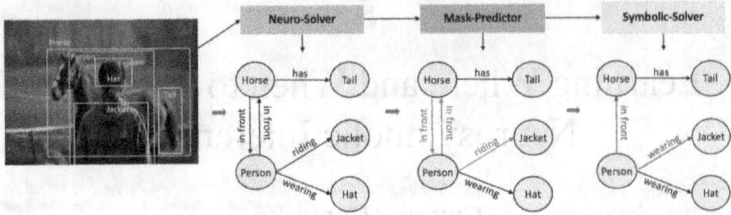

Figure 5.1 Predicate classification in scene graphs. The Neuro-Solver predicts a set of triples between the given objects. The Mask-Predictor identifies the components that potentially violate domain rules (e.g., the predicate *in front* is not symmetric and the object *jacket* is not in the range of the predicate *riding*). Finally, the Symbolic-Solver corrects the predictions.

To tackle this issue, there is a growing interest in neurosymbolic methods aiming to leverage domain knowledge constraints during the learning process [10]. However, existing methods often neglect the crucial aspect of ensuring constraint satisfaction during inference (e.g., modifying the loss function [19, 22] or using adversarial training [4]). While some studies explore constraint enforcement during inference, they often resort to costly reasoning [1, 2, 9, 15, 23] or soft relaxations [7, 8, 18], undermining trust and guarantees.

This chapter introduces a novel neurosymbolic integration approach, the Neural Attention for Symbolic Reasoning (NASR) method, to navigate the trade-off between cost, expressivity, and exactness during inference. NASR combines neural and symbolic approaches by employing an efficient neural solver to solve a task and delegating a symbolic solver to rectify any mistakes. Focusing only on a subset of predictions maintains efficiency, ensuring constraint satisfaction during inference when the facts selected have no false positives. Thus we enjoy most of the benefit of symbolic reasoning about the solution with a cost similar to neural inference.

Our framework[1] is aligned with the "two systems" perspective of [12] and comprises three components: a neural network (Neuro-Solver) trained to solve a given task directly; a hard attention neural network (Mask-Predictor) that determines which predictions are eligible for revision; and a symbolic reasoning engine that revise the identified errors using domain knowledge. The Mask-Predictor learns when and where to reason effectively, facilitating high prediction accuracy and constraint satisfaction with low computation cost.

---

[1] The code is available at: https://github.com/corneliocristina/NASR.

5.2 Method

In this chapter, we consider the type of tasks where multiple interpretable "facts" are predicted by a neural model on which the imposition of hard constraints is desirable. More formally, we consider a set of input data points ($x \in \mathcal{X}$) representing instances to solve (e.g., the picture of a partially filled Sudoku board), and, a set of multidimensional output data points ($y \in \mathcal{Y}$) that correspond to complete interpretable (symbolic) solutions (e.g., the symbolic representation of a completely filled Sudoku board). The collection of N of these pairs of data points will form the *task dataset* $D = \{x^i, y^i\}_{i=1}^N$. Moreover, we require that the task (e.g., completing a partially filled Sudoku board) can be expressed (fully or partially) by a set of rules $\mathcal{R}$ in the form of domain-knowledge constraints (e.g., the rules of the Sudoku game).

The goal is to learn a function $f \colon \mathcal{X} \to \mathcal{Y}$ that associates a solution to a given input instance, *and which further satisfies the rules $\mathcal{R}$*. To solve this class of problems, we propose a neurosymbolic pipeline that integrates three components, the *Neuro-Solver*, the *Mask-Predictor*, and the *Symbolic-Solver* and that works as follows: An input instance is first processed by the Neuro-Solver that outputs an approximate solution. The solution is then analyzed by the Mask-Predictor that has the role of identifying the components of the Neuro-Solver predictions that do not satisfy the set of domain-knowledge constraints/rules $\mathcal{R}$. The masking output of the Mask-Predictor is then combined with the probability distribution predicted by the Neuro-Solver. This is done by deleting the wrong elements of the predictions, leaving the corresponding components "empty" (identified by a masking symbol). This masked probability distribution is then fed to the Symbolic-Solver that fills the gaps with a feasible solution (satisfying the constraints/rules $\mathcal{R}$). In brief, the role of the Symbolic-Solver is to correct the Neuro-Solver prediction errors identified by the Mask-Predictor.

More formally: **(1)** The *Neuro-Solver* is a function $ns(\cdot)$ that that maps an input $x \in \mathcal{X}$ (where $\mathcal{X}$ is the set of all possible inputs for the task under consideration) to a probability distribution over $\mathcal{Y}$ (where $\mathcal{Y}$ is the set of all the possible complete solutions); **(2)** The *Mask-Predictor* is a function $mp(\cdot)$ that takes in input a probability distribution over $\mathcal{Y}$ and produce as output a probability distribution over $\mathcal{Z} = [0, 1]^k$ (where k is the dimension of $y \in \mathcal{Y}$); and **(3)** The *Symbolic Solver* is a function $sb(\cdot)$ that maps $\mathcal{Y}'$ (where $\mathcal{Y}'$ is $\mathcal{Y}$ with an additional class 0, corresponding to a masked solution element) to a probability distribution over $\mathcal{Y}$.

The final hypothesis function f_θ, mapping $\mathcal{X}$ to a probability distribution over $\mathcal{Y}$ and representing the neurosymbolic pipeline approximating the target

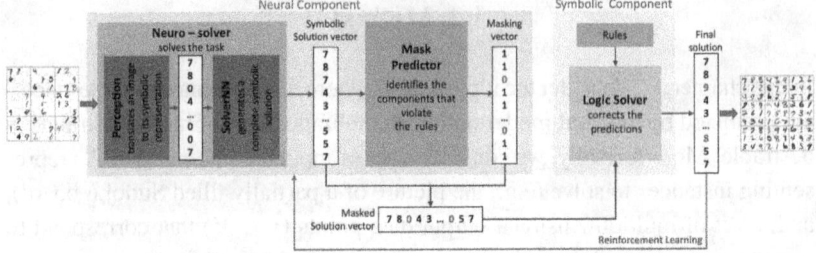

Figure 5.2 Pipeline for solving visual-Sudoku with our Neural Attention for Symbolic Reasoning.

function $f(\cdot)$, is defined as: $f_\theta(x) = sb(\ ns(x) \odot \arg\max(mp(ns(x)))\ ),\ \mathcal{R}\ )$, where $\odot$ is the Hadamard product and θ are the parameters of ns and mp. [2]

5.2.1 Two Example Applications

Visual Sudoku An example application is the visual Sudoku task (Figure 5.2). This task consists of providing a complete Sudoku board $y \in \mathcal{Y}$ corresponding to the solution of an incomplete input board in the form of an image $x \in \mathcal{X}$. $\mathcal{X}$ is defined as $[0, 1]^{252 \times 252}$ and corresponds to the set of the images of a Sudoku board (each Sudoku cell has dimension 28×28). $\mathcal{Y}$ is defined as $\{1, \ldots, 9\}^{81}$ and corresponds to the set of symbolic solutions where each cell is one of the possible nine Sudoku digits. $\mathcal{Z}$ is defined as $\{0, 1\}^{81}$. $\mathcal{Y}'$ is defined as $\{0, \ldots, 9\}^{81}$ and corresponds to the set of symbolic solutions with the nine possible Sudoku digits and the digit 0 indicating empty cells. $\mathcal{R}$ contains the Sudoku rules: each cell needs to be filled with numbers in $\{1, \ldots, 9\}$, without repeating any numbers within the row, column, or block. These can be formalized in different ways, depending on the choice for the symbolic reasoner.

Figure 5.2 shows the pipeline architecture for the Visual Sudoku task.

Scene Graph – PredCl Another example application is the predicate classification (PredCl) task (Figure 5.1). This task consists of predicting the right predicate between a set of objects (given in input in the form of labeled and localized bounding boxes) in an image. $\mathcal{X}$ corresponds to the input images with the set of labeled bounding boxes. This can be vectorized in different ways using appropriate embedding techniques. $\mathcal{Y}$ is defined as $(\mathcal{B} \times \{1, \ldots, m\} \times \mathcal{B})^k$ and corresponds to the set of solutions. A solution is a set of k triples with an object pair and a predicate between them (chosen within m predicates). $\mathcal{B}$ is the space

---

[2] This assumes using "0" as masking symbol. Alternatively, an adapter function can be used: $f_\theta(x) = sb(\ adapt(ns(x), \arg\max(mp(ns(x)))),\ \mathcal{R}\ )$.

of all possible labeled bounding boxes and is defined as $\mathbb{R}^4 \times \{1,\ldots,n\}$ where n is the number of possible objects labels. $\mathcal{Y}'$ is defined as $(\mathcal{B} \times \{0,\ldots,m\} \times \mathcal{B})^k$ and corresponds to the set of symbolic solutions $\mathcal{Y}$ augmented with the class 0 indicating the "empty" predicate (to be filled by the Symbolic-Solver). $\mathcal{Z}$ is defined[3] as $\{0,1\}^k$. $\mathcal{R}$ is an ontology describing the set of object and predicates in the dataset. Some examples are *type*-rules constraining the domain and range of the predicates (e.g., the object *cat* is not in the domain of the predicate *riding*) or the *symmetry/reflexivity* rules (e.g., the predicate *in front of* is asymmetric).

5.2.2 Learning Paradigm

Learning is done in two steps: the Neuro-Solver and the Mask-Predictor are first pre-trained individually in a supervised fashion and then integrated together and refined using reinforcement learning (RL). We refer to our complete pipeline as NASR, and disambiguate ablations where appropriate.

Supervised Learning To train the Neuro-Solver we use the *task dataset* $D = \{x^i, y^i\}_{i=1}^{N}$. For training the Mask-Predictor we generate a synthetic dataset $D_{mp} = \{y_n^i, m^i\}_{i=1}^{N'}$ where: y_n is a symbolic solution instance with the addition of noise that violates the domain-knowledge constraints; and m is the corresponding masking solution. A masking vector m has the same dimension of the input y_n and has a 1 on the components of y_n that do not violate the rules $\mathcal{R}$ and 0 for the components in which noise has been introduced.

The generation of D_{mp} can be done in different ways depending on the type of data we are considering: (1) the input data y_n can be either generated by perturbing the y in D or (2) it can be generated synthetically following a uniform distribution over the possible y_i in D. In the former option, each data point $y_n \in D_{mp}$ has a corresponding data point $y \in D$ of which some components have been modified. The corresponding masking vector m will have a 1 on the components of y_n that has not been modified and 0 for the components in which noise has been introduced. In general, the latter option is not always possible: for example, in the case of the visual Sudoku task, this would require the ability to sample minimal symbolic Sudoku boards uniformly at random, which is still a nontrivial open problem.

Symbolic Solver The Symbolic-Solver will reason about the subset of outputs identified by the Mask-Predictor. The choice of the Symbolic Solver is strongly connected to the type of constraints/rules: For logic-based constraints, classical symbolic reasoners can be used, such as Prolog engines (e.g., SWI-Prolog) or

---

[3] The Hadamard product is intended between the predicate vector $p \in \{1,\ldots,m\}^k$ part of a solution $y \in (\mathcal{B} \times \{1,\ldots,m\} \times \mathcal{B})^k$ and the mask vector $m \in \{0,1\}^k$, since bounding boxes/labels are given in input.

	big_kaggle	minimal_17	multiple_sol*	satnet_data
Symbolic Baseline	74.56	**87.70**	63.50	63.20
SatNet [18]	63.44	0.00	0.00	60.10
SatNet [18] + NASR	69.05	0.02	24.20	**81.40**
NeurASP† [23]	timeout	**89.00**†	timeout	timeout
Our NASR	**84.24**	87.00	**73.00**	82.20

Table 5.1 *Results for the visual Sudoku task: percentage of completely correct solution boards. The best results are in bold font.* * *Multiple_sol dataset: each input board admits more than one solution but only one is provided at training.* † *tested only on 200 of the 5000 test images due to the long run-time.*

probabilistic logic engines (e.g., ProbLog [16], PySwip [17]). For arithmetic constraints, constraints-solvers can be used (e.g., ILP or MILP solvers), general mathematical tools (e.g., Mathematica) or ad-hoc brute force algorithms that exhaustively explore the symbolic solution search space. In this chapter, we mostly consider logic rules and ontologies.

Reinforcement Learning While the Neuro-Solver and Mask-Predictor can be trained independently with supervised learning, the use of RL is necessary for end-to-end learning (since the Symbolic-Solver is not differentiable). End-to-end learning is important so that the neural components can adapt to the expected interventions of the Symbolic-Solver. In this chapter, we use the REINFORCE algorithm [20], with its standard policy loss: $\mathcal{L}(x;\theta) = -r \log P_\theta(m|ns(x))$ where r indicates the RL reward obtained when applying the Symbolic-Solver on the prediction $ns(x)$ masked by m. However, it is possible to use alternative RL algorithms.

5.3 Experimental Results for Visual Sudoku

The main results[4] can be summarized as follows: (1) we outperform the baseline in most of the cases (and never perform worst); (2) we improve the performance of an existing method, by integrating it in our pipeline; (3) we are more efficient, compared to the other methods, in terms computational time vs. performance; and (4) our method is more robust to noise compared to the symbolic baseline.

Baselines We compare NASR with different baselines. As *Symbolic Baseline* we considered the execution of the Symbolic-Solver directly from the output of the Perception module (after applying the arg max operator). Using a

---
[4] For more details on the experimental setup and architectures used, see [6].

probabilistic reasoning engine and using the whole (or partial, e.g., the top k candidates) output distribution of the Perception module as input, is computationally unfeasible. We also compared with two state-of-the-art neurosymbolic methods: *SatNet* [18], a differentiable MAXSAT solver that can be integrated into neural networks; and *NeurASP* [23], an extension of answer set programs (ASPs [5]) that consider a neural network output as the probability distribution over atomic facts in ASPs.

Overall Performance In Table 5.1 we report the performance of our pipeline (with RL) on the different datasets compared with the different baselines. We can see that we outperform all the neurosymbolic methods and in one instance match the Symbolic Baseline (which we outperform in all the other datasets). Note that symbolic baseline fails if even one digit is incorrectly recognized, and thus it is not noise robust. NASR can be integrated with SatNet [18] by replacing our Neuro-Solver with SatNet. The results (SatNet+NASR) show that the soft constraints enforced by SatNet can be improved, sometimes substantially by injection of hard constraints via NASR.

We performed an ablation study to verify the impact of the Mask-Predictor by substituting it with a simple heuristic: We masked the Neuro-Solver output based on its confidence, selecting the threshold by grid search and Bayesian optimization [3]. The results in Table 5.1 show that while this can perform well, it is significantly worse compared to NASR.

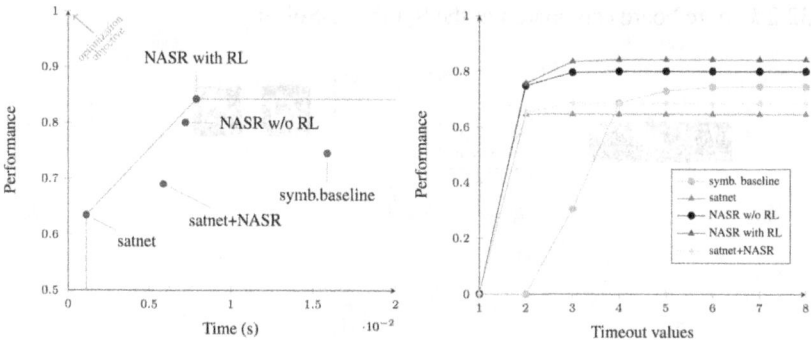

(a) Pareto front (solid line) maximizing the performance (percentage of completely correct boards) and minimizing the computational time. The optimization objective is located on the top left corner of the plot.

(b) Performance analysis for when limiting the computational time of the pipeline. The metric is the percentage of completely correct boards, while increasing the timeout limit.

Figure 5.3 Time efficiency analysis for *big_kaggle* dataset.

Time Efficiency Our system is faster than the symbolic baseline (Perception+Symbolic Solver). This is because our Symbolic Solver needs to fill less empty cells and thus its search space is reduced. In Figure 5.3 we analyze the efficiency in terms of trade-off between the performance (percentage of completely predicted Sudoku boards), and computational time for the *big_kaggle* dataset. Figure 5.3a shows the Pareto front considering the two optimization objectives of minimal computational time and maximal performance. Our method, is always on the Pareto front and usually is the closest to the optimization objective (top left corner of the plot). Figure 5.3b compares the performance of each system when limiting the computation time by different timeout values. We can see that with small timeout limits the neural models behave better compared to the Symbolic-Baseline which requires more time.

Noise Robustness Our pipeline is more robust to noise: in Figure 5.4 we can see the drop in performance of the baseline and our method adding two different type of noise in the input images at inference time. Figure 5.4b shows the results when adding Gaussian blur to the images, while Figure 5.4a when rotating the digits with a random angle in $[-45, 45]$. We can see that our pipeline has a smaller and slower drop in performance compared to the symbolic baseline. These results are less evident in the refined pipeline using RL. We can see that with a high amount of noise the performance gap with the Symbolic-Solver is smaller (e.g., with a rotation between $[-40, 40]$ degrees, our pipeline refined with RL solves 20.3% more boards compared to the Symbolic-Solver), while for a medium amount of noise the gap with the Symbolic-Solver is higher (e.g., with a rotation between $[-25, 25]$ degrees our pipeline refined with RL solves 32.2% more boards compared to the Symbolic-Solver).

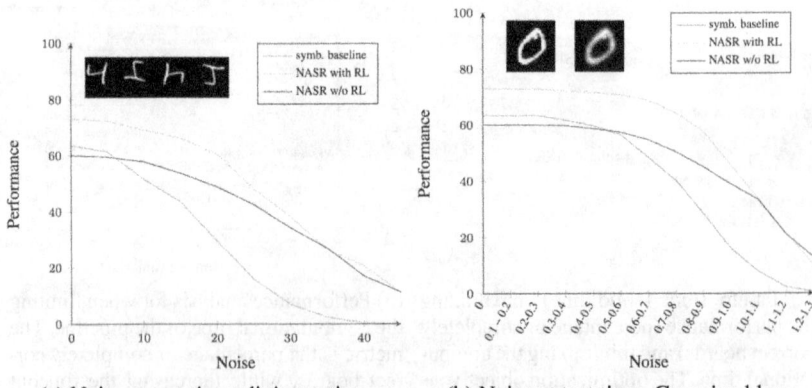

(a) Noise at test time: digit rotation. (b) Noise at test time: Gaussian blur.

Figure 5.4 Performance of the Symbolic Baseline compared to our two pipelines (with or without RL) when adding noise on the dataset *multiple_sol* at test time. Similar results hold for the other datasets.

5.4 Experimental Results on SceneGraph

We evaluate NASR w/o RL on the task of PredCl. Recall that the task consists in providing the right predicate label given the ground truth object labels and bounding boxes.

Dataset and Baseline We consider the GQA dataset introduced by [11]. GQA is a more balanced split of Visual Genome (VG) dataset [14] with cleaner, larger, and more dense scene graphs and with a larger and more balanced variety of objects and predicates (for more details see [11]). It contains 22M pairs of question/answer for real-world visual reasoning and the corresponding scene graphs. We are interested in the latter.

As baseline, we consider the model developed by [13] which is a modified version (that exploits a density-normalized edge loss) of the Message Passing (MP) architecture introduced by [21]. We adopted the data split and evaluation metrics used in their work. For our experiments, we trained their models on the GQA dataset.

Results The results considering a simple domain-range ontology are provided in Table 5.2, using the standard image-level *Recall* metric. We report the results for the Baseline [13] and the improvement over the Baseline of by the Probabilistic Symbolic Baseline (PSB). PSB consists of running the probabilistic symbolic solver directly on the output of the Baseline model. This is computationally very expensive, especially if we consider a slightly more dense ontology. With a more complex ontology this would became computationally intractable. The improvement given by the PSB in the case of PredCl is an upper-bound (Max-improvement) for the performance of NASR. The results of NASR w/o RL are reported as percentage of error correction achieved when compared to the PSB upper bound.

	R@20	R@50	R@100	R@200	R@300
All – shots					
Baseline [13]	29.22	42.35	48.48	50.75	51.11
Max-improvement (PSB)	0.12	0.23	0.32	0.35	0.36
% improvement of NASR w/o RL	99.71	99.58	99.69	99.64	99.64
Zero – shots					
Baseline [13]	16.62	27.65	34.10	37.41	38.11
Max-improvement (PSB)	0.91	1.43	1.93	2.18	2.33
% improvement of NASR w/o RL	100.00	100.00	100.00	100.00	100.00

Table 5.2 *Results on the PredCl task for the GQA dataset using the Recall metric. NASR results are given as percentage of the max achievable improvement under the given ontology, defined by PSB.*

The results show that NASR achieves good performance, and is able to recover the majority of the recoverable errors given the simple domain-range ontology used. This leads, for example, to an improvement between 1% and 2% for the zero-shots predictions. Since we are considering a very simple ontology, the improvement is not as noticeable as in the Visual Sudoku case. However when using a more complex ontology, we expect this difference to become more pronounced.

5.5 Conclusions

To conclude, we presented a neurosymbolic method that aims to efficiently satisfy domain-knowledge constraints at inference, addressing the critical need for robust predictions in the realm of metacognitive AI. This enables a favorable trade-off between accurate predictions, noise robustness, and computation cost. Our framework is generic and can be applied to different types of input (image, text, symbols, etc.) and types of constraints (logic, arithmetic, etc.).

References

[1] Agarwal, Ananye, Shenoy, Pradeep, and Mausam. 2021. End-to-End Neuro-Symbolic Architecture for Image-to-Image Reasoning Tasks. *arXiv preprint arXiv:2106.03121*.

[2] Ahmed, Kareem, Teso, Stefano, Chang, Kai-Wei, Van den Broeck, Guy, and Vergari, Antonio. 2022. Semantic probabilistic layers for neuro-symbolic learning. Pages 29944–29959 of: *Advances in Neural Information Processing Systems*, 22.

[3] Akiba, Takuya, Sano, Shotaro, Yanase, Toshihiko, Ohta, Takeru, and Koyama, Masanori. 2019. Optuna: A next-generation hyperparameter optimization framework. Pages 2623–2631 of: *Proceedings of the 25th ACM SIGKDD International Conference on Knowledge Discovery and Data Mining, Anchorage, AK, USA*.

[4] Ashok, Dhananjay, Scott, Joseph, Wetzel, Sebastian J, Panju, Maysum, and Ganesh, Vijay. 2021. Logic guided genetic algorithms (student abstract). In: *Proceedings of Thirty-Fifth AAAI Conference on Artificial Intelligence, The Eleventh Symposium on Educational Advances in Artificial Intelligence, EAAI*.

[5] Baral, Chitta. 2003. *Knowledge Representation, Reasoning and Declarative Problem Solving*. Cambridge University Press.

[6] Cornelio, Cristina, Stuehmer, Jan, Hu, Shell Xu, and Hospedales, Timothy. 2023. Learning where and when to reason in neuro-symbolic inference. In: *The Eleventh International Conference on Learning Representations (ICLR)*.

[7] Donadello, Ivan, and Serafini, Luciano. 2019. Compensating supervision incompleteness with prior knowledge in semantic image interpretation. In: *The Eleventh International Conference on Learning Representations, Kigali, Rwanda, May 1–5, 2023*. OpenReview.net.

[8] Gan, Leilei, Kuang, Kun, Yang, Yi, and Wu, Fei. 2021. Judgment prediction via injecting legal knowledge into neural networks. In: *Thirty-Fifth AAAI Conference on Artificial Intelligence, Thirty-Third Conference on Innovative Applications of Artificial Intelligence, IAAI 2021, The Eleventh Symposium on Educational Advances in Artificial Intelligence, EAAI 2021, Virtual Event.*

[9] Giunchiglia, Eleonora, and Lukasiewicz, Thomas. 2021. Multi-label classification neural networks with hard logical constraints. *Journal of Artificial Intelligence Research*, **72**(November), 759–818.

[10] Giunchiglia, Eleonora, Stoian, Mihaela Catalina, and Lukasiewicz, Thomas. 2022. Deep learning with logical constraints. Pages 5478–5485 of: *Proceedings of the Thirty-First International Joint Conference on Artificial Intelligence, IJCAI-22.* Survey Track.

[11] Hudson, Drew, and Manning, Christopher. 2019. GQA: A new dataset for real-world visual reasoning and compositional question answering. In: *Proceedings of the IEEE Conference on Computer Vision and Pattern Recognition.*

[12] Kahneman, Daniel. 2011. *Thinking, Fast and Slow.* Farrar, Straus and Giroux.

[13] Knyazev, Boris, de Vries, Harm, Cangea, Catalina, Taylor, Graham W, Courville, Aaron C, and Belilovsky, Eugene. 2020. Graph density-aware losses for novel compositions in scene graph generation. In: *British Machine Vision Conference.*

[14] Krishna, Ranjay, Zhu, Yuke, Groth, Oliver, et al. 2016. Visual genome: Connecting language and vision using crowdsourced dense image annotations. *International Journal of Computer Vision*, **123**, 32–73.

[15] Manhaeve, Robin, Dumancic, Sebastijan, Kimmig, Angelika, Demeester, Thomas, and De Raedt, Luc. 2018. DeepProbLog: Neural probabilistic logic programming. Pages 3753–3763 of: *Advances in Neural Information Processing Systems*, 31.

[16] Raedt, Luc De, Kimmig, Angelika, and Toivonen, Hannu. 2007. ProbLog: A probabilistic prolog and its application in link discovery. In: *International Joint Conference on Artificial Intelligence.*

[17] Tekol, Yüce, and PySwip contributors. 2020. *PySwip v0.2.10.*

[18] Wang, Po-Wei, Donti, Priya L, Wilder, Bryan, and Kolter, Zico. 2019. SATNet: Bridging deep learning and logical reasoning using a differentiable satisfiability solver. *Proceedings of Machine Learning Research*, **97**, 6545–6554.

[19] Wang, Wenya, and Pan, Sinno Jialin. 2020. Integrating deep learning with logic fusion for information extraction. *Proceedings of the AAAI Conference on Artificial Intelligence*, **34**(5), 9225–9232.

[20] Williams, Ronald J. 1992. Simple statistical gradient-following algorithms for connectionist reinforcement learning. *Machine Learning*, **8**(3–4), 229–256.

[21] Xu, Danfei, Zhu, Yuke, Choy, Christopher, and Fei-Fei, Li. 2017. Scene graph generation by iterative message passing. *Computer Vision and Pattern Recognition (CVPR).*

[22] Xu, Jingyi, Zhang, Zilu, Friedman, Tal, Liang, Yitao, and Van den Broeck, Guy. 2018. A semantic loss function for deep learning with symbolic knowledge. *Proceedings of the 35th International Conference on Machine Learning*.

[23] Yang, Zhun, Ishay, Adam, and Lee, Joohyung. 2020. NeurASP: Embracing neural networks into answer set programming. Pages 3097–3106 of: *International Joint Conference on Artificial Intelligence*.

6
Assessment of Competency of Learning Agents via Inference of Temporal Logic Formulas

ZHE XU, NASIM BAHARISANGARI, JEAN-RAPHAËL GAGLIONE, UFUK TOPCU

6.1 Introduction

For enhancing the understanding and collaboration with autonomous agents, there arises a critical need to construct a representation of their task strategies that seamlessly intertwines interpretability, monitoring, and formal reasoning. This representation serves the dual purpose of fostering human comprehension and enabling automated analytical processes. Our approach to achieving this intricate balance involves the formalization of task strategies through the lens of temporal logic formulas. These formulas, resembling natural language and articulating temporal patterns, can provide a comprehensible bridge between the intricacies of autonomous task strategies and human cognition.

In recent years, there has been a surging trend to infer temporal logic formulas from data to explain the behavior of an underlying system and to assess the competency of autonomous learning agents. Our methodology hinges upon a collection of positive examples and negative examples derived from observations of the system. Our goal is to construct a succinct temporal logic formula that is consistent with the provided data. This implies that the model must not only conform to the characteristics outlined by the positive examples but also explicitly deviate from those presented in the negative set, ensuring a consistent and accurate representation of the underlying system.

Existing approaches to temporal logic inference often overlook the prevalence of noise and uncertainties in real-world data, thereby limiting their practicality in deployment scenarios. In response to this challenge, we scrutinize a set of trajectories labeled as positive and negative, acknowledging the existence of noise in the labels, as presented in Gaglione et al. [5]. The primary objective is to infer succinct and interpretable temporal logic formulas that yield minimal *loss* on the data. Here, "loss" denotes the fraction of instances in the labeled trajectories that the inferred formula misclassifies. To confront uncertainties inherent in the data, we extend our focus to labeled *interval trajectories*, where the evolution of values over time is characterized by intervals, effectively capturing the uncertainty, as presented in Baharisangari et al. [1]. Our proposed algorithm strategically maximizes the *worst-case robustness margin*, enhancing the robustness of the inferred formulas in the face of uncertainties. This comprehensive approach aims to not only address noise in the data but also ensure the adaptability and reliability of temporal logic inference methods in real-world applications.

6.2 Preliminaries

In this section, we set up definitions and notations used throughout this chapter.

Finite Trajectories We can describe the state of an underlying system by a vector $x = [x^1, x^2, \ldots, x^n]$, where n is a non-negative integer (the superscript i in x^i refers to the ith dimension). The domain of x is denoted by $\mathbb{X} = \mathbb{X}^1 \times \mathbb{X}^2 \times \cdots \times \mathbb{X}^n$, where each $\mathbb{X}^i$ is a subset of $\mathbb{R}$. The evolution of the underlying system within a finite time horizon is defined in the discrete time domain $\mathbb{T} = \{t_0, t_1, \ldots, t_J\}$, where J is a non-negative integer. We define a finite *trajectory* describing the evolution of the underlying system as a function $\zeta : \mathbb{T} \to \mathbb{X}$. We use $\zeta_j \triangleq x(t_j)$ to denote the value of ζ at time-step t_j.

Intervals and Interval Trajectories An *interval*, denoted by $[\underline{a}, \overline{a}]$, is defined as $[\underline{a}, \overline{a}] := \{a \in \mathbb{R}^n | \underline{a}^i \leq a^i \leq \overline{a}^i, i = 1, \ldots, n\}$, where $\underline{a}, \overline{a} \in \mathbb{R}^n$, and $\underline{a}^i \leq \overline{a}^i$ holds true for all i. The superscript i refers to the ith dimension. For the purpose of this work, we introduce *interval trajectories*. We define an *interval trajectory* $[\underline{\zeta}, \overline{\zeta}]$ as a set of trajectories such that for any $\zeta \in [\underline{\zeta}, \overline{\zeta}]$, we have $\zeta_j \in [\underline{\zeta}_j, \overline{\zeta}_j]$ for all $t_j \in \mathbb{T}$ [9]. We know that the time length of a trajectory $\zeta \in [\underline{\zeta}, \overline{\zeta}]$ is equal to the time length of an interval trajectory $[\underline{\zeta}, \overline{\zeta}]$; thus, we can denote the time length of an interval trajectory $[\underline{\zeta}, \overline{\zeta}]$ with $|\zeta|$.

6.2.1 Signal Temporal Logic

We first briefly review the signal temporal logic (STL). We start with the Boolean semantics of STL. The domain $\mathbb{B} = \{True, False\}$ is the Boolean domain. Moreover, we introduce a set $\Pi = \{\pi_1, \pi_2, \ldots, \pi_n\}$ which is a set of predefined *atomic predicates*. Each of these predicates can hold values *True* or *False*. The syntax of STL is defined recursively as follows.

$$\varphi := \top \mid \pi \mid \neg \varphi \mid \varphi_1 \wedge \varphi_2 \mid \varphi_1 \vee \varphi_2 \mid \varphi_1 \mathbf{U}_I \varphi_2,$$

where ⊤ stands for the Boolean constant *True*, π is an atomic predicate in the form of an inequality $f(x) > 0$ where f is some real-valued function. $\neg$ (negation), $\wedge$ (conjunction), $\vee$ (disjunction) are standard Boolean connectives, and "**U**" is the temporal operator "until." We add syntactic sugar, and introduce the temporal operators "**F**" and "**G**" representing "eventually" and "always," respectively. I is a time interval of the form $I = [a, b)$, where $a < b$, and they are non-negative integers.

Definition 6.1 The Boolean semantics of an STL formula ϕ, for a trajectory ζ with the time length of $|\zeta|$ at time-step t is defined recursively as follows.

$$(\zeta, t) \models \pi \text{ if and only if } t \leq T \text{ and } f(\zeta(t)) > 0,$$
$$(\zeta, t) \models \neg \phi \text{ if and only if } (\zeta, t) \not\models \phi,$$
$$(\zeta, t) \models \phi_1 \wedge \phi_2 \text{ if and only if } (\zeta, t) \models \phi_1 \text{ and } (\zeta, t) \models \phi_2,$$
$$(\zeta, t_j) \models \varphi_1 \mathbf{U}_{[a,b)} \varphi_2 \text{ if and only if } \exists j' \in [j+a, j+b),$$
$$(\zeta, t_{j'}) \models \varphi_2 \text{ and } \forall j'' \in [j+a, j'), (\zeta, t_{j''}) \models \varphi_1.$$

Definition 6.2 Robust semantics quantifies the margin at which a certain trajectory satisfies or violates an STL formula ϕ at time-step t. The robustness margin of a trajectory ζ with respect to an STL formula ϕ at time-step t is given by $r(\zeta, \phi, t)$, where $r(\zeta, \phi, t)$ can be calculated recursively via the robust semantics [4].

$$r(\zeta, \pi, t) = f(\zeta(j)),$$
$$r(\zeta, \neg \phi, t) = -r(\zeta, \phi, t),$$
$$r(\zeta, \phi_1 \wedge \phi_2, t) = \min(r(\zeta, \phi_1, t), r(\zeta, \phi_2, t)),$$
$$r(\zeta, \varphi_1 \mathbf{U}_{[a,b)} \varphi_2, t_j) = \max_{j+a \leq j' < j+b} (\min(r(\zeta, \varphi_2, t_{j'}),$$
$$\min_{j+a \leq j'' < j'} r(\zeta, \varphi_1, t_{j''}))).$$

We can define the robustness margin of an interval trajectory in two views: *worst-case* and *best-case*. The worst-case view chooses the trajectory with the minimum corresponding robustness within an interval trajectory (Eq. (6.1)). The best-case view chooses the trajectory with maximum corresponding robustness within an interval trajectory (Eq. (6.2)); thus, we define the robustness margin of an interval trajectory in two views, as follows.

$$\underline{r}([\underline{\zeta}, \overline{\zeta}], \varphi, t_j) = \min_{\zeta \in [\underline{\zeta}, \overline{\zeta}]} r(\zeta, \varphi, t_j), \tag{6.1}$$

$$\overline{r}([\underline{\zeta}, \overline{\zeta}], \varphi, t_j) = \max_{\zeta \in [\underline{\zeta}, \overline{\zeta}]} r(\zeta, \varphi, t_j). \tag{6.2}$$

Syntax DAG Any STL formula can be represented as a syntax-directed acyclic graph, i.e., syntax DAG. In a syntax DAG, the nodes are labeled with atomic predicates or temporal operators that form an STL formula [7]. For instance, Figure 6.1a shows the unique syntax DAG of the formula $(\pi_1 \, \mathbf{U} \, \pi_2) \wedge \mathbf{G}(\pi_1 \vee \pi_2)$, in which the subformula π_2 is shared. Figure 6.1b shows the arrangement of the identifiers of each node in the syntax DAG ($i \in \{1, \ldots, 7\}$).

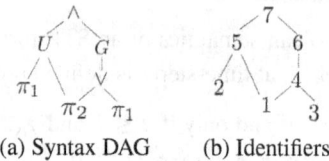

(a) Syntax DAG (b) Identifiers

Figure 6.1 Syntax DAG and identifier of syntax DAG of the formula $(\pi_1 \, \mathbf{U} \, \pi_2) \wedge \mathbf{G}(\pi_1 \vee \pi_2)$.

6.3 Problem Formulation

Problem Input As the input of this problem, in addition to a sample $Z \subset (\mathbb{R}^m)^* \times \{0, 1\}$ consisting of labeled signals, we have a finite set of predicates Π. The set of predicates consists of the atoms for the prospective STL formulas.

We define a *loss* function which assigns a real value to a given sample Z and an STL formula φ. Intuitively, the function evaluates how "well" the STL formula φ classifies a sample. While there are numerous ways of defining it (e.g., quadratic loss function, regret, etc.), we use the definition:

$$l(Z, \varphi) = \sum_{(\zeta, l) \in Z} \frac{|V(\varphi, \zeta) - l|}{|Z|}, \qquad (6.3)$$

which calculates the fraction of traces in Z which the STL formula φ misclassified.

Having defined the setting, we now formally describe the problem we solve:

Problem 6.3 Given Z, Π, find a minimal STL formula φ using predicates from Π such that $l(Z, \varphi) \leq \kappa$.

Intuitively, the margin on the achieved loss κ allows for a bounded fraction of the traces to be considered as noise.

Definition 6.4 Given a labeled set of interval trajectories $\mathcal{D}_{unc} = \{([\underline{\zeta}, \overline{\zeta}]^i, l_i)\}_{i=1}^{N_D}$, $l_i = +1$ represents the desired behavior and $l_i = -1$ represents the undesired behavior, an STL formula φ, which is evaluated at time t_0, perfectly classifies the desired behaviors and the undesired behaviors if the following condition is satisfied.

If $l_i = +1$, then $\forall \zeta \in [\underline{\zeta}, \overline{\zeta}]^i$, we have $(\zeta, t_0) \models \varphi$; if $l_i = -1$, then $\forall \zeta \in [\underline{\zeta}, \overline{\zeta}]^i$, we have $(\zeta, t_0) \models \neg \varphi$.

Definition 6.5 We define that two interval trajectories $[\underline{\zeta}, \overline{\zeta}]$ and $[\underline{\zeta}, \overline{\zeta}]'$ are separable if there exists at least one time-step t_j and one dimension k such that the two intervals $[\underline{\zeta}_j^k, \overline{\zeta}_j^k]$ and $[\underline{\zeta}_j^k, \overline{\zeta}_j^k]'$ do not intersect, i.e., $[\underline{\zeta}_j^k, \overline{\zeta}_j^k] \cap [\underline{\zeta}_j^k, \overline{\zeta}_j^k]' = \emptyset$.

Definition 6.6 We define that two finite sets of interval trajectories $\mathcal{Z}$ and $\mathcal{Z}'$ are separable if all pairs of interval trajectories $[\underline{\zeta}, \overline{\zeta}] \in \mathcal{Z}$ and $[\underline{\zeta}, \overline{\zeta}]' \in \mathcal{Z}'$ are separable.

By extension, we write that a labeled set of interval trajectories $\mathcal{D}_{unc} = \{([\underline{\zeta}, \overline{\zeta}]^i, l_i)\}_{i=1}^{N_D}$ is separable if $\left\{[\underline{\zeta}, \overline{\zeta}]^i \middle| l_i = +1\right\}$ and $\left\{[\underline{\zeta}, \overline{\zeta}]^i \middle| l_i = -1\right\}$ are separable.

Problem 6.7 Given a possibly non-separable set of labeled interval trajectories $\mathcal{D}_{unc} = \{([\underline{\zeta}, \overline{\zeta}]^i, l_i)\}_{i=1}^{N_D}$, compute an STL formula φ that maximizes $F(\mathcal{D}_{unc}, \varphi)$ such that $|\varphi| \leq N$, where N is a predetermined positive integer.

Theorem 6.8 *If a given labeled set of interval trajectories $\mathcal{D}_{unc} = \{([\underline{\zeta}, \overline{\zeta}]^i, l_i)\}_{i=1}^{N_D}$ is separable, then there exists at least one STL formula φ that perfectly classifies $\mathcal{D}_{unc}$.*

Given a set of N_D labeled interval trajectories $\mathcal{D}_{unc} = \{([\underline{\zeta}, \overline{\zeta}]^i, l_i)\}_{i=1}^{N_D}$, we define in Eq. (6.4) a function $\tilde{F}$ that gives the worst-case robustness margin of an interval trajectory $[\underline{\zeta}, \overline{\zeta}]^i$ with respect to φ or $\neg \varphi$ if $l_i = +1$ or $l_i = -1$, respectively.

$$\tilde{F}([\underline{\zeta}, \overline{\zeta}]^i, l_i, \varphi) := \begin{cases} \underline{r}([\underline{\zeta}, \overline{\zeta}]^i, \varphi, t_0), & \text{if } l_i = +1, \\ \underline{r}([\underline{\zeta}, \overline{\zeta}]^i, \neg \varphi, t_0), & \text{if } l_i = -1. \end{cases} \quad (6.4)$$

We then construct in Eq. (6.5) our objective function F. If we consider the STL formula for perfect classification of $\mathcal{D}_{unc}$: $\bigwedge_{\zeta \in [\underline{\zeta}, \overline{\zeta}]^i, l_i=+1} (\zeta \models_S \varphi) \wedge \bigwedge_{\zeta \in [\underline{\zeta}, \overline{\zeta}]^i, l_i=-1} (\zeta \models_S \neg \varphi)$, F would be the worst-case robustness margin of it. Hence, F represents the lower worst-case robustness margin amongst all the interval trajectories.

$$F(\mathcal{D}_{unc}, \varphi) := \min_{i=1,\ldots,N_D} \tilde{F}([\underline{\zeta}, \overline{\zeta}]^i, l_i, \varphi). \quad (6.5)$$

6.4 Inferring STL Formulas from Noisy and Uncertain Data

For solving Problems 6.3 and 6.7, we devise an algorithm based on ideas from the learning algorithm of Neider and Gavran [7] for inferring linear temporal logic (LTL) formulas that perfectly classify a sample. The satisfiability problem involves determining whether a logical formula is satisfiable or not. For the continuous domain's satisfiability problem where an explicit objective function is given, we can use *optimizing satisfiability modulo theories* (optSMT) solvers [2].

Propositional Logic Formula We utilize propositional formulas in the satisfiability problem to infer STL formulas from data with noise/uncertainty. If $\mathcal{P}$ is a set of propositional variables in the Boolean domain, then a propositional variable $p \in \mathcal{P}$ is a propositional formula. Moreover, if Φ and Ψ are propositional formulas, then $\neg \Phi$ and $\Phi \vee \Psi$ are propositional formulas as well. Consequently, we define a model of a propositional formula as a mapping $v \colon \mathcal{P} \to \mathbb{B}$. The semantics of this propositional valuation are given by a satisfaction relation defined as follows: $v \models p$ if and only if $v(p) = 1$, $v \models \neg \Phi$ if and only if $v \not\models \Phi$, $v \models \Phi \vee \Psi$ if and only if $v \models \Phi$ or $v \models \Psi$. If $v \models \Phi$, then we introduce v as a *model* of Φ. If such a model v exists, the propositional formula Φ is satisfiable. This model provides sufficient information to construct an STL formula based on that model.

MaxSMT Unlike SAT problems, SMT (Satisfiability Modulo Theories) deals with the satisfiability of first-order formulas over background theories. Similar to MaxSAT, MaxSMT is the problem of finding models that maximize the number of satisfiable clauses [8]. The formal problem definition remains the same as in the case of MaxSAT. For our algorithm, we will exploit the Partial Weighted MaxSMT for the theory of Linear Real Arithmetic (LRA). Standard SMT solvers like Z3 [3] can handle such problems.

6.4.1 MaxSMT-based Algorithm

Given that we are using MaxSAT solvers that possess the capability of handling Partial Weighted MaxSAT problems, we can solve a stronger version of Problem 6.3. In this stronger version, the loss based on which we search for STL formulas takes the following form:

$$wl(Z, \varphi, \Omega) = \sum_{(\zeta, l) \in Z} \Omega(\zeta) |V(\varphi, \zeta) - l|,$$

Algorithm 2: Learning algorithm based on maximum satisfiability
Input: A sample Z, Ω function, Threshold κ
1 $n \leftarrow 0$
2 **repeat**
3 $\quad$ $n \leftarrow n + 1$
4 $\quad$ Construct formula $\Phi_n^Z = \Phi_n^{str} \wedge \Phi_n^{stf}$
5 $\quad$ Assign weights to soft constraints in Φ_n^Z:
6 $\quad\quad$ $w(y_{n,0}^\zeta) = \Omega(\zeta)$ for $(\zeta, 1) \in Z$, and $w(\neg y_{n,0}^\zeta) = \Omega(\zeta)$ for $(\zeta, 0) \in Z$
7 $\quad$ Find model v using MaxSAT solver
8 **until** *Sum of weights of soft constraints* $\geq 1 - \kappa$
9 **return** φ_v

where Ω is a function that assigns a positive real-valued weight to each ζ in the sample in such a way that $\sum_{(\zeta,l)\in Z}\Omega(\zeta) = 1$.

Following the algorithm from [7], we translate the problem of inferring STL formulas into problems in Partial Weighted MaxSAT and then use an optimized MaxSAT solver to find a solution. More precisely, we construct a propositional formula Φ_n^Z and assign weights to its clauses in such a way that a model v of Φ_n^Z that satisfies all the hard constraints, satisfies two properties:

(i) Φ_n^Z contains sufficient information to extract an STL formula φ_v of size n, and

(ii) the sum of weights of the soft constraints satisfied by it is equal to $1 - wl(Z, \varphi_v, \Omega)$.

To obtain a complete algorithm (Algorithm 2), we increase the value of n (starting from 1) until we find a model v of Φ_n^Z that satisfies the hard constraints and ensures that the sum of weights of the soft constraints is greater than $1 - \kappa$.

Structural Constraints To include the features of STL in the structure of the syntax DAG, we introduce the following additional variables: $a_i \in \mathbb{N}$ and $b_i \in \mathbb{N}$ for $i \in \{1, \ldots, n\}$, which encode that the temporal bounds of Node i is $[a_i, b_i)$ when the operator labeling Node i uses temporal bounds (i.e., is $\mathbf{U}_I$), and $\theta_i \in \mathbb{R}$ for $i \in \{1, \ldots, n\}$, which encode the value of the parameterized threshold of Node i when a predicate is labeling Node i.

The formula Φ_n^{str} constrains the variables $x_{i,\lambda}$, $l_{i,j}$, $r_{i,j}$, a_i, b_i, and θ_i to encode a valid syntax DAG, such that a valuation v of these variables satisfying Φ_n^{str} describes an STL formula φ_v. A unique φ_v can be extracted from v as for STL, where we also assign interval $[a_p, b_p)$ and parameter θ_p to Node p when labeled with some λ that expect respectively an interval and a parameter.

Semantic Constraints We define Φ_u^n, which tracks the valuation of the STL formula encoded by Φ_n^{str} on u, as the conjunction of Formulas 6.6 to 6.9.

$$\bigwedge_{1\le i\le n}\bigwedge_{\pi\in\Pi} x_{i,\pi} \to \left[\bigwedge_{0\le\tau<|u|} y_{i,\tau}^u \leftrightarrow f_\pi(u_\tau) \ge \theta_i\right], \tag{6.6}$$

$$\bigwedge_{\substack{1\le i\le n \\ 1\le j<i}} x_{i,\neg} \wedge l_{i,j} \to \left[\bigwedge_{0\le\tau<|u|} \left[y_{i,\tau}^u \leftrightarrow \neg y_{j,\tau}^u\right]\right], \tag{6.7}$$

$$\bigwedge_{\substack{1\le i\le n \\ 1\le j,j'<i}} x_{i,\vee} \wedge l_{i,j} \wedge r_{i,j'} \to \left[\bigwedge_{0\le\tau<|u|} \left[y_{i,\tau}^u \leftrightarrow y_{j,\tau}^u \vee y_{j',\tau}^u\right]\right] \tag{6.8}$$

$$\bigwedge_{\substack{1\le i\le n \\ 1\le j,j'<i}} x_{i,\mathbf{U}_I} \wedge l_{i,j} \wedge r_{i,j'} \to$$
$$\left[\bigwedge_{0\le\tau<|u|}\left[y_{i,\tau}^u \leftrightarrow \bigvee_{\tau+a_i\le\tau'<\min(\tau+b_i,|u|)} \left[y_{j',\tau'}^u \wedge \bigwedge_{\tau+a_i\le t<\tau'} y_{j,t}^u\right]\right]\right]. \tag{6.9}$$

The correctness of the algorithm adapted to learn STL formulas follows from the correctness of the formula Φ_n^Z.

Theorem 6.9 *Given a sample Z, predicates Π and threshold $\kappa \in \mathbb{R}$, the MaxSMT-based STL learning algorithm terminates and outputs an STL formula φ that has $wl(Z,\varphi,\Omega) \le \kappa$ and is the minimal in size among all STL formulas that have predicates in Π and $wl(Z,\varphi,\Omega) \le \kappa$.*

6.4.2 Uncertainty-aware SMT-based Algorithm

In this section, we explain the algorithm we use to infer uncertainty-aware STL formulas for a given set of finite interval trajectories $\mathcal{D}_{unc}$ consisting of two labeled sets P_{unc} and N_{unc} where P_{unc} contains the interval trajectories with desired property (or behavior) and N_{unc} contains the interval trajectories with the undesired property (or behavior). For an arbitrary STL formula ϕ, we express that an interval trajectory ζ is *consistent* with ϕ if the interval trajectories in the positive set $\mathcal{DSTL}^{pos}$ satisfy ϕ, and the interval trajectories in the negative set $\mathcal{DSTL}^{neg}$ violate ϕ.

Algorithm 3 outlines the framework used for learning uncertainty-aware STL formulas, where the upper bound on the size of the true STL formula, denoted by N and the minimum robustness margin is given. In Algorithm 3, we convert the task of inferring STL formulas from interval trajectories to a satisfiability problem in propositional logic. In using propositional formulas in inferring uncertainty-aware STL formulas, we encode the robust semantics of the STL formulas as constraints such that the robust margin of the inferred formula maximizes the objective function in Eq. (6.5). To do so, we introduce the following variables to track the best-case and worst-case robustness margins of an interval trajectory (Eqs. (6.10) and (6.11), respectively). Then, we define the objective function in Eq. (6.12) using these variables.

$$\underline{y}^\zeta_{i,j} = \min_{\zeta \in [\underline{\zeta},\overline{\zeta}]} r(\zeta, \varphi_i, t_j), \quad (6.10)$$

$$\overline{y}^\zeta_{i,j} = \max_{\zeta \in [\underline{\zeta},\overline{\zeta}]} r(\zeta, \varphi_i, t_j), \quad (6.11)$$

$$Y^\zeta := \min_{i=1,\ldots,N_D} \begin{cases} +\underline{y}^\zeta_{i,0}, & \text{if } l_i = +1, \\ -\overline{y}^\zeta_{i,0}, & \text{if } l_i = -1. \end{cases} \quad (6.12)$$

Uncertainty-Aware Temporal Logic Inference algorithm (*TLI-UA*) Algorithm 3 outlines the steps for *TLI-UA*. We gradually increase the size of the sought-after formula n (starting from 1) until one of the specified stopping criteria (explained later) is met. In each iteration, we first create the formula of the structural constraints of the DAG (represented as Φ^{str}_n) at line 4.

On top of these constraints, we set the objective function Y^ζ (defined in Eq. (6.12)) at line 5. We then use OptSMT to obtain a model v of Φ^{str}_n that maximizes Y^ζ (line 6). We reconstruct the inferred formula and evaluate the achieved objective function value (line 7).

The first stopping criterion activates when the maximum iteration $N \in \mathbb{N}^+$ (a given parameter) is reached, resulting in a formula of maximum size N.

The second stopping criterion activates when the robustness margin threshold $R \in \mathbb{R}$ (another given parameter) is reached.

To address Problem 6.7, one can set N to the specified positive integer and $R = +\infty$ to disregard the second stopping criterion. With only N as the stopping criterion, the loop of the algorithm could be skipped, and we could directly start at $n = N$. In this scenario, Algorithm 3 returns one of the formulas of size N that maximizes $F(\mathcal{D}_{unc}, \varphi)$ (such a formula is not unique).

When a finite R is defined, Algorithm 3 returns an STL formula with a size possibly less than N but with $F(\mathcal{D}_{unc}, \varphi_v) \geq R$. This is particularly valuable when the expected size of the STL formula is unknown, and $N = +\infty$.

6.5 Numerical Evaluation

6.5.1 STL Inference from Noisy Data

In this section, we propose a case study and evaluate the performance of Algorithm 2.

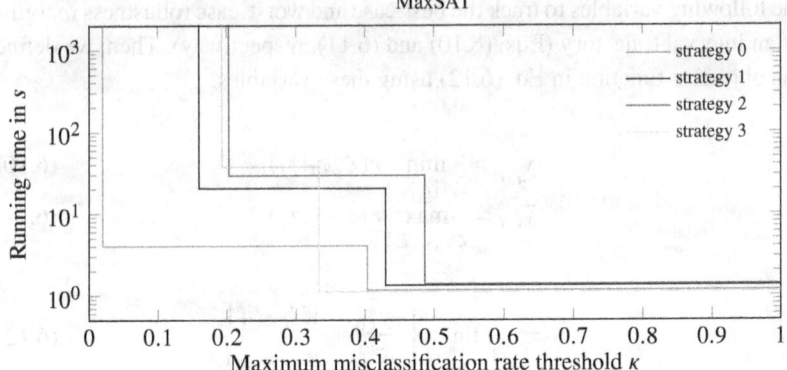

Figure 6.2 Impact of the threshold κ on the running time of MaxSAT, represented as a step function, for each strategy. Each step corresponds to a certain number of iterations in Algorithm 2, i.e., to an inferred STL formula of a certain size, with a misclassification rate lower than or equal to κ.

Our samples consist of traces generated by policies learned from reinforcement learning (RL) using *model-based reinforcement learning* (MBRL) algorithm [6]. These traces describe a Pusher-robot that interacts with a ball and a wall. The states of the system are composed of seven features in total, with their corresponding predicates: two Boolean features with corresponding predicates in the form $u_j = \theta$ for $j \in \{1, 2\}$ (for example, $u_1 = 1$ when the ball is in contact with the robot) and five continuous features with corresponding predicates in the form $u_j > \theta$ for $j \in \{3, \ldots, 7\}$ (for example, u_4 represents the total upper arm movement of the Pusher-robot). We note that this system is hybrid, but we simply consider Boolean features as continuous features.

We consider a total of four samples, each of them corresponding to an identified strategy of the Pusher-robot we would like to explain with an STL formula. Each sample contains 300 traces: 150 positive traces from the current strategy, and 150 negative traces from the other three strategies. We set a timeout of $900s$ on each run.

Figure 6.2 shows the running time of MaxSAT for different numbers of iterations in Algorithm 2, presented by the misclassification rate. For example, on the strategy 3 sample, we could infer the formula $\mathbf{F}_{[1,3)} s_0 = 0$ of size 2 with a misclassification rate of 19.33% (any $\kappa \in [0.1933, 0.3333)$ would have the same effect), with a runtime of 37 seconds. On the same sample, we could infer the formula $(s_5 > 0.003) \mathbf{U}_{[1,3)} (s_0 = 0)$ of size 3 with a misclassification rate of 15.67%, with a runtime of 38 minutes (which exceeds the chosen timeout but is a good example of non-trivial inferred STL formula).

Algorithm 3: *TLI-UA*

Input: Sample $\mathcal{D}_{unc} = \{([\underline{\zeta}, \overline{\zeta}]^i, l_i)\}_{i=1}^{N_D}$, Maximum iteration $N \in \mathbb{N}^+$, Minimum robustness margin $R \in \mathbb{R}$

1    $n \leftarrow 0$
2   **repeat**
3        $n \leftarrow n + 1$
4   Construct formula Φ_n^{str}
5   Assign objective function Y^ζ to be maximized
6   Find model ν using OptSMT solver
7   Construct φ_ν and evaluate $r \leftarrow F(\mathcal{D}_{unc}, \varphi_\nu)$
8   **until** $r \geq R$ or $n > N$
9   **return** φ_ν

6.5.2 STL Inference from Data with Uncertainties

In this subsection, we evaluate the performance of Algorithm 3. To do so, we generate 4 datasets non-separable datasets each containing up to 3 interval trajectories with a time length of up to 10. As a comparison reference, we use a MaxSMT-based algorithm on finitely many randomly sampled trajectories within the interval trajectories to infer STL formulas where we sample 200 trajectories from each interval trajectory in each dataset. We choose 1000 seconds for the timeout on each execution. Figure 6.3 illustrates the results where the execution time of *TLI-UA* is at most $1/100$ of the execution time of *TLI-RS* (for a dataset with 800 sampled trajectories in total). The obtained formulas φ with their corresponding optimal robustness margin $\underline{r}^*([\underline{\zeta}, \overline{\zeta}], \varphi, t_0)$ for the five datasets are respectively as follows. (1) $(x^1 < 10) \wedge (x^1 - x^2 > 8.9)$ and -10, (2) $\mathbf{G}_{[9,10)}(x^1 - x^2 > 4.5)$ and -5, (3) $\mathbf{F}_{[0,2)}(x^1 + x^2 > 8) \rightarrow (x^1 + x^2 > 8)$ and -4, and (4) $\mathbf{F}_{[1,10)}(\neg x^2 < 4.5)$ and -0.5.

6.6 Conclusion

In this chapter, we proposed two frameworks for inferring STL formulas from noisy data and data with uncertainties. We demonstrated with our first framework that allowing for a certain threshold of misclassified data when these classes are themselves subject to noise allows for faster computation of smaller STL formulas overall, making our method more robust to noise and less subject to overfitting. Our findings with our second framework indicated that incorporating awareness of uncertainties in STL inference accelerates the overall inference process when uncertainties are present. Leveraging uncertainty-aware

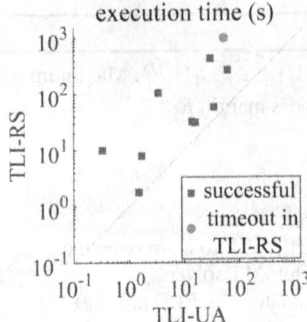

Figure 6.3 The comparison of the execution time between *TLI-UA* and *TLI-RS*, where the execution time of *TLI-UA* is at most 1/100 of the execution time of *TLI-RS* (for a dataset with 800 sampled trajectories in total).

STL inference to improve reinforcement learning represents a potential avenue for future research.

6.7 Acknowledgments

This work is partially supported by NSF CNS 2304863, CNS 2339774, ONR N00014-23-1-2505, DARPA HR001120C0032, ARL W911NF2020132, ACC-APG-RTP W911NF, NSF 1646522, and DFG 434592664.

References

[1] Baharisangari, Nasim, Gaglione, Jean-Raphaël, Neider, Daniel, Topcu, Ufuk, and Xu, Zhe. 2022. Uncertainty-Aware Signal Temporal Logic Inference. Pages 61–85 of: Bloem, Roderick, Dimitrova, Rayna, Fan, Chuchu, and Sharygina, Natasha (eds), *Software verification*. Springer International Publishing.

[2] Bjørner, Nikolaj, Phan, Anh-Dung, and Fleckenstein, Lars. 2015. νZ – An Optimizing SMT Solver. Pages 194–199 of: Baier, Christel, and Tinelli, Cesare (eds), *Tools and Algorithms for the Construction and Analysis of Systems – 21st International Conference, TACAS 2015, Held as Part of the European Joint Conferences on Theory and Practice of Software, ETAPS 2015, London, UK, April 11–18, 2015. Proceedings*. Lecture Notes in Computer Science, vol. 9035. Springer.

[3] de Moura, Leonardo Mendonça, and Bjørner, Nikolaj. 2008. Z3: An efficient SMT Solver. Pages 337–340 of: *TACAS*. Lecture Notes in Computer Science, vol. 4963. Springer.

[4] Fainekos, Georgios E, and Pappas, George J. 2009. Robustness of temporal logic specifications for continuous-time signals. *Theoretical Computer Science*, **410**(42), 4262–4291.

[5] Gaglione, Jean-Raphaël, Neider, Daniel, Roy, Rajarshi, Topcu, Ufuk, and Xu, Zhe. 2022. MaxSAT-based temporal logic inference from noisy data. *Innovations in Systems and Software Engineering*, **18**(3), 427–442.

[6] Nagabandi, Anusha, Konoglie, Kurt, Levine, Sergey, and Kumar, Vikash. 2019. Deep dynamics models for learning dexterous manipulation. Pages 1–12 of: *Third Conference on Robot Learning*.

[7] Neider, Daniel, and Gavran, Ivan. 2019. Learning linear temporal properties. Pages 148–157 of: *Proceedings of the Eighteenth Conference on Formal Methods in Computer-Aided Design, FMCAD 2018*.

[8] Sebastiani, Roberto, and Trentin, Patrick. 2017. On optimization modulo theories, MaxSMT and sorting networks. *CoRR*, abs/1702.02385.

[9] Xu, Zhe, and Duan, Xiaoming. 2021. Robust pandemic control synthesis with formal specifications: A case study on COVID-19 pandemic. In: *60th IEEE Conference on Decision and Control*.

PART IV

Metacognition with LLMS

7
Metacognitive Intervention for Accountable LLMs through Sparsity

TIANLONG CHEN

In recent years, the field of Natural Language Processing (NLP) has witnessed a significant leap forward, largely propelled by the development of Large Language Models (LLMs) [11, 12, 16]. Such advances have even made the arrival of superhuman AI plausible. However, despite their impressive capabilities, LLMs are not without flaws. One of the most critical challenges they face is the tendency toward "hallucination" or generating misleading or dishonest information [9]. This issue is particularly concerning in high-stakes domains such as medical diagnostics, where accuracy is paramount [10].

Currently, there is a gap in the literature regarding effective post-deployment interventions for LLMs to address these errors. Existing methods like few-shot or zero-shot prompting show promise but lack certainty in post-prompting performance and heavily rely on human expertise for error detection and prompt crafting [11, 15]. Other approaches, such as fine-tuning LLM parameters, come with risks of overfitting and forgetting previously learned information [4]. Furthermore, techniques like activation-level intervention may lead to prohibitive inference latency due to their repetitive nature [8]. Against this backdrop, we trifurcate the challenges for LLM intervention into three folds. ❶ Firstly, the "black-box" nature of LLMs obscures the malfunction source within the multitude of parameters, complicating targeted intervention. ❷ Secondly, rectification typically depends on domain experts to identify errors, hindering scalability and automation. ❸ Thirdly, the architectural complexity and sheer size of LLMs render pinpointed intervention an overwhelmingly daunting task.

Here we call for a novel paradigm for LLM intervention inspired by cognitive science principles. This paradigm, termed the Concept-Learning-enabled SparsE metAcognitive inteRvention (CLEAR), aims to equip LLMs with self-awareness in error identification and correction, emulating human cognitive efficiency. As shown in Figure 15.3, CLEAR would enable LLMs to form transparent decision-making pathways guided by human-comprehensible concepts, allowing for precise model intervention. This framework ensures more effective resource allocation for challenging instances, addressing the crucial need for honesty and reliability in superhuman AI systems (Figure 7.1).

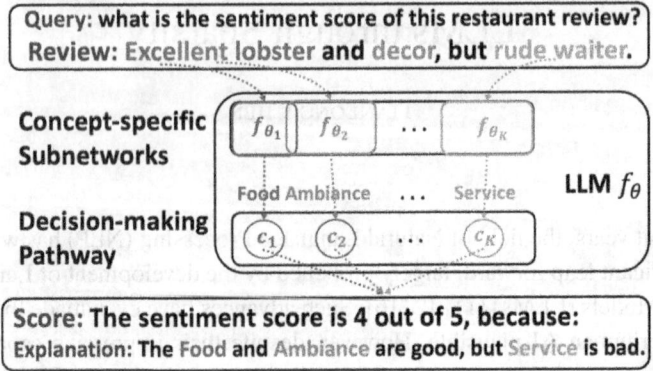

Figure 7.1 The illustration of an LLM f_θ, learned through *CLEAR*, outlines a transparent decision pathway for each input, seamlessly progressing from tokens, via pertinent subnetworks and concepts, to the final output. This pathway provides a unique interface for debugging and intervening on the LLM for erroneous samples.

Specifically, this chapter investigates the following two major research goals:

- *Sparsity-based Concept Learning*, which aims at learning transparent concept-specific sparse subnetworks.
- *Metacognitive Intervention*, which involves autonomous and efficient error identification and rectification.

7.1 Preliminary: Enable Concept Bottlenecks for LLMs

We focus on interpreting the predictions of fine-tuned LLMs for both classification and regression tasks. Given data $\mathcal{D} = \{(x^{(i)}, y^{(i)}, c^{(i)})_{i=1}^{n}\}$, where $x \in \mathbb{R}^d$ is the original text input, $y \in \mathbb{R}$ is the target label or $y \in \mathbb{R}^{d'}$ is the target-generated text, and $c \in \mathbb{R}^k$ is a vector of k concepts from the concept set C with $|C| = k$. We consider an LLM f_θ encoder that embeds an input text $x \in \mathbb{R}^d$ into its latent representation $z \in \mathbb{R}^e$. Vanilla fine-tuning strategy can be abstracted as $x \to z \to y$.

Concept-Bottleneck-Enabled Large Language Models The original concept bottlenecks in CBMs [6] come from resizing one of the layers in the CNN encoder to match the number of concepts. However, since LLM encoders typically provide text representations with much higher dimensions than the number of concepts, directly reducing the neurons in the layer would significantly impact the quality of learned text representation. To address this issue, we instead add a linear layer with the sigmoid activation, denoted as p_ψ, that projects the learned latent representation $z \in \mathbb{R}^e$ into the concept space $c \in \mathbb{R}^k$. This process can be represented as $x \to z \to c \to y$. Note that, unlike the previous works for image classification, each concept here does not need to be binary (i.e., present or not). We allow multi-class concepts, e.g., the concept "Food" in a restaurant review, which can be positive, negative, or unknown. We refer to the LLM and the projector (f_θ, p_ψ) together as the *concept encoder* and the complete model $(f_\theta, p_\psi, g_\phi)$ as *Concept-Bottleneck-Enabled Pre-trained Language Models* (CBE-LLMs). During training, CBE-LLMs seek to achieve two goals: (1) align concept prediction $\hat{c} = p_\psi(f_\theta(x))$ to x's ground-truth concept labels c and (2) align label prediction $\hat{y} = g_\phi(p_\psi(f_\theta(x)))$ to ground-truth task labels y. We jointly training LLM with the concept and task labels entails learning the concept encoder and label predictor via a weighted sum, $\mathcal{L}_{joint}$, of the two objectives:

$$\begin{aligned}\theta^*, \psi^*, \phi^* &= \arg\min_{\theta,\psi,\phi} \mathcal{L}_{joint}(x, c, y) \\ &= \arg\min_{\theta,\psi,\phi} [\mathcal{L}_{CE}(g_\phi(p_\psi(f_\theta(x)), y) + \gamma \mathcal{L}_{CE}(p_\psi(f_\theta(x)), c)].\end{aligned} \quad (7.1)$$

It's worth noting that the LLM-CBMs trained jointly are sensitive to the loss of weight γ. We set the default value for γ as 5.0 for its better performance.

7.2 Sparsity-based Concept Learning

This section first constructs concept-specific sparse subnetworks. To achieve this, we introduce *SparseCBM*, which surpasses traditional concept bottleneck models (CBMs) in both concept and task label prediction, offering insights into key neuron activations and their roles in learning specific concepts. Our framework starts with decomposing the joint optimization defined in Eq. (7.1.) according to each concept c_k, which is formulated as follows:

$$\begin{aligned}\theta^*, \psi^*, \phi^* &= \{(\theta_k^*)_{k=1}^K\}, \{(\psi_k^*)_{k=1}^K\}, \{(\phi_k^*)_{k=1}^K\} = \arg\min_{\theta,\psi,\phi} \sum_{k=1}^K \mathcal{L}_{joint}(x, c_k, y) \\ &= \arg\min_{\theta,\psi,\phi} \sum_{k=1}^K [\mathcal{L}_{CE}(g_{\phi_k}(p_{\psi_k}(f_\theta(x)), y) + \gamma \mathcal{L}_{CE}(p_{\psi_k}(f_\theta(x)), c_k)],\end{aligned}$$
$$(7.2)$$

where ϕ_k, ψ_k are the weights of the kth parameter of the projector and classifier, and θ_k is the subnetwork specific for the concept c_k, which is explained later. Since both of them are comprised of a single linear layer (with or without the activation function), the involved parameters for c_k can be directly indexed from these models and are self-interpretable [1, 6].

We further enhance interpretability by introducing 0/1 weight masks for each subnetwork, allowing for a clear understanding of the importance of weight in concept learning. These masks, represented as $\theta_{M_k} = M_k \odot \theta^*$, facilitate a more transparent decision-making pathway during inference. With well-optimized $\{(M)_{k=1}^K\}$, during inference, the decision-making pathway can be represented as:

$$\hat{y} = \sum_{k=1}^{K} \phi_k^* \cdot \sigma(\psi_k^* \cdot f_{\theta_{M_k}}(x)) = \sum_{k=1}^{K} \phi_k^* \cdot \sigma(\psi_k^* \cdot f_{M_k \odot \theta^*}(x)), \quad (7.3)$$

where $\sigma(\cdot)$ is the sigmoid activation function of the projector. This decision-making pathway defined in Eq. (7.3) factorizes the parameters of the SparseCBM, and can be optimized through one backward pass of the discomposed joint loss defined in Eq. (7.2) with $\theta_k^* = \theta_{M_k}$.

Those sparsity masks are calculated through a second-order unstructured pruning [5, 7] for LLMs. Initially, the joint loss $\mathcal{L}$ (we omit the subscript *joint* for brevity in subsequent equations) can be expanded at the weights of subnetwork θ_{M_k} via Taylor expansion:

$$\mathcal{L}(\theta_{M_k}) \simeq \mathcal{L}(\theta^*) + (\theta_{M_k} - \theta^*)^\top \nabla \mathcal{L}(\theta^*) + \frac{1}{2}(\theta_{M_k} - \theta^*)^\top H_{\mathcal{L}}(\theta^*)(\theta_{M_k} - \theta^*), \quad (7.4)$$

where $H_{\mathcal{L}}(\theta^*)$ stands for the Hessian matrix of the decomposed joint loss at θ^*. Since θ^* is well-optimized, we assume $\nabla \mathcal{L}(\theta^*) \approx 0$ as the common practice [5, 7]. Then, the change in loss after pruning is:

$$\Delta \mathcal{L}(\Delta \theta) = \mathcal{L}(\theta_{M_k}) - \mathcal{L}(\theta^*) \simeq \frac{1}{2} \Delta \theta^\top H_{\mathcal{L}} \Delta \theta, \quad (7.5)$$

where, $\Delta \theta = \theta_{M_k} - \theta^*$ signifies the change in LLM weights, that is, pruned parameters. Given a target sparsity $s \in [0, 1)$, we seek the minimum loss change incurred by pruning. Then, the problem of computing the sparsity masks can be formulated as a constrained optimization task:

$$\min_{\Delta \theta} \frac{1}{2} \Delta \theta^\top H_{\mathcal{L}}(\theta^*) \Delta \theta, \; s.t. \; e_b^\top \Delta \theta + \theta_b = 0, \quad \forall b \in Q, \quad (7.6)$$

where e_b denotes the bth canonical basis vector of the block of weights Q to be pruned. This optimization can be solved by approximating the Hessian at θ^*

via the dampened empirical Fisher information matrix [5, 7]. Hence, we can derive the optimized concept-specific masks $\{(M_k)_{k=1}^K\}$.

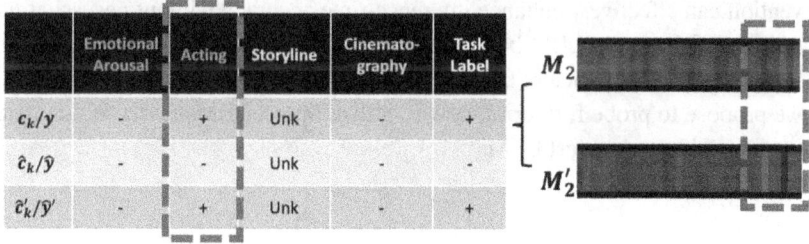

Figure 7.2 Illustration of the explainable prediction for a real-world example from the IMDB-C dataset using OPT as the backbone. The brown boxes with dashed lines indicate the test-time intervention on corresponding concepts by modulating the corresponding mask. M_2 and M_2' denote the parameter masks for the second concept, "Acting," before and after the intervention, respectively. We visualize M_2' after seeing all test samples.

Uncertainty-based Inference-time Intervention SparseCBMs exhibit the capability to allow inference-time concept intervention, which can elicit more honest knowledge from LLMs to mitigate hallucination. The core idea is to subtly modify the concept-specific masks for the LLM backbone when a mispredicted concept is detected. Specifically, parameters of the LLM backbone f_θ, projector p_ψ, and the classifier g_ϕ are frozen, while the concept-specific masks $\{(M_k)_{k=1}^K\}$ is kept trainable. During the test phase, if a concept prediction $\hat{c}_k$ for an input text x is incorrect, we acquire the gradient $\mathcal{G}_k(x)$ for the corresponding subnetwork $f_{\theta_{M_k}}$, and modulate the learned mask M_k accordingly. Inspired by [2, 14], we define the uncertainty scores for LLM parameters by the l_2-norm of the product of the gradient of the mask and the parameter weights: $\mathcal{S} = \|\mathcal{G}_k(x) \cdot \theta^*\|$. Then, we perform the following two operations based on the uncertainty scores:

(1) **Drop** a proportion of r unpruned weights with the lowest uncertainty scores: $\text{argmin}_m^{r \cdot |\theta|} S_m, \forall m \in |\theta_{M_k}|$;
(2) **Grow** a proportion of r pruned weights with the highest uncertainty scores: $\text{argmax}_m^{r \cdot |\theta|} S_m, \forall m \in |\theta \setminus \theta_{M_k}|$. Here m refers to the parameter index of the LLM backbone.

By dropping and growing an equal number of parameters, the overall sparsity s of the LLM backbone remains unchanged. This mask-level intervention is further optimized through the decomposed joint loss $\mathcal{L}_{joint}$ defined in Eq. (7.2). Note that r is set as a relatively small value (e.g., 0.01) to compel the model to retain the overall performance while learning from the mistake. As shown in Figure 7.2, our experiments validate that the proposed sparsity-based intervention can effectively enhance inference-time accuracy without necessitating training of the entire LLM backbone. Also, the intervened parameters provide insight into the parameters that contributed to each misprediction. Therefore, we propose to probe further on how to utilize SparseCBMs better to facilitate more reliable and honest LLMs.

7.3 Metacognitive Intervention

Another key goal of this chapter focuses on investigating if LLMs have the inherent capability to identify and rectify potentially erroneous samples, aiming to preempt initial mispredictions. Under this goal, we introduce the concept of Building **Concept-Specific Sparse Subnetworks through the Mixture of Concept Experts (MoCE)**. This goal involves the development of a MoCE framework, an innovative method for constructing pathways that are specifically anchored to certain concepts, thereby improving the precision of targeted interventions. Our approach is inspired by the mixture-of-expert (MoE) paradigms, notably those presented by [13]. The MoCE framework distinguishes itself by dynamically activating unique subsets of the network tailored to each input and emphasizes concept-driven processing. This results in the creation of sparse, efficient modules that are adept at fine-tuning the encoding of text inputs based on their intrinsic concepts.

The experts can be symbolized as $\{e_m\}_{m=1}^M$, where m signifies the expert index and M is the total count of experts. For each concept c_k, an auxiliary routing mechanism, dubbed $r_k(\cdot)$, is deployed. This mechanism identifies the top-T experts based on peak scores $r_k(x)_m$, with x representing the present intermediate input embedding. Generally, T is much smaller than N, which underscores the sparse activations among modules of the LLM backbone, making

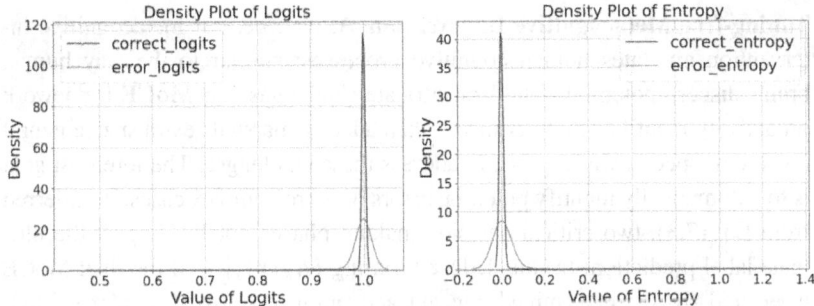

Figure 7.3 Logit entropy scrutiny. It can be observed that predictions with errors tend to demonstrate lower confidence and larger entropy.

the inference of the model more efficient. The output, x', emanating from the expert layer is:

$$x' = \sum_{k=1}^{K} \sum_{m=1}^{T} r_k(x)_m \cdot e_m(x); \quad r_k(x) = \text{top-T}(\text{softmax}(\zeta(x)), T), \quad (7.7)$$

where ζ is a shallow MLP representing learnable routers [3]. For the kth concept, the expert $e_t(\cdot)$ initially processes the given features, after which the router amplifies it using coefficient $r_k(x)_t$. The combined embeddings across concepts yield the output x'. The top-T operation retains the top T values, nullifying the others. Typically, a balancing mechanism, such as load or importance balancing loss [13], is implemented to avert the risk of representation collapse, preventing the system from repetitively selecting the same experts across diverse inputs. Transitioning to matrix representation for all MoE layers in the LLM structure, we derive

$$\hat{y} = \sum_{k=1}^{K} \phi_k \cdot \sigma(\psi_k \cdot f_{\theta_k}(x)) = \sum_{k=1}^{K} \phi_k \cdot \sigma\left(\psi_k \cdot \sum_{m=1}^{T} R_k(x)_m \cdot E_m(x)\right), \quad (7.8)$$

where $\sigma(\cdot)$ is the sigmoid projector's activation function, with $R(\cdot)$ and $E(\cdot)$ symbolizing matrix incarnations of all expert layer routers and experts. Crucially, Eq. (7.8) portrays a factorized decision trajectory, streamlining the classification framework. This can be optimized through a single backward iteration of the composite loss as outlined in Eq. (7.7). Note that Eq. (7.8) accomplishes a **core objective**: during inference, the LLM backbone's final classifications intrinsically rely on the learned routing policies, the chosen experts, and the perceived concepts. This unique accountability offers an interface for precise error identification and interventions.

Tuning-free Metacognitive Intervention At its core, our metacognitive intervention emulates human cognitive processes: similar to the way human brains discern potential pitfalls or intricate challenges, our **MoCE** framework proactively identifies these issues. It then adeptly marshals extra sparse neural resources, specifically experts, to address these challenges. The foremost goal is to automatically identify potential errors or more complex cases. As inferred from Eq. (7.3), two critical decision-making phases notably impact the ultimate label prediction: (a) the deduced routing $\{R_k(x)\}_{k=1}^{K}$ of the final MoCE layer, and (b) the determined concept activation $\hat{a} = \{\hat{a}_k\}_{k=1}^{K} = \psi \cdot f_\theta(x)$. Intuitively, an elevated entropy of predictive logits denotes a more dispersed distribution over experts or concept options, signifying lower model confidence and pinpointing instances that deserve additional attention. For this purpose, the Shannon entropy is utilized for logits within the routine and concept activation: $H(p) = -\sum_{j=1} \texttt{softmax}(l_j) \log(\texttt{softmax}(l_j))$.

For illustration, the distributions of logits and entropy for concept prediction are depicted using kernel density estimation in Figure 7.3. It is evident that predictions with errors tend to demonstrate lower confidence and augmented entropy, reinforcing our premise. For automation, as we iterate through the concepts, K-Means clustering is employed to divide confidence levels into two clusters ($K = 2$). The subset with lower confidence is considered to stem from the more challenging instances. K-Means offers the advantage of determining thresholds dynamically, eliminating human involvement. If, for a single concept prediction relating to an instance, the confidence levels of both the routine and concept activation surpass the corresponding thresholds, we tag this concept prediction as potentially erroneous.

Once an erroneous prediction is identified, we allocate augmented computational resources to secure a more reliable prediction. This operation can be easily achieved by setting the maximum expert number from T to a larger number T' for the router ζ as $r_k(x) = \texttt{top-T}(\texttt{softmax}(\zeta(x)), T')$. Note that this operation is very efficient since no extra parameter tuning is involved.

▷ *Case Study*: To further illustrate, we present a detailed case study of the metacognitive intervention process in Figure 7.4. This depiction illuminates the transition of the predicted label for the concept "Cinematography" from incorrect "−" to correct "+," subsequently refining the final task label. Texts highlighted in red indicate the clues overlooked by insufficient experts. Moreover, by analyzing expert and concept activations before and after the intervention, we reveal the neural mechanics underpinning the intervention strategy at the subnetwork level, offering additional real-world implications. For instance, we can compute the influence I of each concept c_k to the final decision by the product of the concept activation $\hat{a}_k$ and the corresponding weight w_k in the

Metacognitive Intervention for Accountable LLMs 119

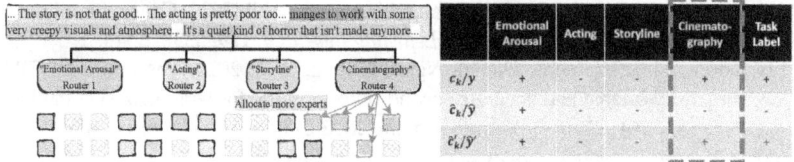

Figure 7.4 Illustration of an case study for the accountable metacognitive intervention.

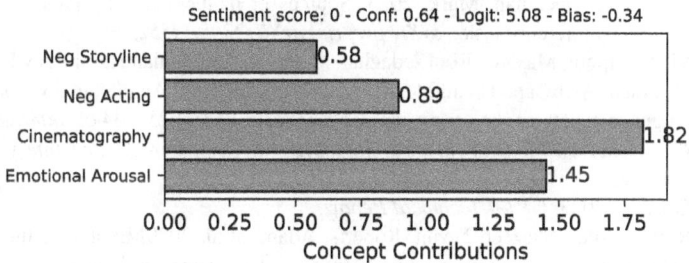

Figure 7.5 Contributions of Concepts.

linear classifier: $I(c_k) = \hat{a}_k \cdot w_k$. The results are visualized in Figure 7.5. This capability to correct and interpret the underlying causes for prediction errors further boosts the model's overall trustworthiness and usability.

References

[1] Tan, Zhen, Cheng, Lu, Wang, Song, Yuan, Bo, Li, Jundong and Liu, Huan. 2023. Interpreting pretrained language models via concept bottlenecks. *Advances in Knowledge Discovery and Data Mining: 28th Pacific-Asia Conference on Knowledge Discovery and Data Mining, PAKDD 2024, Taipei, Taiwan.*

[2] Evci, Utku, Gale, Trevor, Menick, Jacob, Castro, Pablo Samuel, and Elsen, Erich. 2020. Rigging the lottery: Making all tickets winners. Pages 2943–2952 of: *Proceedings of the 37th International Conference on Machine Learning.*

[3] Fedus, William, Zoph, Barret, and Shazeer, Noam. 2022. Switch transformers: Scaling to trillion parameter models with simple and efficient sparsity. *Journal of Machine Learning Research*, **23**(1), 5232–5270.

[4] Hardt, Moritz, and Sun, Yu. 2023. Test-time training on nearest neighbors for large language models. *arXiv preprint arXiv:2305.18466.*

[5] Hassibi, Babak, and Stork, David. 1992. Second order derivatives for network pruning: Optimal brain surgeon. Pages 164–171 of: *Proceedings of the 6th International Conference on Neural Information Processing Systems.*

[6] Koh, Pang Wei, Nguyen, Thao, Tang, Yew Siang, et al. 2020. Concept bottleneck models. *Proceedings of Machine Learning Research*, **119**, 5338–5348.
[7] Kurtic, Eldar, Campos, Daniel, Nguyen, Tuan, et al. 2022. The optimal BERT surgeon: Scalable and accurate second-order pruning for large language models. Pages 4163–4181 of: *Proceedings of the 2022 Conference on Empirical Methods in Natural Language Processing*.
[8] Li, Kenneth, Patel, Oam, Viégas, Fernanda, Pfister, Hanspeter, and Wattenberg, Martin. 2023. Inference-time intervention: Eliciting truthful answers from a language model. *arXiv preprint arXiv:2306.03341*.
[9] McKenna, Nick, Li, Tianyi, Cheng, Liang, Hosseini, Mohammad Javad, Johnson, Mark, and Steedman, Mark. 2023. Sources of hallucination by large language models on inference tasks. *arXiv preprint arXiv:2305.14552*.
[10] Monajatipoor, Masoud, Rouhsedaghat, Mozhdeh, Li, Liunian Harold, Jay Kuo, C-C, Chien, Aichi, and Chang, Kai-Wei. 2022. BERTHop: An effective vision-and-language model for chest x-ray disease diagnosis. Pages 725–734 of: *International Conference on Medical Image Computing and Computer-Assisted Intervention*. Springer.
[11] OpenAI. 2023. *GPT-4 Technical Report*.
[12] Raffel, Colin, Shazeer, Noam, Roberts, Adam, et al. 2020. Exploring the limits of transfer learning with a unified text-to-text transformer. *Journal of Machine Learning Research*, **21**(1), 5485–5551.
[13] Shazeer, Noam, Mirhoseini, Azalia, Maziarz, Krzysztof, et al. 2017. Outrageously large neural networks: The sparsely-gated mixture-of-experts layer. *arXiv preprint arXiv:1701.06538*.
[14] Sun, Mingjie, Liu, Zhuang, Bair, Anna, and Kolter, J Zico. 2023. A simple and effective pruning approach for large language models. *arXiv preprint arXiv:2306.11695*.
[15] Wei, Jason, Wang, Xuezhi, Schuurmans, Dale, et al. 2022. Chain-of-thought prompting elicits reasoning in large language models. Pages 24824–24837 of: *Advances in Neural Information Processing Systems*, 35.
[16] Zhou, Yongchao, Muresanu, Andrei Ioan, Han, Ziwen, et al. 2022. Large language models are human-level prompt engineers. In: *The Eleventh International Conference on Learning Representations*.

8
Metacognitive Insights into ChatGPT's Arithmetic Reasoning

NOEL NGU, PAULO SHAKARIAN, ABHINAV KOYYALAMUDI,
LAKSHMIVIHARI MAREEDU

We study the performance of a commercially available large language model (LLM) known as ChatGPT on math word problems (MWPs) from the dataset DRAW-1K. To our knowledge, this is the first independent evaluation of Chat-GPT. We found that ChatGPT's performance changes dramatically based on the requirement to show its work, failing 20% of the time when it provides work compared with 84% when it does not. Further, several factors about MWPs relate to the number of unknowns and number of operations that lead to a higher probability of failure when compared with the prior, specifically noting (across all experiments) that the probability of failure increases linearly with the number of addition and subtraction operations. We also have released the dataset of ChatGPT's responses to the MWPs to support further work on the characterization of LLM performance and present baseline machine learning models to predict if ChatGPT can correctly answer an MWP.

8.1 Introduction

The emergence of large language models (LLM) has gained much popularity in recent years. At the time of this writing, some consider OpenAI's GPT 3.5 series models as the state of the art [1]. In particular, a variant tuned for natural dialogue known as ChatGPT [2], released in November 2022 by OpenAI, has gathered much popular interest, gaining over one million users in a single week [3]. However, in terms of accuracy, LLMs are known to have performance issues, specifically when reasoning tasks are involved [1, 4]. This issue, combined with the ubiquity of such models, has led to work on prompt generation and other aspects of the input [5, 6]. Other areas of machine learning, such as meta-learning [7, 8] and introspection [9, 10], attempt to predict when a model will succeed or fail for a given input. An introspective tool, especially for certain tasks, could serve as a frontend to an LLM in a given application.

As a step toward such a tool, we investigate aspects of math word problems (MWPs) that can indicate the success or failure of ChatGPT on such problems. We found that ChatGPT's performance changes dramatically based on the requirement to show its work, failing 20% of the time when it provides work compared with 84% when it does not. Further several factors about MWPs can lead to a higher probability of failure when compared with the prior, specifically noting that the probability of failure increases linearly with the number of addition and subtraction operations (across all experiments). We also have released the dataset of ChatGPT's responses to the MWPs to support further work on the characterization of LLM performance. While there has been previous work examining the LLM performance on MWPs [4], such work did not investigate specific aspects that increase MWP difficulty nor did it examine performance on ChatGPT in particular.

The remainder of this chapter proceeds as follows. In Section 8.2, we describe our methodology. Then we describe our results in Section 8.3. Using these intuitions, we present baseline models to predict the performance of ChatGPT in Section 8.4. This is followed by a discussion of related work (Section 8.5) and future work (Section 8.7).

8.2 Methodology

MWP Dataset In our study, we employed the DRAW-1K dataset [11–13], which not only includes 1000 MWPs with associated answers but also template algebraic equations that one would use to solve such a word problem. As a running example, consider the following MWP.

One whole number is three times a second. If 20 is added to the smaller number, the result is 6 more than the larger.

We show ChatGPT's (incorrect) response to this MWP in Figure 8.1. The DRAW-1K dataset not only includes the correct answer, which in this case is 12 and 7 but also includes template equations used to solve the problem. For our running example, this consists of the equations $m - n = a - b$ and $c \times m - n = 0$. This information represents a symbolic representation of the problem, which can potentially be used to identify aspects that make such problems more difficult.

Entering Problems into ChatGPT at Scale At the time of our study, OpenAI, the maker of ChatGPT had not released an API. However, using the ChatGPT

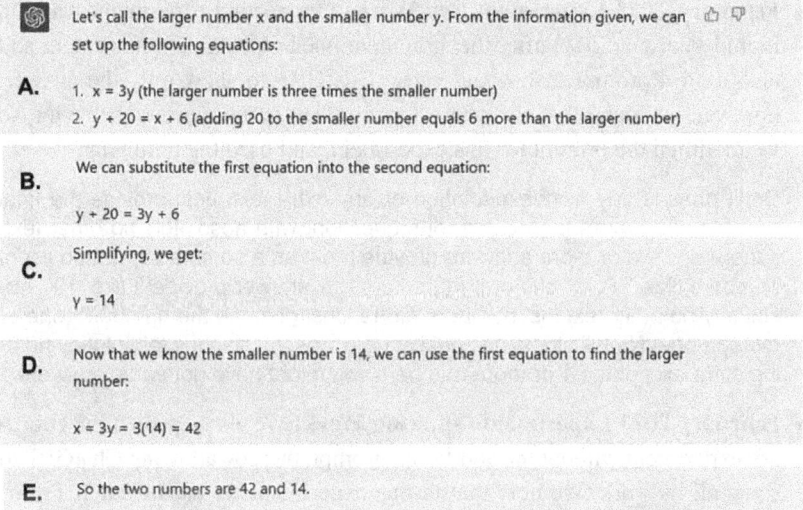

Figure 8.1 ChatGPT's response (Jan. 24, 2023) to MWP *One whole number is three times a second. If 20 is added to the smaller number, the result is 6 more than the larger.* In Step A it correctly identifies the set of equations needed to solve the problem and correctly simplifies it in Step B. However, it fails to correctly perform the algebraic operation in Step C (it should state $2y = 14$). This leads ChatGPT to obtain an incorrect result, returning 42 and 14 instead of 21 and 7.

CLI Python Wrapper[1] we interfaced with ChatGPT allowing us to enter the MWP's at scale. For the first two experiments, we would add additional phrases to force ChatGPT to show only the final answer. We developed these additions to the prompt based on queries to ChatGPT to generate the most appropriate phrase. However, we found in our third experiment that this addition impacted results. We ran multiple experiments to test ChatGPT's ability with these problems.

- **January 2023 Experiment (No Work)** Our first experiment was run in the early January 2023 prior to OpenAI's announcement of improved performance on mathematical tasks on January 30, 2023[2] and in this experiment we included the following statement as part of the prompt.

  Don't provide any work/explanation or any extra text. Just provide the final number of answers for the previous question, with absolutely no other text. If there are two or more answers provide them as a comma-separated list of numbers.

[1] We used ChatGPT CLI Python Wrapper by Mahmoud Mabrouk, see https://github.com/mmabrouk/chatgpt-wrapper
[2] https://help.openai.com/en/articles/6825453-chatgpt-release-notes

- **February 2023 Experiment (No Work)** Our second experiment was run in mid-February 2023 after the aforementioned OpenAI announcement and also used a prompt that would cause ChatGPT to show only the answer; however, we found that our original prompt led to more erratic behavior, so we modified the prompt for this experiment, and used the following.

  Don't provide any work/explanation or any extra text. Just provide the final number of answers for the previous question, with absolutely no other text. if there are two or more answers provide them as a comma separated list of numbers like: '10, 3,' etc; or if there is only 1 answer provide it like '10'. Absolutely no other text just numbers alone. Just give me the numbers (one or more) alone. No full stops, no spaces, no words, no slashes, absolutely nothing extra except the 1 or more numbers you might have gotten as answers.

- **February 2023 Experiment (Showing Work)** We also repeated the February experiment without the additional prompt, thereby allowing ChatGPT to show all its work. We note that in this experiment we used ChatGPT Plus which allowed for faster response. At the time of this writing, ChatGPT Plus is only thought to be an improvement to accessibility and not a different model.[3]

8.3 Results

The key results of this study are as follows: (1) the creation of a dataset consisting of ChatGPT responses to the MWPs, (2) identification of ChatGPT failure rates (84% for January and February experiments with no work and 20% for the February experiment with work), (3) identification of several factors about MWPs relating to the number of unknowns and number of operations that lead to a higher probability of failure when compared with the prior (Figure 8.3), (4) identification that the probability of failure increases linearly with the number of addition and subtraction operations (Figure 8.5), and (5) identification of a strong linear relationship between the number of multiplication and division operations and the probability of failure in the case where ChatGPT shows its work.

Dataset We have released ChatGPT's responses to the 1000 DRAW-1K MWP's for general use at **https://github.com/lab-v2/ChatGPT_MWP_eval**. We believe that researchers studying this dataset can work to develop models that can combine variables operate directly on the symbolic template, or even identify

---

[3] https://openai.com/blog/chatgpt-plus/

aspects of the template from the problem itself in order to predict LLM performance. We note that at the time of this writing, collecting data at scale from ChatGPT is a barrier to such work as APIs are not currently directly accessible, so this dataset can facilitate such ongoing research without the overhead of data collection.

Overall Performance of ChatGPT on DRAW-1K As DRAW-1K provides precise can complete answers for each problem, we classified ChatGPT responses in several different ways and the percentage of responses in each case is shown in Figure 8.2.

(i) *Returns all answers correctly.* Here ChatGPT returned all answers to the MWP (though it may round sometimes).
(ii) *Returns some answers correctly, but not all values.* Here the MWP called for more than one value, but ChatGPT only returned some of those values.
(iii) *Returns "No Solution."* Here ChatGPT claims there was no solution to the problem. This was not true for any of the problems.
(iv) *Returns answers, but none are correct.* Here ChatGPT returned no correct answers (e.g., see Figure 8.1).

Throughout this chapter, we shall refer to the probability of failure as the probability of cases 3 and 4 above (considered together). In our February experiment, we found that when ChatGPT omitted work, the percentages, as reported in Figure 8.2 remained the same, though they differed significantly when work was included. We also report actual numbers for all experiments in Table 8.1. We note that the probability of failure increases significantly when the work is not shown. However, when the work is included, ChatGPT obtains performance in line with state-of-the-art models (i.e., EPT [16, 18]) which has a reported 59% accuracy while ChatGPT (when work is shown) has fully a correct (or rounded) answers 51% of the time, but can be viewed as high as 80% if partially correct answers are included.

Factors Leading to Incorrect Responses We studied various factors from the templated solutions provided for the MWP in the DRAW-1K dataset and these included number of equations, number of unknowns, number of division and multiplication operations, number of addition and subtraction operations, and other variants derived from the metadata in the DRAW-1K dataset. We identified several factors that, when present, cause ChatGPT to fail with a probability greater than the prior (when considering the lower bound of a 95% confidence interval). These results are shown in Figure 8.3. One interesting aspect we noticed is that when the system would be required to show its work, the number of unknowns present no longer seems to increase the probability

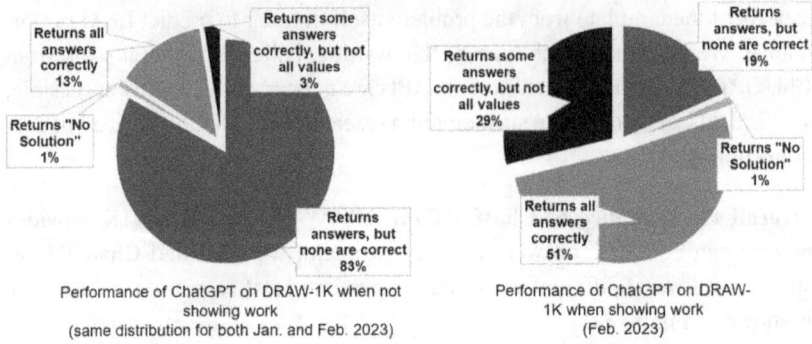

Performance of ChatGPT on DRAW-1K when not showing work (same distribution for both Jan. and Feb. 2023)

Performance of ChatGPT on DRAW-1K when showing work (Feb. 2023)

Figure 8.2 Overall results on the 1000 MWPs in DRAW-1K based on ChatGPT's response.

Response Type	Jan. 2023 (No work)	Feb. 2023 (No work)	Feb. 2023 (Showing work)
Returns answers, but none are correct	831	830	186
Returns "No Solution"	9	10	14
Returns all answers correctly	135	134	513
Returns some answers correctly but not all values	25	26	287

Table 8.1 *Number of responses for each ChatGPT variant.*

of failure (this was true for all quantities of unknowns in addition to what is shown in Figure 8.3). Additionally, the number of multiplication and division operations, while increasing the probability of failure greater than the prior in the January experiment was not significant (based on 95% confidence intervals) in the February experiment (when work was not shown) – possibly a result of OpenAI's improvements made at the end of January. However, there was a significant relationship between the number of multiplication and division operations and failure when work was shown. In fact, we found a strong linear relationship ($R^2 = 0.802$) for this relationship in the case where work was shown.

Correlation of Failure with Additions and Subtractions Previous work has remarked on the failure of LLMs in multi-step reasoning [1, 4]. In our study, we identified evidence of this phenomenon. Specifically, we found a strong linear relationship between the number of addition and subtraction operations with

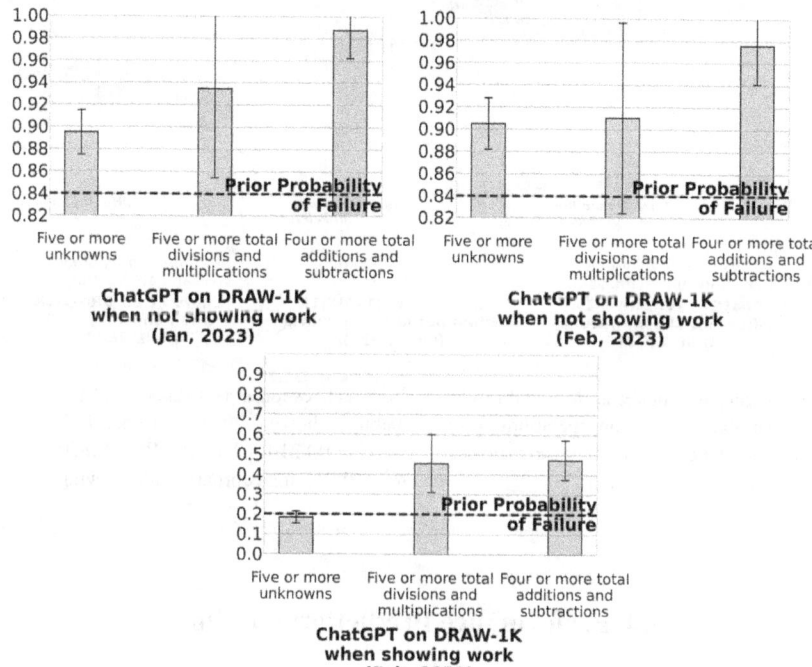

Figure 8.3 Aspects of MWPs that led to ChatGPT failure more often than the prior (95% confidence intervals shown).

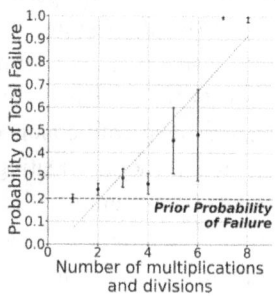

Figure 8.4 Additional finding specific to the February, 2023 experiment where ChatGPT displayed its work relating number of multiplications to probability of failure, $R^2 = 0.802$, 95% confidence intervals.

the probability of failure ($R^2 = 0.821$ for the January experiment, $R^2 = 0.870$ for the February experiment and $R^2 = 0.915$ when work was shown). We show this result in Figure 8.5. It is noteworthy that the relationship existed in all of our experiments, and seemed to be strengthened when ChatGPT included work in the result.

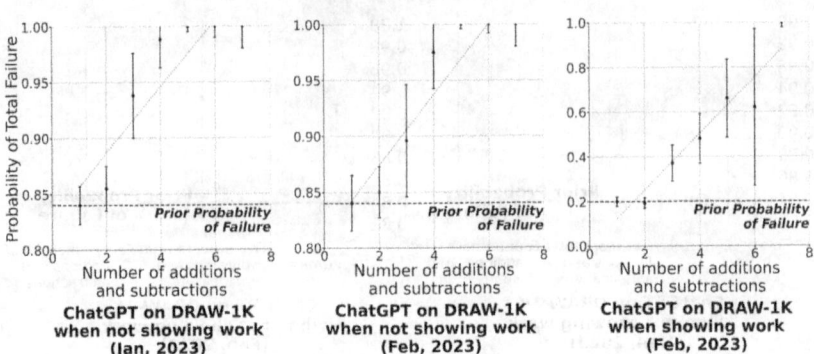

Figure 8.5 Increase in probability of an incorrect response as a function of the number of addition operations (prior probability shown with dashed line, 95% confidence intervals, linear regression with $R^2 = 0.821$ for January, $R^2 = 0.870$ for February without showing work and $R^2 = 0.915$ for February with showing work).

8.4 Performance Prediction Baselines

The results of the previous section, in particular, the factors indicating a greater probability of failure (e.g., Figures 8.3–8.5), may indicate that the performance of ChatGPT can be predicted. In this section, we use features obtained from the equations associated with the MWPs to predict performance. Note that here we use ground-truth equations to derive the features, so the models presented in this section are essentially using an oracle – we leave extracting such features from equations returned by ChatGPT or another tool (e.g., EPT [18]) to future work. That said, as these features deal with counts of operations, unknowns, and equations, a high degree of accuracy in creating the equations would not be required to faithfully generate such features.

Following the ideas of machine learning introspection [9, 10], we created performance prediction models using random forest and XGBoost. We utilized scikit-learn 1.0.2 and XGBoost 1.6.2, respectively. In our experiments, we evaluated each model on each dataset using a five-fold cross-validation and report average precision and recall in Table 8.2 (along with F1 computed based on those averages). In general, our models were able to provide higher precision than random on predicting incorrect answers for both classifiers. Further, XGBoost was shown to be able to provide high recall for predicting correct responses. While these results are likely not suitable for practical use, they do demonstrate that the features extracted provide some amount of signal to predict performance and provide a baseline for further study.

Version of ChatGPT	Model Type	Incorr. Prec.	Incorr. Recall	Incorr. F1
Jan. (No work)	RF	0.90	0.88	0.89
	XGBoost	0.95	0.22	0.36
Feb. (No work)	RF	0.94	0.89	0.91
	XGBoost	0.98	0.35	0.51
Feb. (Showing work)	RF	0.78	0.69	0.73
	XGBoost	0.77	0.59	0.67

Table 8.2 *Performance Prediction baseline models using ground truth equations.*

8.5 Related Work

The goal of this challenge dataset is to develop methods to introspect a given MWP in order to identify how an LLM (in this case ChatGPT) will perform. Recent research in this area has examined MWPs that can be solved by providing a step-by-step derivation [14–17]. While these approaches provide insight into potential errors that can lead to incorrect results, this has not been studied in this prior work. Further, the methods of the aforementioned research are specific to the algorithmic approach. Work resulting from the use of our challenge dataset could lead to solutions that are agnostic to the underlying MWP solver – as we treat ChatGPT as a black box. We also note that, if such efforts to introspect MWPs are successful, it would likely complement a line of work dealing with "chain of thought reasoning" for LLMs [5, 6] which may inform better ways to generate MWP input into an LLM (e.g., an MWP with fewer additions may be decomposed into smaller problems). While some of this work also studied LLM performance on Math Word Problems (MWPs), it only looked at how various prompting techniques could improve performance rather than underlying characteristics of the MWP that leads to degraded performance of the LLM.

8.6 Future Work

Understanding the performance of commercial black-box LLMs will be an important topic as they will likely become widely used for both commercial and research purposes. Further future directions would also include an examination of ChatGPT performance on datasets of other MWPs [13], investigating ChatGPT's nondeterminism, and exploring these studies on upcoming commercial LLM's to be released by companies such as Alphabet and Meta.

8.7 Acknowledgments

Some of the authors have been funded by the ASU Fulton Schools of Engineering.

References

[1] Fu, Yao. 2022. How does GPT obtain its ability? Tracing emergent abilities of language models to their sources. https://yaofu.notion.site.

[2] ChatGPT: Optimizing language models for dialogue. https://openai.com/blog/chatgpt/.

[3] Mollman, Steve. 2022. ChatGPT gained 1 million users in under a week. Here's why the AI chatbot is primed to disrupt search as we know it. www.yahoo.com/video/chatgpt-gained-1-million-followers-224523258.html.

[4] Hoffmann, Jordan, Borgeaud, Sebastian, Mensch, Arthur, et al. 2022. Training compute-optimal large language models. *arXiv preprint arXiv:2203.15556 [cs]*.

[5] Wei, Jason, Wang, Xuezhi, Schuurmans, Dale et al. 2022. Chain-of-thought prompting elicits reasoning in large language models. Pages 24824–2483 of: *Proceedings of the Thirty-Sixth International Conference on Neural Information Processing Systems*.

[6] Wang, Xuezhi, Wei, Jason, Schuurmans, Dale, et al 2023. Self-consistency improves chain of thought reasoning in language models. *arXiv preprint arXiv:2203.11171*.

[7] Hospedales, Timothy, Antoniou, Antreas, Micaelli, Paul, and Storkey, Amos. 2022. Meta-learning in neural networks: A survey. *IEEE Transactions on Pattern Analysis and Machine Intelligence*, **44**(9), 5149–5169.

[8] Zhou, Kaiyang, Liu, Ziwei, Qiao, Yu, Xiang, Tao, and Loy, Chen Change. 2023. Domain generalization: A survey. *IEEE Transactions on Pattern Analysis and Machine Intelligence*, **45**(4), 4396–4415

[9] Daftry, Shreyansh, Zeng, Sam, Bagnell, J Andrew, and Hebert, Martial. 2016. Introspective perception: Learning to predict failures in vision systems. *arXiv preprint arXiv:1607.08665 [cs]*.

[10] Ramanagopal, Manikandasriram Srinivasan, Anderson, Cyrus, Vasudevan, Ram, and Johnson-Roberson, Matthew. 2017. Failing to learn: Autonomously identifying perception failures for self-driving cars. *arXiv preprint arXiv:1707.00051 [cs]*.

[11] Upadhyay, Shyam, Chang, Ming-Wei, Chang, Kai-Wei, and Yih, Wen-tau. 2016. Learning from explicit and implicit supervision jointly for algebra word problems. Pages 297–306 of: *Proceedings of the 2016 Conference on Empirical Methods in Natural Language Processing*.

[12] Upadhyay, Shyam and Chang, Ming-Wei. 2017. Annotating derivations: A new evaluation strategy and dataset for algebra word problems. *arXiv preprint 07197. arXiv.1609.07197*.

[13] Lan, Yihuai, Wang, Lei, Zhang, Qiyuan, et al. MWPToolkit: An open-source framework for deep learning-based math word problem solvers. https://ojs.aaai.org/index.php/AAAI/article/view/21723. doi:10.1609/aaai.v36i11.21723, number: 11.

[14] Gong, Zheng, Zhou, Kun, Zhao, Xin, Sha, Jing, Wang, Shijin, and Wen, Ji-Rong. 2022. Continual pre-training of language models for math problem understanding with syntax-aware memory network. Pages 5923–5933 of : *Proceedings of the 60th Annual Meeting of the Association for Computational Linguistics (volume 1: Long papers)*.

[15] Ki, Kyung Seo, Lee, Donggeon, Kim, Bugeun, and Gweon, Gahgene. 2020. Generating equation by utilizing operators: GEO model. Pages 426–436 of: *Proceedings of the 28th International Conference on Computational Linguistics*.

[16] Kim, Bugeun, Ki, Kyung Seo, Rhim, Sangkyu, and Gweon, Gahgene. 2022. EPT-x: An expression-pointer transformer model that generates explanations for numbers. Pages 4442–4458 of: *Proceedings of the Sixtieth Annual Meeting of the Association for Computational Linguistics (volume 1: Long Papers)*.

[17] Xia, Yuancheng, Li, Feng, Liu, Qing, et al. 2023. ReasonFuse: Reason path driven and global–local fusion network for numerical table-text question answering. *Neurocomputing*, **516**, 169–181.

[18] Kim, Bugeun, Ki, Kyung Seo, Lee, Donggeon, and Gweon, Gahgene. 2020. Point to the expression: Solving algebraic word problems using the expression-pointer transformer model. Pages 3768–3779 of: *Proceedings of the 2020 Conference on Empirical Methods in Natural Language Processing (EMNLP)*.

PART V

Metacognition in Learning Agents

9

Uncertainty Quantification's Role in Metacognition

GAVIN STRUNK

9.1 Introduction

It is no secret that AI has a significant impact on our daily lives, from making recommendations for what to watch on TV to answering our questions about almost anything. These areas have shown a huge leap in performance with the progress of AI technologies, but these systems do make mistakes. Generally, the consequences are low if an AI system recommends a TV show or product that does not pertain to your interest. However, more recently, these technologies are being applied in applications where the consequences for incorrect results or information are more severe. For example, if an autonomous car incorrectly perceives an obstacle, this could result in a crash that damages the occupants [8].

There is value in designing systems with the highest possible performance, but in complex applications, like driving a car, it is nearly impossible to simulate all possible scenarios an autonomous car will encounter. It is equally important to support robust operation when the AI system is not completely certain, and ideally, continue learning during operation to improve performance when the situation is encountered in the future. This adaptation requires an accurate estimation of the prediction uncertainty, which is not commonly implemented in deep neural networks. Figure 9.1 shows an example of a traditional deep neural

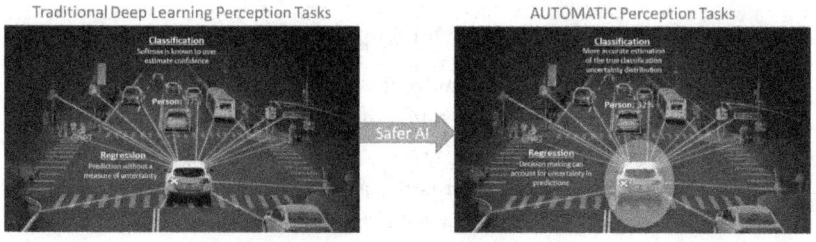

Figure 9.1 Autonomous driving example.

network (DNN) processing approach compared to an approach incorporating uncertainty. For classification tasks, it is common to use the confidence estimate from the final softmax layer as the uncertainty. This approach is notoriously unreliable due to domain shift, uncalibrated networks, and sensitivity to noise [5]. For regression tasks, common approaches provide a prediction of the desired continuous variables but do not provide a prediction of the uncertainty. This leaves a downstream fusion or decision-making algorithm to rely on the prediction with absolute certainty, which leads to an overly confident perception. The right-hand side of Figure 1.1 demonstrates the same scenario incorporating uncertainty estimation techniques. In this case, classification networks provide a more accurate uncertainty measure that is robust to out-of-distribution samples, domain shifts, and other sources of error. In the regression case, an uncertainty representation, such as normal distribution, is provided with the prediction. These techniques enable the use of a wealth of research in decision-making and data fusion algorithms that have been developed. However, being robust to training data is only the first part of the story to enable learning agents to function safely in highly dynamic environments around and with humans.

Adapting behavior based on your confidence is a natural response for humans. Consider the task of parking a car. If there is a large space and no cars, humans are typically able to park a car with a relatively quick maneuver. However, if the space is small and cars are either side are parked close to the line, typically the parking maneuver is executed with significantly less speed to account for the reduced margin for error. This could also be thought of as reduced certainty in our ability to successfully perform the maneuver at a higher speed. As a person performs the parking maneuver more times, they become increasingly proficient and can increase their speed. This is the behavior one would expect to achieve with a metacognitive learning agent.

9.2 Preliminaries

9.2.1 Metacognition

Metacognition is defined as the awareness and understanding of one's own thought process [4], but the interpretation of this definition has some ambiguity when applied to learning systems. For our purposes, we adopt the following modified definition: metacognition is the ability of an AI system to become aware of unknown objects or environments and understand how to improve its response to them. Metacognitive abilities allow an agent to continuously learn during its operations and improve its performance, as well as adapt to unforeseen objects or decisions that were not explicitly encountered during training.

A metacognitive agent is realized in two steps as shown in Figure 9.2. The agent must recognize high-uncertainty objects in the environment and probe the environment to gather information until the performance converges.

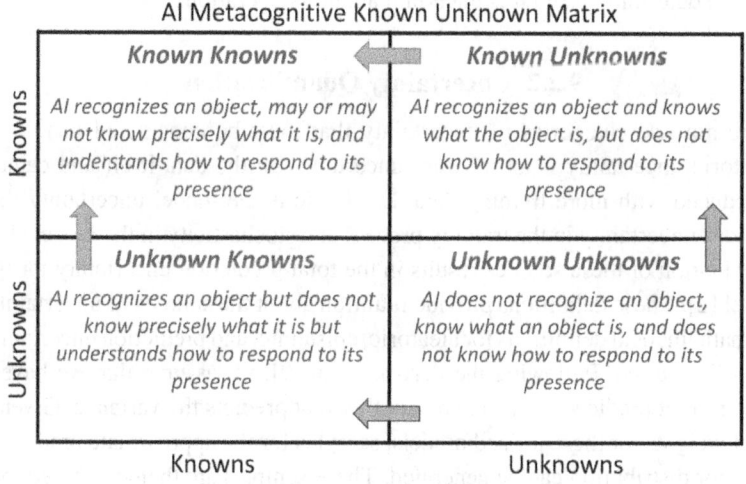

Figure 9.2 AI metacognition known unknown matrix.

In the first step, the AI senses or detects something that has a high uncertainty, which represents an unknown unknown. As the agent begins to track this object, there are two paths that can be taken. First, the agent could use unsupervised learning to recognize the object without knowing the specific class or object. In this case, the object becomes known as unknown, and the agent is able to recognize it without knowing exactly what it is. The second option is to identify the specific object class using a mechanism such as human feedback. In this case, the class is provided to the agent, and it learns to recognize the specific object. The second step involves a continual learning framework to choose safe actions that probe the environment to measure the performance under different decisions. This is challenging to do in the general sense while not exposing the system to risk. If we assume the system can safely probe the environment, then after sufficient interactions the agent can learn to perform well with this new object, which moves it from either an unknown known or known unknown to a known known. The agent does not necessarily know what this object is at this point, but it has learned how to respond when it observes it. This process could also involve a human to identify an object for the agent or identify key characteristics (e.g., friend or foe) to improve sample efficiency, but the principles are the same regardless. With these steps in mind, this chapter is primarily focused on the first step of detecting an unknown object, which is formally defined as an object that the agent has high uncertainty about. This highlights the importance of accurate uncertainty quantification for the learning agent to determine when it is uncertain about its operation.

9.2.2 Uncertainty Quantification

There are two categories of uncertainty that must be considered in DNNs. Aleatoric uncertainty is an inherent uncertainty in the data itself and cannot be reduced with more training data. Epistemic is the model uncertainty that includes uncertainty in the training process and stochasticity in the model. The combination of these sources results in the total prediction uncertainty for the model [2]. These definitions provide intuition about the sources of uncertainty, but mathematical definitions for aleatoric, epistemic, and prediction uncertainty are still required. Following the derivation in [9], we assume that we have a network that predicts the mean and another that predicts the variance. Given a set of weights for these trained models, samples for the approximate predictive posterior distribution can be generated. These samples are then combined into a Gaussian mixture conditional distribution:

$$p(y \mid x) \sim \mathcal{N}(\mu(x), \sigma^2(x)).$$

Focusing on the variance model, it follows that

$$\sigma^2(x) = M^{-1} \sum_i (\sigma_i^2(x) + \mu_i^2(x)) - \mu^2(x).$$

Separating the terms and replacing the summations with the expectation results in the following prediction uncertainty (PU):

$$\sigma^2(x) = \mathbb{E}_i[\sigma_i^2(x)] + \text{Var}_i[\mu_i(x)].$$

The prediction uncertainty is the sum of the aleatoric and epistemic uncertainties:

$$PU = AU + EU.$$

Therefore, the aleatoric uncertainty is defined as:

$$AU = \mathbb{E}_i[\sigma_i^2(x)],$$

while the epistemic uncertainty is defined as:

$$EU = \text{Var}_i[\mu_i(x)].$$

9.2.3 Related Work

We provide a brief summary of the uncertainty quantification approaches that are relevant to this chapter. However, this is a highly studied area so we recommend he interested readers to read [1, 5] for a more comprehensive review of uncertainty quantification approaches in deep learning.

Russell [9] quantified multivariate uncertainty for a regression task using a Gaussian density loss function to retrain the network with a covariance output. Next, they retrained the network with a Kalman Filter in the loop and demonstrated improved performance in their uncertainty accuracy by including the filter in the training process. Their approach is advantageous in regression applications where a Gaussian uncertainty assumption is reasonable. However, retraining the entire network can be prohibitive for large models and the approach is not readily extensible to different representations of uncertainty.

Another approach to DNN uncertainty quantification is evidential learning, which chooses an evidential prior distribution, over the likelihood function, as the uncertainty representation. Training samples are then used to learn the parameters of the prior distribution that best fits the "evidence" or samples. The prior distribution can be selected to provide various types of uncertainty (e.g. aleatoric, epistemic, out of distribution, etc.), making the approach flexible for a variety of tasks [2, 3, 10, 12]. Evidential learning typically uses the mean prediction to compute the loss function, which can be less accurate if complex state-dependent noise is present in the data because the uncertainty is not conditioned on the input as well.

Shen et al. [11] developed a post-hoc method for detecting out-of-distribution samples, which they called Meta Modeling. Their method of augmenting existing trained models with uncertainty prediction capability has computational advantages compared to retraining the entire network to add an uncertainty output. Our work in this chapter extends the concepts developed in this chapter to apply to uncertainty representations that are relevant to regression tasks.

The uncertainty quantification approach for a metacognitive agent needs to be a post-hoc method that does not require retraining the entire network. In addition to the computational benefits, the agent will be learning online or potentially from human feedback to reduce this uncertainty. Therefore, the agent needs to also update the uncertainty estimation to accurately reflect the new knowledge. A post-hoc approach allows these processes to occur independently. Next, we will discuss our post-hoc approach that generalizes to a variety of tasks by allowing a choice of suitable uncertainty distribution.

9.3 Meta Modeling

9.3.1 Overview

A Meta Model is an external network that is trained to output the parameters of a distribution of interest using the intermediate layers of a base network, as shown in Figure 9.3. The Meta Models can be trained to predict a variety of quantities including: aleatoric uncertainty, epistemic uncertainty, prediction uncertainty, and out-of-distribution (OOD) samples. The original authors in [11] studied OOD detection for classification on the Modified National Institute of Standards and Technology (MNIST) handwritten digits and fashion datasets (FashionMNIST). They studied several base network architectures to predict numbers from the MNIST dataset, and the Meta Model was trained to learn the parameters of a Dirichlet distribution. They demonstrated the Meta Model had a sharp distribution for in-distribution samples (MNIST) and more uniform behavior for OOD samples (FashionMNIST), and therefore they were able to successfully detect OOD samples by thresholding the sharpness of the Dirichlet distribution.

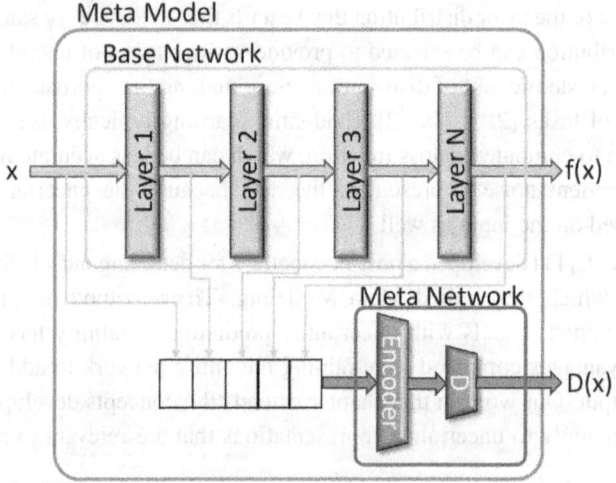

Figure 9.3 Meta model architecture overview.

To create the Meta Model, the intermediate outputs of the base network are used as inputs to an encoder network comprised of linear layers. The encoder network extracts meta-features, which is essentially a lower-dimensional latent space that learns which parameters in intermediate layers are the most critical to predict the desired output quantity. These meta-features are then used as input to an additional distribution network comprised of small linear layers that are designed to learn the distribution parameters that represent the quantity of interest. For example, in classification, the conjugate prior Dirichlet distribution is chosen. The Meta Model is then trained using the distribution loss between the predicted distribution and the true output distribution using Kullback–Leibler (KL) divergence. It is also worth noting that the Meta Model can provide the desired output directly or can be processed through an additional function. For example, the Dirichlet distribution does not directly output information about the sample being out-of-distribution, and therefore the distribution needs to be processed through a function that measures distribution "sharpness" to determine if the sample is OOD.

9.3.2 Meta Modeling for Regression Tasks

As previously mentioned, the choice of the statistical model depends on the objective of the Meta Model, and we are primarily interested in extending this concept to estimate prediction uncertainty in regression tasks. For regression tasks, a natural distribution choice is the multivariate normal distribution because the covariance representation lends itself well to combining multiple measurements with Kalman filtering methods. However, using a single Gaussian is known to poorly fit complex multimodal distributions that can arise in real sensor data uncertainty. If this occurs, other distributions can be selected (Laplace, Gaussian Mixture Model, etc.) and the method will apply similarly.

Once a suitable distribution has been chosen, the next step is to choose the intermediate layers of the base network that will be used to estimate the uncertainty. For most models, it is acceptable to use all intermediate layers because the Meta network is relatively small. However, for very large models, such as large language models (LLMs), it is likely that a subset of the layers should be chosen to reduce the computational requirement. The Meta feature networks are constructed with a single linear layer with a ReLU activation function. The input size of the Meta feature network is matched to the base network's layer output size, and the output is the number of parameters in the chosen distribution. Finally, we choose the architecture for the distribution network as a single linear layer and ReLU activation. Now we have constructed the full Meta Model for the desired base network.

There are a couple of options for training the Meta Model which both fall under a supervised learning approach. The first is to use the maximum likelihood for the loss function, and the second is a distribution-to-distribution metric, such as KL Divergence or Wasserstein distance. The benefit of using the maximum likelihood estimator (MLE) approach for a normal distribution is that it does not require the ground truth covariance for the loss function, which is normally not available in real-world datasets. However, the disadvantage is that the MLE has to be formulated for each desired output distribution, making it less generic. A distribution-to-distribution metric can be used to directly optimize the parameters to predict the truth parameters, and this approach works regardless of the chosen distribution. The downside is that the ground truth uncertainty representation needs to be known to successfully train.

Since we are selecting a normal distribution, we also need to ensure valid covariance matrix constraints are enforced during the Meta Model training and prediction. This is handled by first performing an LDL decomposition of the covariance matrix. We then apply exponential activation to the diagonal matrix D to ensure positive values. Finally, we only predict the upper triangular parameters and then complete the covariance matrix using a copy for the lower triangular terms. This ensures the covariance is symmetric. With the constraints satisfied, we can train with a traditional supervised learning approach with either the dataset that was used to train the base network or a dataset from the same distribution.

9.4 Experiments

9.4.1 Thermistor Experiment

The first experiment models the behavior of a thermistor probe measuring temperature and outputting resistance, shown in Figure 9.4. This single input single output example allows us to not only validate the accuracy of the Meta Model, but also inject different sources of uncertainty and compare against known ground truth uncertainties.

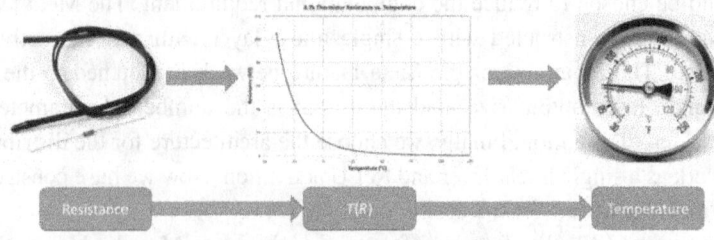

Figure 9.4 Thermistor example application.

Thermistors are known to have a nonlinear relationship between their output resistance and the temperature of the object it is measuring. Normally, this nonlinear relationship is calibrated by collecting a dataset of at least three resistance measurements of an object with a known temperature. The calibration data is then fitted with the Steinhart–Hart (SH) equation, shown in Eq. (9.1), to determine the parameters K_a, K_b, K_c.

$$T(R) = \frac{1}{K_a + K_b \ln R + K_c (\ln R)^3}. \tag{9.1}$$

The SH equation is then used as the transformation from resistance to temperature for that specific thermistor probe. For our example, we use the SH equation as the ground truth model for the resistance-to-temperature relationship and train a DNN to predict the temperature given resistance as an input.

The SH model parameters used to generate the thermistor model are shown in Table 9.1. The thermistor model also assumes the resistance values are between 100 and 15,000 Ohms. By limiting the resistance range of the training data, it is also possible to investigate cases when the measurement is out-of-distribution (OOD). Various types of noise are added to this ground truth model to create the training and test datasets.

Model Parameters	Value
K_a	1.283e3
K_b	2.362e4
K_c	9.285e8

Table 9.1 *SH model parameters.*

Two base networks are constructed to learn the mapping between resistance and temperature: deterministic and stochastic. Both cases are studied because the deterministic network is expected to have zero epistemic uncertainty because prediction uncertainty is conditioned on a fixed set of model parameters. As a result, the prediction uncertainty is equal to the aleatoric uncertainty. The stochastic network contains randomness in the model parameters, so the prediction uncertainty is not conditioned on a fixed set of weights and therefore epistemic uncertainty is non-zero. The base deterministic DNN structure is comprised of linear layers with ReLU activation in the inner layers, and the stochastic network is similarly constructed with the addition of a dropout layer added to the end of the ReLU layers. The dropout layer is also active during training and inference to produce stochastic outputs.

Meta Models were trained to predict normal distribution parameters and the results are shown in Figure 9.5. The Meta Models in both cases predicted the true variance with less than 7% error. The deterministic network case, shown in the top four figures, also confirms the Meta Model predicts zero epistemic

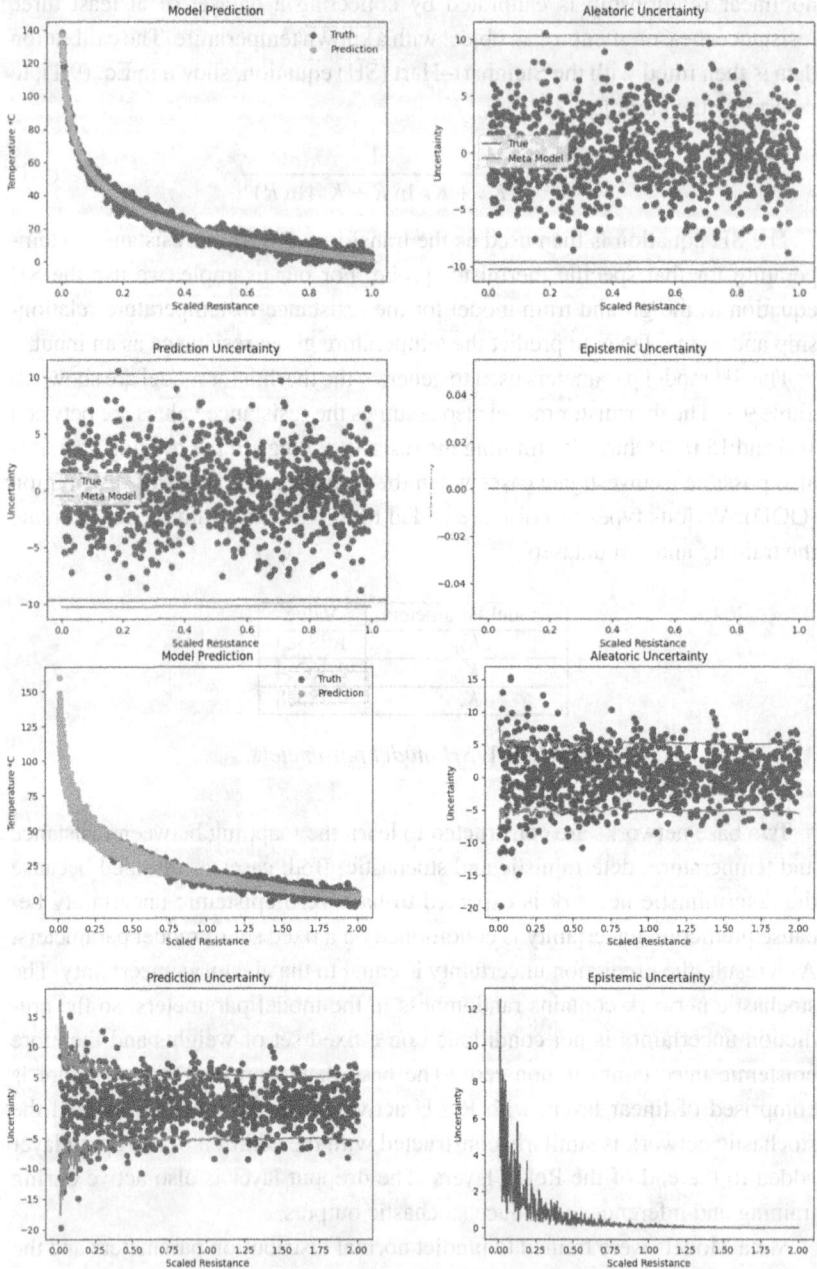

Figure 9.5 Results of Meta Models on uncertainty estimation.

uncertainty. The stochastic case shows the Meta Model also accurately captures the epistemic uncertainty. This example demonstrates our approach is able to separate the sources of uncertainty and provide accurate uncertainty estimations.

9.4.2 KITTI Dataset Experiment

The second experiment was performed on the real-world autonomous car dataset, KITTI [6]. In real-world data, it is typical that the ground truth uncertainty is not known. Therefore, we evaluate the accuracy of our uncertainty estimation indirectly by tracking the performance of a task that uses the uncertainty. The task for this experiment is pose estimation, which is performed by fusing data from the inertial measurement unit (IMU) and the LIDAR measurements. The pretrained LoRCoN-LO DNN [7] was used to predict the relative 6D pose of the vehicle from sequential LIDAR measurements, and a static covariance was computed for the IMU.

To evaluate the pose estimation task, the task metrics are a combination of both absolute and relative errors between the predicted and ground truth poses. The absolute trajectory metric provides a measure of overall accuracy throughout the entire run, whereas the relative metrics evaluate the performance of individual predictions. The definitions of the metrics are as follows:

- **Average Trajectory Error (ATE):** average distance error between the predicted trajectory and the ground truth trajectory
- **Relative Pose Translation Error (RPTE):** average error between the relative translation error computed over the entire trajectory
- **Relative Pose Rotation Error (RPRE):** average error between the relative angular error computed over the entire trajectory

We compared three cases to evaluate the performance of our Meta Models. First, we look at the IMU only to determine the baseline performance before incorporating the processed LIDAR data. Second, we fused the LIDAR with the IMU data by computing a static covariance from the LIDAR pose prediction error across the entire dataset. The third case combines the IMU and LIDAR data, but uses the Meta Model covariance prediction at each measurement update. We then compare the fusion results using pose metrics.

Figure 9.6 shows the pose trajectory results for KITTI sequence 02. It is seen in the left figure that the IMU only poses are noisy and drift significantly. This is expected because the global positioning system (GPS) measurement was not used, so there was no absolute measurement update to correct the drift. The center image shows the inclusion of the LIDAR measurements using

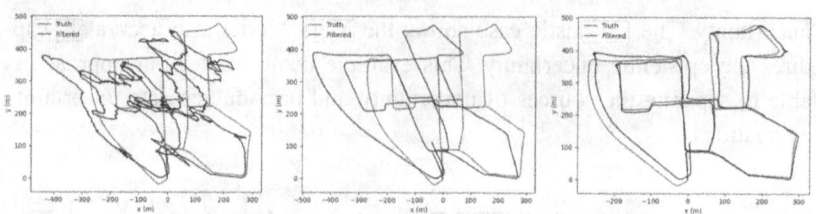

Figure 9.6 LoRCoN-LO results.

the static covariance dramatically smooths the trajectory, but the drift is still substantial. The image on the right shows the Meta Model fusion was able to smooth the trajectory, but also significantly reduce the drift. Table 9.2 shows the quantitative results which confirm both absolute and relative trajectory errors improved with the Meta-Model approach. Since the measurements were filtered with a Kalman Filter and the pose predictions used the same LoRCoN-LO base network, the improvement in the trajectory error is due to an improvement in the measurement uncertainty accuracy used to combine the measurements.

Metric	IMU	Static	Meta Model
ATE	123.3	19.38	17.64
RPTE	1.12	0.97	0.93
RPRE	0.92	0.818	0.78

Table 9.2 *KITTI dataset performance metrics.*

9.5 Discussion and Conclusion

The first step toward achieving a metacognitive agent is to determine objects or decisions with high uncertainty. To do this, the agent requires an accurate uncertainty estimate in a representation that can be combined with multiple sources of data. This chapter has shown a methodology to achieve this component using Meta Models.

9.6 Acknowledgments

This research was, in part, funded by the Defense Advanced Research Projects Agency (DARPA) under Agreement No. HR00112290109. The views, opinions, and/or findings expressed are those of the author(s) and should not be interpreted as representing the official policies, either expressed or implied, of the U.S. Government.

References

[1] Abdar, Moloud, Pourpanah, Farhad, Hussain, Sadiq, et al. 2021. A review of uncertainty quantification in deep learning: Techniques, applications and challenges. *Information Fusion*, **76**(December), 243–297.

[2] Amini, Alexander, Schwarting, Wilko, Soleimany, Ava, and Rus, Daniela. 2020. Deep Evidential Regression. Pages 14927–14937 of: *Proceedings of the 34th International Conference on Neural Information Processing Systems*.

[3] Duan, Ruxiao, Caffo, Brian, Bai, Harrison X, Sair, Haris I, and Jones, Craig. Evidential Uncertainty Quantification: A Variance-Based Perspective. Pages 2121–2130 of: *IEEE/CVF Winter Conference on Applications of Computer Vision*.

[4] Dunlosky, John, and Metcalfe, Janet. 2008. *Metacognition*. SAGE Publications.

[5] Gawlikowski, Jakob, Tassi, Cedrique Rovile Njieutcheu, Ali, Mohsin, et al. 2022 (January). A Survey of uncertainty in deep neural networks. *arXiv preprint arXiv:2107.03342* [cs, stat].

[6] Geiger, Andreas, Lenz, Philip, and Urtasun, Raquel. 2012 (June). Are we ready for autonomous driving? The KITTI vision benchmark suite. Pages 3354–3361 of: *2012 IEEE Conference on Computer Vision and Pattern Recognition, Providence, RI, USA*.

[7] Jung, Donghwi, Cho, Jae-Kyung, Jung, Younghwa, Shin, Soohyun, and Kim, Seong-Woo. 2023 (Mar.). LoRCoN-LO: Long-term recurrent convolutional network-based LiDAR odometry. *arXiv preprint arXiv:2303.11853* [cs].

[8] McCarthy, Roger L. 2021. Autonomous vehicle accident data analysis: California OL 316 reports: 2015–2020. *ASCE-ASME Journal of Risk and Uncertainty in Engineering Systems, Part B: Mechanical Engineering*, **8**(034502).

[9] Russell, Rebecca L, and Reale, Christopher. 2021 (June). Multivariate uncertainty in deep learning. *arXiv preprint arXiv:1910.14215* [cs, stat].

[10] Sensoy, Murat, Kaplan, Lance, and Kandemir, Melih. 2018. Evidential deep learning to quantify classification uncertainty. In: *Thirty-Second Conference on Neural Information Processing Systems*.

[11] Shen, Maohao, Bu, Yuheng, Sattigeri, Prasanna, Ghosh, Soumya, Das, Subhro, and Wornell, Gregory. 2022 (December). Post-hoc uncertainty learning using a Dirichlet meta-model. *arXiv preprint arXiv:2212.07359* [cs].

[12] Ye, Kai, Chen, Tiejin, Wei, Hua, and Zhan, Liang. 2024 (January). Uncertainty regularized evidential regression. *arXiv preprint arXiv:2401.01484* [cs].

10

The Role of Predictive Uncertainty and Diversity in Embodied AI and Robot Learning

RANSALU SENANAYAKE

10.1 Why Do We Need (or Not Need) Uncertainty?

We want our robots to *efficiently learn* how to *robustly act* in the unrestrained physical world. A robot's ability to experience and think through *diverse* possible scenarios – whether in perception or action – and their consequences helps enhance robustness and generalizability. In principle, countless conceivable scenarios can exist. However, learning all possible scenarios or gaining a complete understanding of them is not only impossible but often redundant. Instead, empowering the robots with the ability to discern what is important and what is not facilitates efficient learning and scrutiny of their own decisions. Having a sense of the *likelihood* of these scenarios aids in prioritizing their importance.

Characterizing the likelihood of events and actions about robots and the physical world they operate in inherently involves uncertainty, whether represented as probabilities, sets, or any other form. However, some techniques for handling uncertainty tend to be computationally expensive and sometimes even inaccurate, thereby defeating the purpose of working with uncertainty for efficient learning and enhancing robustness. This issue becomes particularly pronounced as models grow larger, especially with limitations of hardware in

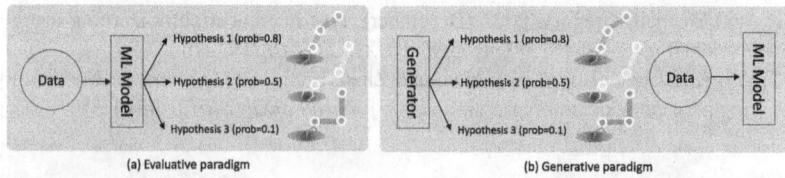

Figure 10.1 Two paradigms of diversity. (a) In the *evaluative paradigm*, the machine learning model provides various hypotheses with associated likelihoods based on data it has seen. (b) In the *generative paradigm*, we need to generate hypothetical outcomes.

The Role of Predictive Uncertainty and Diversity 149

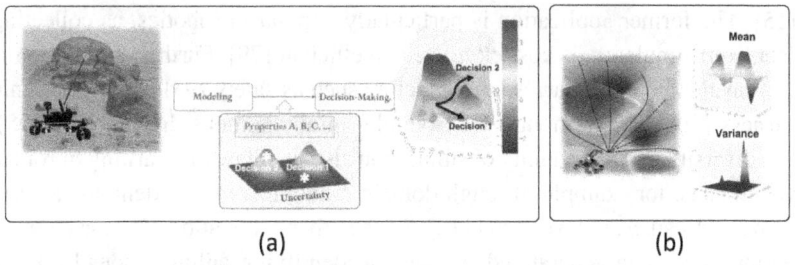

Figure 10.2 An example of epistemic uncertainty. (a) A rover builds an elevation map of Mars [1] to make decisions. If we have a distribution of maps instead of a single map, we can have different decision options, for instance to minimize risk, power consumption, etc. (b) The pits behind the boulders are not visible from the rover's front view. However, if the rover quantifies the *epistemic uncertainty*, then it knows how to make safe decisions. To represent this uncertainty, in its simplest form, we need the mean and variance of elevation [118].

embodied agents. Therefore, in any given application, it is crucial to balance the trade-off among accuracy, uncertainty, and computational complexity.

As shown in Figure 10.1, there are at least two main paradigms where the diversity and uncertainty of predictions prove beneficial: *evaluative paradigm* and *generative paradigm*. In the evaluative paradigm, the robot learns a model that captures its uncertainty about the world. For instance, consider the mapping example illustrated in Figure 10.2. If the robot cannot observe a particular area of the environment, say, due to occlusion, the model should quantify the uncertainty of that area as high. With this uncertainty, we can generate multiple maps with varying degrees of likelihood. In probabilistic terms, if we have a map that represents mean and variance, finitely many maps can be sampled from that map representation. Given the geography, a 10 km deep pit behind the boulder is unlikely. However, the presence of flat ground or even a 2 m deep pit behind the boulder is more plausible. Thus, access to uncertainty enables the generation of many such hypotheses. Having access to such diverse hypotheses aids in devising various decision options, which can then be evaluated to make the optimal decision. When the robot is capable of quantifying the uncertainty of the world, it can make risk-averse or risk-seeking decisions [97], which in turn helps with certifying robots and building trustworthiness.

In the generative paradigm, our objective is to generate diverse worlds, scenarios, or data. Generation can be performed using a machine learning (ML) model, such as text2video [18], or a digit twin, such as a physics-based simulator [73, 131] or virtual reality [130]. The generated cases can either be used to learn an ML model [68, 144] or to test an existing model [26,

115]. The former application is particularly popular in robotics, as collecting data from simulators is cost-effective and efficient [79]. Further, simulation is also invaluable in scenarios where factors such as safety [93] and algorithmic fairness [105] is paramount, for example, in autonomous driving [27, 105]. The diversity and uncertainty of simulation also facilitate the learning of robust ML models, for example, through domain randomization, as demonstrated in frameworks such as BayesSim [112]. As the other application of the generative paradigm, we can generate edge cases for identifying failure modes [26] and out-of-distribution scenarios [86, 98?].

10.2 When Do We Need (or Not Need) Uncertainty?

We already discussed that uncertainty helps with making more informed decisions, learning robust models, and testing existing pre-trained models. Uncertainty, also known as ambiguity, stems from the stochasticity of the world. Stochasticity, also known as randomness, is a comfort term we often use when we do not know how to model the reality[1] or we do not care about perfectly modeling the reality. Attempting to deal with all sources of stochasticity is often wasteful in real-time applications such as robotics. Therefore, it is important to understand the potential sources of uncertainty in embodied AI agents.

10.2.1 The Sources of Uncertainty

As also illustrated in Figure 10.3, broadly speaking, uncertainty could be due to internal sources or external sources:

(i) Physical limitations: Uncertainty can arise from errors in measurement devices, actuators, and human inputs. All measurement devices exhibit systematic, random, and gross errors. These errors can compound, exacerbating the uncertainty. For example, consider an autonomous vehicle that attempts to localize a pedestrian using a pre-computed map. The vehicle's LIDAR system introduces measurement error when determining the distance to the pedestrian. Additionally, the vehicle's precise location may be uncertain due to poor GPS signals around high-rise buildings, further increasing the uncertainty of the pedestrian's location estimation in the map. As another example, robot actuators, such as stepper motors, may not execute commands with perfect accuracy. This inaccuracy can worsen

---

[1] This statement is partially biased toward the philosophy of *determinism* though there are many arguments that supports the philosophy of *indeterminism*.

due to mechanical wear and tear, operation outside specified conditions (such as overloading or extreme temperatures), and calibration issues. Uncertainty can also stem from the way humans abstract their knowledge through models. When humans provide prior knowledge to models, in terms of preferences [16, 103] or Bayesian priors [43], this information may be incomplete and prone to errors.

(ii) Model limitations: Machine learning models are increasing in size, yet the constraints of limited onboard hardware and the need for real-time inference necessitate the use of smaller models. However, opting for simpler models or quantizing larger models to reduce their size invariably leads to errors and high uncertainty [85].

(iii) Partial observability: Occlusion [118], environmental factors such as fog or ambient darkness [105], and clutter [71] can introduce uncertainty into a robot's observations. Employing multiple sensors or sensor modalities can sometimes mitigate this uncertainty [83]. Factors such as human intentions may also be considered partially observable [78, 152]. Partial observations can arise from hardware limitations or a model's inability to process them, representing more internal sources of uncertainty. Conversely, changes in the operating environment can be viewed as an external source of uncertainty. Partial observability is explicitly addressed in certain decision-making models, such as Partially Observable Markov Decision Processes (POMDPs) [62, 64].

(iv) Environment dynamics: The environment in which a robot operates changes over space and time. For instance, objects on a tabletop might move and people in a house or street may walk. Such changes invariably result in uncertainty [57, 120, 141].

(v) Domain shifts: When a robot encounters situations not covered during the training of its neural networks, the robot becomes more uncertain. For instance, if an autonomous vehicle is trained exclusively with data from Arizona, it may struggle to operate in Boston, where snow is common [26, 98, 105], or in Australia, where it might encounter unfamiliar animals such as kangaroos [25].

10.2.2 A Few Examples of Uncertainty in Embodied AI

Let us now discuss a few examples of how these various sources of uncertainty affect three robotics applications.

Vision-language-based navigation: With the advancement of multi-modal ML models, the development of mobile robots capable of interacting with and

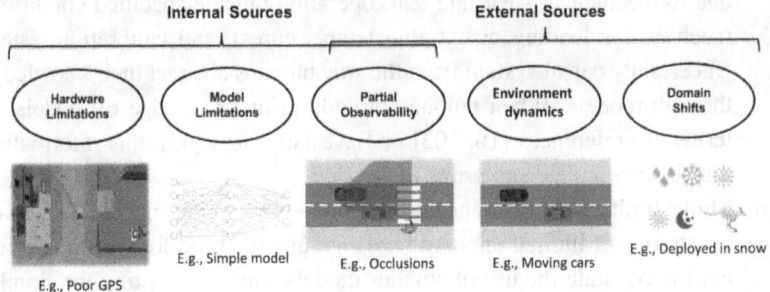

Figure 10.3 Sources of uncertainty.

assisting humans is becoming increasingly feasible. Currently, preliminary versions of such robots are already prevalent, serving purposes such as cleaning and providing aid to humans, exemplified by devices such as Amazon Astro and Roomba connected to Alexa. Consider a robot designed to navigate within homes. In order to navigate, it first needs to construct a map of the house. Traditionally, this involves creating an occupancy map, a process that requires exploring uncertain (i.e., unseen) areas of the house. In a house, walls are typically fixed in place, furniture is semi-permanent, and humans and pets are the most dynamic elements. Robots may encounter challenges such as becoming stuck, performing unsafe actions, or facing tasks that are beyond their physical capabilities – for instance, a Roomba cannot open doors, whereas a Spot robot can. Moreover, these robots will need to incorporate large language models (LLMs) and vision-language models (VLMs) to interact with human to understand their unknown intents and behaviors. Further, these large models needs to be small enough, by construction or quantization, to fit in limited hardware. However, reducing the model size to accommodate computing constraints compromises model performance and increase in uncertainty of estimates [133].

Manipulation Many robotics tasks in the future will require manipulation. Robots need to plan how to move the end effector of a robot arm from place A to place B, and then grasp it. Uncertainty in motion planning could be due to unknown or occluded objects along the robot's path. When grasping an object, objects can be occluded or the object shape can be uncertain [74]. Further, the deformability, type of material, unknown mass and friction, all can contribute to uncertainty. Even if these factors are known, the pose can be uncertain [54, 128], especially for mobile manipulators such as a manipulator on a quadruped. Multiple sensor modalities such as vision and touch helps to reduce the uncertainty. Uncertainty can also arise when the object is dynamic, for instance when picked up an object from a conveyor belt or from another robot or human [17].

Field Robotics Field robots are primarily designed for outdoor operations. They take various forms, such as drones, copters, rovers, boats, and legged robots with arms, autonomous underwater vehicles (AUVs), and other specialized configurations. These robots find applications in a range of activities, including surveillance, environmental monitoring, mining, agriculture, and exploration in underwater or space environments. The unstructured nature of these environments introduces significant uncertainty into their operation. Further, in environments like mines, underwater, or space, GPS signals are unavailable, necessitating the estimation of the robot's location using its trajectory or nearby landmarks through techniques such as Simultaneous Localization and Mapping (SLAM) [2]. However, SLAM does not typically account for uncertainties caused by occlusions and domain shifts. As another source of uncertainty, dynamic elements such as moving people and objects can make field robotics tasks challenging, especially in urban environments. Control also becomes particularly difficult in conditions where control commands may not be executed as expected due to air or water turbulence, or slippery surfaces caused by snow.

10.3 How Do We Quantify Uncertainty?

In embodied agents, we obtain data and then fit an ML model. Irrespective of the sources discussed in Section 10.2.1 and how well the model is trained, no model is perfect. Therefore, we need to quantify the uncertainty of the predictions. However, how we can quantify this uncertainty depends on the type of uncertainty.

10.3.1 Types of Uncertainties: The Known Unknowns and Unknown Unknowns

Uncertainty can be categorized into two types:

(i) *Aleatoric uncertainty (known unknowns)*: This uncertainty is also known as statistical uncertainty. It represents the inherent randomness of a system and cannot be reduced with more data. Aleatoric uncertainty is widely studied in probabilistic robotics [134] and is less challenging to work with.

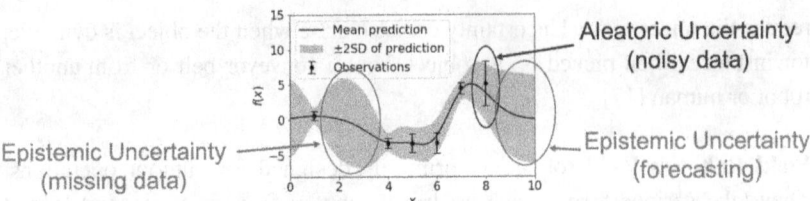

Figure 10.4 Types of uncertainty. Aleatoric uncertainty (known unknowns) is due to the inherent randomness of data whereas epistemic uncertainty (unknown unknowns) is due to lack of data. If we do not have data in a particular region of the input space or if we try to predict the future, the epistemic uncertainty is high.

(ii) *Epistemic uncertainty (unknown unknowns)*: This uncertainty is also known as model uncertainty or systematic uncertainty. Since it stems from the lack of knowledge about the world, the more data we collect, the more we can reduce this uncertainty. Epistemic uncertainty is not straightforward to estimate.

To explain the distinction between the two types of uncertainties more pictorially, consider Figure 10.4. When we take measurements for a given x, the measurements vary slightly each time. This randomness, depicted by error bars in the plot, is known as aleatoric uncertainty as it is the inherent noise. In regions where we lack data, (x, y), whether we are interpolating or extrapolating, the epistemic uncertainty is high. The further we are from data, the higher the uncertainty becomes. Estimating aleatoric uncertainty is relatively straightforward, and most frequentist statistical methods address it. However, quantifying epistemic uncertainty involves representing multiple models through a set, an ensemble, or a probability distribution. In deep ensembles [70] and Monte Carlo dropout, it is obvious that there are multiple models because they consider different weight combinations of the neural network. When a probability distribution is introduced overweights of a neural network, as in Bayesian methods, it implicitly creates an infinite number of neural networks where some models are more likely than others.

The term *Bayesian* is typically an overloaded term in robotics. In some classical application areas of robotics such as filtering, the term Bayesian is used whenever the Bayes theorem is applied. In more learning tasks, a distribution over the parameters of the ML model is introduced and the Bayes theorem is used to estimate the parameter distribution given data. In some sub-communities of Bayesian statistics, the term Bayesian is used only when an approximate Bayesian inference techniques such as Markov chain Monte Carlo or variational inference is used to solve a complex problem with many priors and hyperpriors. Under such nomenclature, even vanilla Gaussian processes (GPs), which are considered as a Bayesian nonparamatric technique in statistical ML, are not Bayesian enough as they do not introduce parameters over the hyperparameters of the GP kernels.

10.3.2 Measuring Uncertainty: Metrics

Historically, most uncertainty estimates are represented as probabilities though other forms exist. From a *measure theory* perspective, a probability measure μ is a real-valued function defined in a σ-algebra, while satisfying Kolmogorov axioms of (1) non-negativity, (2) unit measure, and (3) countable additivity [13]. In other words, a probability measure assigns a real number in the range [0, 1] to each event in the σ-algebra, such that the measure of the entire sample space is 1. These probabilities can be used to develop metrics to measure uncertainty in different ways.

Variance It measures the *uncertainty of a random variable*. A high variance indicates high uncertainty as it represents the dispersion (i.e., how far) from the mean. For a discrete random variable X with outcomes $\{x_1, x_2, ..., x_N\}$, associated probabilities $\{p_1, p_2, ..., p_N\}$, and mean $\mu = \sum_{n=1}^{N} p_n x_n$, the variance can be computed as

$$\sigma^2 = \sum_{n=1}^{N} p_n (x_n - \mu)^2. \tag{10.1}$$

Entropy It measures the *uncertainty of the possible outcomes of a random variable or a system*. As the entropy represents the expected amount of information that is needed to describe the random variable X, it can be computed as

$$H(X) = -\sum_{n=1}^{N} p_n \log p_n. \tag{10.2}$$

Entropy only considers the probability of outcomes rather than outcomes themselves. Entropy is measured in bits for the logarithmic base of 2. The more uncertain it is, the higher the entropy is. In other words, if every outcome is equally probable, then the entropy is high. For instance, for a binary classifier with the probability of the positive class p, the entropy can be computed as $-(p \log p + (1 - p) \log(1 - p))$. This entropy is 0 when the probability is 1 or 0 (i.e., very much certain) while it achieves its highest when the probability is 0.5 (i.e., most uncertain).

Negative Log Probability The negative log-likelihood (NLL) measures the *correctness of a classifier's prediction compared to the ground truth*. For a softmax output p of the ground truth class, the negative log-likelihood (NLL)[2] can be computed as $-\log p$. If an ML model outputs a high probability to the correct class, then the NLL value will be closer to 0. Conversely, if it outputs a low probability to the correct class, the NLL value will be high. Therefore, the lower the NLL, the better the model is.

The prediction of an ML model can be a probability distribution rather than a scalar. In such cases, NLL can be used to measure the *correctness of a classifier's predictive distribution compared to the ground truth*. If the predictive distribution of an ML model is a Gaussian with mean y_* and standard deviation σ_*, the negative log probability [117] can be computed as

$$-\log p(y_*|y_t) = \frac{1}{2}\log(2\pi\sigma_*^2) + \frac{(y_* - y_t)^2}{2\sigma_*^2} \qquad (10.3)$$

for ground truth value y_t.

Mahalanobis Distance The Mahalabonis distance measures the *distance between a point and a probability distribution*. It is defined as

$$\sqrt{(\mathbf{x} - \mu)^\top \Sigma^{-1}(\mathbf{x} - \mu)} \qquad (10.4)$$

for a point $\mathbf{x} \in \mathbb{R}^d$ from a multivariate distribution with mean $\mu \in \mathbb{R}^d$ and covariance matrix $\Sigma^2 \in \mathbb{R}^{d \times d}$. When $d = 1$, the distance is $||x - \mu||_2/\sigma$. In regression, the Mahalnobis distance can be used for measuring the distance between the ground truth value and prediction represented by the mean and variance. It is also used in out-of-distribution detection [98].

f-**divergences** In probability theory, the *difference between two probability distributions* can be measured by f-divergences. Special cases of these divergences include total variation distance, Hellinger distance, and α-divergences. The total variation distance is simply the half the absolute area difference between the two probability density or mass functions. The Hellinger distance for discrete probability distributions $p = \{p_1, p_2, ..., p_N\}$ and $q = \{q_1, q_2, ..., q_N\}$ is defined as

---

[2] In classification settings with softmax outputs, the equation of NLL is similar to that of cross entropy.

Figure 10.5 A summary of Section 10.3

> **Types of uncertainty**: Aleatoric, epistemic
>
> **Representing uncertainty**: Probabilities, sets/intervals, polytopes, functions
>
> **Assessing uncertainty**: Variance, entropy, NLL, Mahalanobis distance, total variation distance, Hellinger distance, α-divergences, Wasserstein distance
>
> **Uncertainty quantification methods**: Ensembles, MC dropout, Laplace approximation, variational inference, MCMC, conformal prediction, prior/-posterior networks, epistemic neural networks
>
> **Assessing uncertainty calibration**: Entropy, NLL, Brier score, confidence plots, ECE, ACE
>
> **Uncertainty calibration methods**: Temperature scaling, histogram bining, Platt scaling, isotonic regression, Beta calibration, BBQ

$$\frac{1}{\sqrt{2}}\sqrt{\sum_{n=1}^{N}(\sqrt{p_n} - \sqrt{q_n})^2}. \tag{10.5}$$

Squaring this metric and adding 1 gives the Bhattacharyya distance, $\sum_{n=1}^{N}\sqrt{p_n q_n}$. α-divergences play a significant role in many uncertainty quantification techniques such as variational inference. A special case is KL divergence, defined as

$$\mathbb{KL}[p\|q] = \sum_{n=1}^{N} p_n \log\left(\frac{p_n}{q_n}\right). \tag{10.6}$$

Since the smaller the divergence is, the more similar the distributions are, as a non-negative metric, $\mathbb{KL}[p\|q] = 0$ divergence indicates perfect similarity between the two distributions. Note that $\mathbb{KL}[p\|q] \neq \mathbb{KL}[q\|p]$ (Eq. 10.5).

Wasserstein Distance In the theory of optimal transport (OT), the Wasserstein distance measures the *difference between two probability distributions*. Defined by the Monge–Kantorovich theorem [139]. Wasserstein distance takes into account the geometry of the space.[3] Wasserstein distance can be defined for

---

[3] Wasserstein is proposed in Differential Geometry whereas KL divergence is proposed in Information Theory.

any metric space. 1-Wasserstein distance, also known as the Earth mover's distance (EMD), and 2-Wasserstein are more popular. For points x_i in p and x_j in q, the latter is defined as

$$W_2(p,q) = \min_M \sum_{i,j} M_{i,j} \|x_i - x_j\|_2^2 \qquad (10.7a)$$

subject to the constraints

$$M.\mathbb{1} = p, \qquad (10.7b)$$

$$M^\top.\mathbb{1} = q, \qquad (10.7c)$$

$$M \geq 0 \qquad (10.7d)$$

for a coupling matrix M. Constraints ensure that the probability of a sample in p matches with all other points in q. Intuitively, the OT problem determines the optimal way to move the probability distribution p to another q.

When the distributions are Gaussian or uniform distributions, the 2-Wasserstein distance has a closed form solution.[4] 1-Wasserstein distance also has a closed-form solution for one dimensional distributions. Whenever such simplifications are not available for the integer program that is expensive to solve, the Sinkhorn algorithm can be used [23] by introducing an entropic regularization[5] term,

$$\sum_{i,j} M_{i,j} \|x_i - x_j\|_2^2 - \lambda^{-1} \underbrace{\left(- \sum_{i,j} M_{i,j} \log M_{i,j} \right)}_{\text{entropy}}. \qquad (10.8)$$

10.3.3 Quantifying Uncertainty: Techniques

Our objective is to quantify uncertainty arising from various sources discussed in Section 10.2.1. Rather than representing the predictions as a single output, we want diverse outputs to represent the uncertainty. This predictive uncertainty can be represented probabilistically or by other means.[6]

---

[4] OT algorithms are implemented in POT: Python Optimal Transport library [37].
[5] λ can be set to a large number depending on the machine precision of the computer. The larger it is, the more accurate it would be but the slower the convergence is and more prone to computation errors.
[6] Historically, from the days of Jacob Bernoulli and Thomas Bayes in the 1700s, uncertainty estimations were treated in a more Bayesian sense. While the efforts by Pierre-Simon Laplace and Adolphe Quetelet by applying probability in physics and social sciences in 1800s helped to disseminate statistics in various fields and Andrey Kolmogorov's work helped with mathematical formalization of these ideas [66]. With the successful applications of probability in Biometrics and the establishment of journals such as Biometrika, frequentist statistics became more popular in early 1900s. Since many methods were inefficient, Arthur P.

Central to most probability-based uncertainty techniques is the Bayes theorem given by

$$\underbrace{q(z)}_{\text{approximate posterior}} \approx \underbrace{p(z|y)}_{\text{posterior}} = \frac{\overbrace{p(y,z)}^{\text{joint dist.}}}{\underbrace{p(y)}_{\text{evidence}}} = \frac{\overbrace{p(y|z)}^{\text{likelihood}}\overbrace{p(z)}^{\text{prior}}}{\underbrace{\int p(y|z)p(z)\mathrm{d}z}_{\text{evidence}}}. \tag{10.9}$$

The prior distribution indicates our prior belief. For instance, if a robot tries to estimate where it is in a room, it can have a rough estimate, say, represented as a Gaussian distribution, of its location first. The likelihood is how we attempt to represent our data, i.e., this is data y, given the location z. By multiplying with the likelihood with prior, it accounts for all potential locations and provides the location estimate given data, $p(z|y)$. To make the posterior a probability distribution, we have to divide this product with the evidence, also known as marginal likelihood.

In general, the prior can come from domain knowledge, rough estimates, or previous estimates. The latter is specially common in many sequential or iterative robotics estimation problems where the posterior at time t can be used as the prior in the next time step $t+1$. If we do not know anything about the prior, we can choose a non-informative prior. This is typically a uniform distribution, or more commonly in practice, it is an exponential distribution such as a Gaussian distribution with a large standard deviation. If we do not know about the prior, we can also introduce a probability distribution over the parameters of the prior which is called a hyper-prior $p(s)$, resulting in $p(y|z)p(z|s)p(s)$. The integral in the denominator of the Bayes theorem makes posterior estimates intractable for most choices of the prior and likelihood. The class of priors known as conjugate priors make the posterior estimation problem simpler. The choice of prior is sometimes an art; it should be good enough to make the posterior tractable but also good enough to represent the reality.

As shown in Figure 10.6, in most ML models, we introduce a prior distribution over the parameters of the model, $p(w)$, and estimate the posterior

Dempster and others were focusing on developing alternative techniques [143] to represent believes and uncertainty. With the advancement of computers in the 1980s, modern Bayesian inference techniques such as MCMC were developed. They became popular, especially after the application of Gibbs sampling in image processing [44] and they were the most successful way to train neural networks back then [96]. While MCMC techniques continued to grow, in the late 1990s and 2000s Michael I. Jordan and others popularized variational inference techniques as a tractable way to estimate Bayesian posterior probabilities. With the advancement of large neural networks, Bayesian inference techniques have also been struggling. By the time of writing this article, having foundations in probability theory, conformal predictions are gaining their popularity as they are straightforward to use in deep neural networks [7] though they do not provide precise probabilities.

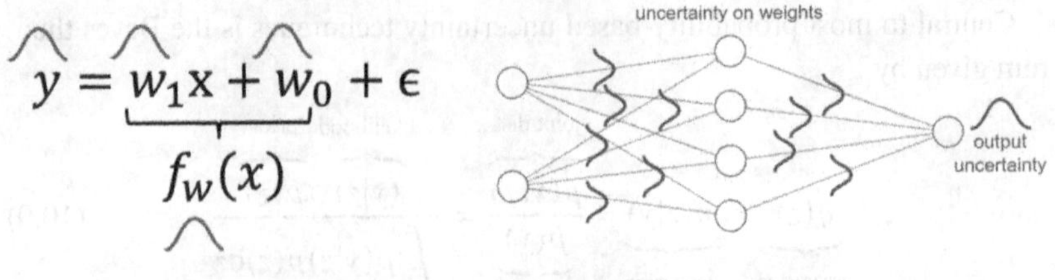

$$y = \underbrace{w_1 x + w_0}_{f_w(x)} + \epsilon$$

Figure 10.6 We need many models to quantify the epistemic uncertainty. This can be done explicitly by having multiple models as in ensembles or implicitly by introducing probability distributions over the parameters (weights of the neural network) of the ML model.

parameters given data, $p(w|\mathcal{D})$, using Bayes' theorem. Once we know this posterior distribution, the *predictive probability distribution* for a new input y_* can be obtained by integrating over posterior parameter estimates:

$$p(y_*|\mathcal{D}) = \int \underbrace{p(y_*|w)}_{\text{likelih.}} \underbrace{p(w|\mathcal{D})}_{\text{poster.}} dw. \qquad (10.10)$$

This predictive distribution can also be imperially approximated by sampling from the posterior distribution, $\{w_k \sim p(w|\mathcal{D})\}_{k=1}^K$, and computing the likelihood values, $\{p(y_*|w_k)\}_{k=1}^K$. The mean and variance can be computed from this set of predictive likelihood values.

Since posterior estimation becomes quickly intractable for complex models, mainly because of the integral in Eq. (10.9), we often have to resort to approximation techniques. Some of these approximation techniques such as variational inference and MCMC are explicit and derived from the first principle whereas some others are somewhat implicit such as MC dropout and ensembles. In what follows, we present a number of popular probabilistic and non-probabilistic techniques to quantify uncertainty in ML models.

Ensembles A collection of NNs can be independently trained, ideally in parallel for computational efficiency, with different weight initializations or other changes. During test time, the mean and variance can be calculated using the outputs of different NNs. If an ensemble was performed on NNs that are adversarially trained on common NN loss functions,[7] then the predictive uncertainties are more accurate [58, 70].

Monte Carlo Dropout To calculate uncertainty from MC dropout, the same input is passed through the network multiple times but a certain percentage

---

[7] To be precise, the loss should be a *proper scoring rule*.

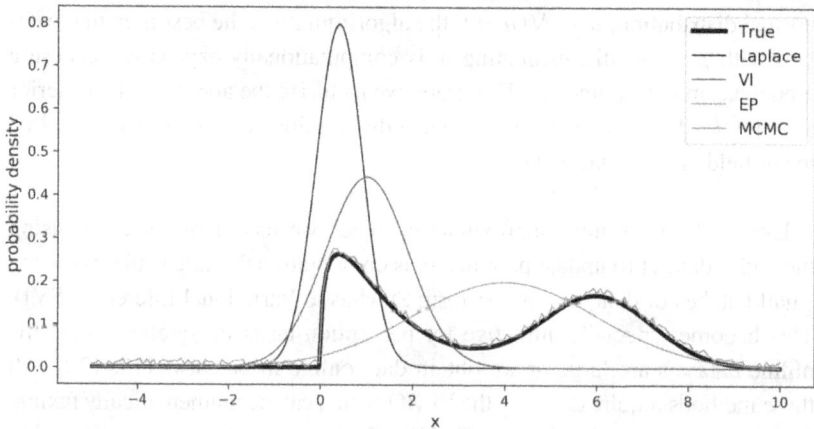

Figure 10.7 Approximate Bayesian inference techniques. The Laplace approximation aligns with the peak. Variational inference (VI) typically aligns with one of the modes of distribution. By contrast, expectation propagation (EP) considers the average over all modes, resulting in a more dispersed distribution. Markov Chain Monte Carlo (MCMC) can theoretically obtain the exact distribution if finitely many samples are taken.

of neurons are randomly disabled every time.[8] Since this process implicitly induces a distribution over neural network weights, the multiple output represents epistemic uncertainty. This process can also be thought as variational inference [40], however with a rather weak approximate posterior [38].

Laplace Approximation (LA) The posterior is approximated with a unimodal distribution, typically a Gaussian. By using the second-order Taylor approximation of the *maximum a posteriori* (MAP) estimate, it places the Gaussian around the mode of the true posterior. Since this is simply a mode-matching technique, it is simple and faster but less accurate.

Variational Inference (VI) The objective of variational inference is to learn a parameterized distribution $q(w)$ that matches with the true posterior distribution $p(w|y)$ by minimizing $\mathbb{KL}[q(w)||p(w|y)]$ for model parameters w. Since we do not know the true posterior, we have to derive a lower bound – typically known as the Evidence Lower Bound (ELBO) – that does not depend on the true posterior. By doing so, we convert the inference problem into an optimization problem. For instance, if the parameters of the approximate posterior follow a

---

[8] Dropout and MC dropout have two different objectives and work differently. The former runs in training time and works as a regularizer. The latter runs at inference time and acts as an uncertainty estimator.

normal distribution, $w \sim \mathcal{N}(\mu, \sigma)$, the algorithm finds the best μ, σ that overlaps with p. Typically, estimating w is computationally expensive because a model has many parameters. Therefore, we factorize the approximate posterior as $q(w) = \prod_i q_i(w_i)$ and estimate each distribution separately. This is called mean-field approximation [15].

Even with mean-field approximation, when we have large datasets, using the entire dataset to update parameters is computationally infeasible. By using small batches of data, we can perform Stochastic Variational Inference (SVI). This becomes specially attractive for perception tasks in robotics where the offline datasets are large or we obtain data online in batches [118, 120]. All these methods require deriving the ELBO which can be mathematically taxing. Black-box variational inference (BBVI) alleviates this. Rather than obtaining an analytical form, it uses automatic differentiation.[9] This allows learning complex distributions required for real-world robotics tasks [121]. We also often see amortized variational inference in variational autoencoders (VAEs) in which a neural network is used to estimate the parameters of the variational distribution. However, the objective in VAEs is not obtaining outputs with uncertainty. As a separate note, when the p and q terms of the divergence term of VI are swapped,[10] $\mathbb{KL}[p(w|y)||q(w)]$, we call this Expectation Propagation (EP).

Markov Chain Monte Carlo (MCMC) In VI, we assume a particular distribution for the posterior. Although this assumption makes computations tractable and faster, it also limits the representation power of the posterior distribution. For instance, if a unimodal Gaussian is used for q, then it will not appropriately cover the true bimodal distribution p. Instead of assuming a parametric distribution, we can represent a distribution using samples.[11] Stein's variational inference maintains a set of particles to represent the distribution – thus a non-parametric technique – and performs variational inference using them [77].

In Monte Carlo (MC) techniques, we often assume a proposal distribution q that we know of and can query. The proposal is used merely to guide sampling and needs not to match with the true distribution, although it should, ideally, encapsulate the true distribution. In rejection sampling, one of the MC techniques, we evaluate the probability density of proposal distribution q and true distribution p at a sampled position i. If $k \times q[i] < p[i]$ for some k, then we reject the ith sample, otherwise keep it. The collection of kept samples represents

---

[9] Read more about Probabilistic Programming Languages (PPL) such as Pyro, Edward (now part of TensorFlow Probability), Stan, WebPPL, and Turing.jl.
[10] Note that KL is not a symmetric metric.
[11] An infinite amount of samples represent the true distribution.

a distribution. In rejection sampling, samples are drawn independently. This uninformed sampling technique quickly becomes inefficient as it results in a lot of samples being rejected, especially when we try to estimate multimodal or high-dimensional distributions. Therefore, in techniques such as Metropolis or Metropolis–Hastings [52], we sample in such a way that the next sampling step depends on the previous sample – from a Markov chain. Gibbs sampling [44] samples from one or a few dimensions at a time, making it an efficient variation on Metropolis–Hastings (MH) algorithm for high-dimensional distributions. Hamiltonian Monte Carlo (HMC) allows larger jumps in MH, which results in a few samples to represent a complex distribution. No-U-Turn Sampler (NUTS) automatically fine-tunes hyperparameters in HMC [19].[12]

Langevin Monte Carlo (LMC) such as Metropolis-adjusted Langevin Algorithm (MALA) [114], takes only a single leapfrog step in HMC. Therefore, although HMC is more suitable for high-dimensional spaces, LMC is simpler to implement and easier to use. Since traditional MCMC algorithms fall short when it comes to large datasets as the entire dataset is used to update, in Stochastic Gradient Langevin Dynamics (SGLD), parameters are updated using SGD with minibatches [146].

Conformal Prediction (CP) Similar to MC dropout, conformal prediction is a post-hoc uncertainty estimation technique as it estimates the uncertainty of a pre-trained model using a calibration dataset and a quantile value. Conformal prediction aims at representing uncertainty with sets (or intervals). For image classification, it predicts a possible set of classes (e.g., {cat, tiger}). The more complicated the image is or the poorer the model is, it adds more element into the set as any outcome becomes possible (e.g., {cat, tiger, jaguar, lion}). In regression tasks, the uncertainty is represented as an interval. Given the simplicity, conformal predictions are also well-suited for generative models such as large-language models [90]. They are model-agnostic and provide theoretical guarantees. They have many applications in robotics as well [113, 129, 129, 142, 145]. CP, in its vanilla form, assumes *exchangeability* [124], making it less straightforward to apply in certain time-dependent tasks such as robot control, though new treatments exists [11]. Angelopoulos and Bates [7] provide an excellent introduction for practitioners.

Direct Uncertainty Estimation Techniques such as Prior Networks (PNs) [80] and Posterior Networks (PostNets) [20] try to estimate the probability estimates directly while trying to maintain the properties of a predictive distribution

---

[12] These techniques are implemented in Pyro and Stan probabilistic programming language packages.

(i.e., higher uncertainty away from data). Especially, since the PostNets use normalizing flow density estimators, they can represent complex probability distributions. These techniques are inspired from the ideas of evidential deep learning [6, 122][13] and have been used in applications such as autonomous driving [56]. As another line of work, Epistemic Neural Networks (ENNs), presented as a generalized representation of other epistemic models such as Bayesian neural networks and ensembles, model the epistemic uncertainly using a small additional network called Epinet [102].

Other Methods and Representations There are many other works on using alternative techniques for representing and estimating uncertainty and diversity. In work such as deep kernel learning (DKL) [148], Gaussian processes are used to estimate the uncertainty. However, rather than using a standard kernel, a deep neural network is used to capture nonlinear patterns in data. Kernel learning techniques have proven to be useful in many domains [55, 137, 147] and they are useful in robotics problems where using large neural networks is prohibitive. As another popular technique, particle filtering, also known as sequential Monte Carlo (SMC), is one of the most commonly used methods that helps with maintaining diverse solutions in robotics [134]. All these are standard Bayesian techniques that were not discussed under previous methods. In addition to them, various Covariance Matrix Adaptation (CMA) techniques have been used in robot control to maintain diverse solutions [8, 136]. Belief functions [123] in DST, probability box (p-box) [32, 35] in engineering analysis, fuzzy sets [3, 155] all deal with alternative uncertainty specifications and have applications to robotics [9, 32, 56, 132].

10.3.4 Calibration: Are Our Uncertainties Correct?

How do we know if the probabilities estimated by various techniques discussed in Section 10.3.3 are indeed accurate? Incorrect probabilities can lead to incorrect believes about safety. While various visualizations and metrics (see Section 10.3.2) such as accuracy vs. confidence plots, NLL, entropy, average discrepancy between class probability and ground truth (i.e., Brier score), etc. can be used to assess the correctness of uncertainty [104].

Considering the robustness and flexibility, Expected Calibration Error (ECE) [49] can be considered as the most popular method to assess calibration. In ECE, N predictions are arranged into M bins each with size $1/M$ according to predicted label confidences. If average accuracy and average confidence

---

[13] They are based on Dempster–Shafer Theory (DST).

within each bin is similar, then the model is well-calibrated. This notion can be formulated as

$$\text{ECE} = \sum_{m=1}^{M} \frac{|B_m|}{N} \left| \underbrace{\text{acc}(B_m)}_{\frac{1}{B_m} \sum_{i \in B_m} \mathbb{1}(\hat{y}_i = y_i)} - \underbrace{\text{conf}(B_m)}_{\frac{1}{B_m} \sum_{i \in B_m} \hat{p}_i} \right|, \quad (10.11)$$

where $y_i, \hat{y}_i$ and $\hat{p}_i$ are the ground truth label, predicted label, and predicted label confidence, respectively, for sample i in bin B_m. The lower the ECE, the better calibrated the model is. Adaptive Calibration Error (ACE) [99] has been proposed as an improvement to ECE. Ovadia et al. [104] conducted a study on the quality of uncertainty on out-of-distribution samples for various epistemic uncertainty estimation techniques. They concluded that the quality of uncertainty lowers with data shift and deep ensembles and stochastic variational inference techniques are more promising.

If a model is miscalibrated, it needs to be calibrated. Temperature scaling and histogram binning are simple post-hoc calibration techniques. The former is parametric and the latter is nonparametric. Temperature scaling of probabilities, softmax(logits/T), has a single parameter T that can be tuned to minimize the ECE based on a validation dataset. A more generalized version of temperature scaling is Platt scaling, where a logistic regression model is fitted [107]. Isotonic regression involves fitting a piece-wise-constant nonparametric model. Bayesian Binning into Quantiles (BBQ) [94] is another calibration technique.[14]

10.4 How Do We Leverage Uncertainty?

Traditionally, robotics systems are designed to be modular for reasons such as interpretability and ease of designing and debugging. Nowadays, especially in research settings, the robots are also trained in an end-to-end fashion. Either way, our objective is to control the robot to achieve our objective[15] by observing a limited amount of uncertain data.

10.4.1 Uncertainty in Perceiving and Representing the World

In order for the robot to get an idea about the world, it performs various tasks.

---

[14] net:cal – Uncertainty Calibration Python library implements a number of these techniques.
[15] Our objective itself can also be uncertain.

Figure 10.8 Uncertainty in object detection in autonomous driving. Uncertainty can be associated with the detected object class or bounding boxes.

Mapping the Environment One of the most basic requirements for a robot to act in its environment is a map. Occupancy grid maps are one of the commonly used metric mapping techniques in mobile robotics [31, 135]. Considering their assumptions about discretization and inability to quantify epistemic uncertainty due to occlusions, Gaussian process-based occupancy mapping techniques were introduced [100, 118]. However, considering their prohibitive computational complexity on robotic hardware, especially when the robot gathers more and more data, Bayesian Hilbert mapping (BHM) – a scalable technique that uses variational inference to estimate a distribution of maps – has been proposed [120]. These techniques can estimate both aleatoric and epistemic and uncertainty (e.g., uncertainty due to lack of data or unseen areas). There are also techniques to only estimate the aleotoric uncertainty (e.g., sensor noise) by maximizing the maximum likelihood of a Bernoulli distribution [111] or Gaussian mixture [46]. When the occupancy is represented as a deep neural network, DST has been used to quantify uncertainty [56, 138]. More recently, variational inference [125] and post-hoc Laplace approximations [47] have been used to represent uncertainty in Neural radiance fields (NeRF). A single-layer NeRF functions as a BHM with random Fourier features. There have been applications of environment uncertainty quantification techniques for modeling table top objects for manipulation [149], tactile localization and mapping [61], mapping dynamic environments [48, 57, 87, 138], etc.

Localization and Tracking The exact location of a robot or an external agent with respect to an origin in the environment it is in is difficult to determine because of sensor limitations (e.g., low GPS signal levels), limited landmarks, occlusions, dynamics, etc. Particle filters, a sequential Monte Carlo (SMC) estimation technique, and various Kalman filters such as Extended Kalman filters (EKF), Unscented Kalman filters (UKF) have been widely used to estimate the location of the robot [29, 134]. They have also been used for object tracking – determining the position over time. All these techniques focus only on aleatoric uncertainty. Also, in addition to this MC technique, variational inference has been used for localization [88, 89]. There are also tracking techniques that use uncertainty in deep neural networks [156, 158].

Object Detection and Pose Estimation An embodied agent typically needs to identify various objects around it to act intelligently. An object in a scene is represented by a bounding box – origin, width and height – and the associated object class. Therefore, quantifying the uncertainty of object detection requires, computing the uncertainty of the bounding box parameters and assigned class, for instance using MC dropout [127]. Feng et al. [34] survey various probabilistic object detection techniques. More recently, conformal prediction has been used as a post-hoc uncertainty estimation technique for object detection [24]. For most robotics tasks, merely detecting objects is not sufficient, the pose of the object needs to be estimated [151]. Estimating the poses typically requires estimating the (x, y, z) position and (roll, pitch, yaw) orientation, making it an estimation problem in $\mathbb{R}^6$. Ensemble techniques [150] as well as directional statistics have been used in pose estimation. In particular, probability distributions such as the von Mises-Fisher [109, 157] and Bingham [45, 101] can represent uncertainty.

Semantic Segmentation Semantic segmentation is often useful in robotics tasks for delineating the foreground from the background, determining the road for driving AVs, etc. Semantic uncertainty can be related to the entire class (e.g., the entire sky is identified as a river) or the class boundaries (e.g., a few pixels in the border between road and trees are incorrectly assigned). Especially, the latter case is unavoidable. Most loss functions and evaluation metrics used in semantic segmentation focus on the pixel-level semantic assignment of the entire image rather than the semantic boundaries. However, for safety-critical tasks such as AVs, knowing the exact road boundaries is important. There is very high epistemic and aleatoric uncertainty in these boundaries due to ML model limitations, occlusions, etc. Similar to the mapping techniques discussed earlier in this subsection, semantic uncertainty has been quantified using relevance vector machines [41], MC dropout [92], etc.

Imagination and Future Prediction The human ability to imagine helps us understand potential risks and efficient decisions. The action of "pretend play" in toddlers is considered to help with cognition, language, social, and emotional development [51]. Similarly, training robots in diverse simulations and through domain randomization helps learn robust models [112]. The progress in generative AI has greatly simplified and improved the process of imagination.

One special case of imagination is predicting future outcomes, which indeed is highly uncertain. Similar to humans' tendency to anticipate the future before making-decisions, robots can predict the future state of any of the tasks discussed thus far. As examples, Gaussian processes [119], filtering [48, 57], and convolutional LSTMs [82, 138] have been used to predict how the occupancy changes in space and time for the next few time steps. The somewhat similar problem of video prediction is also useful in robotics, for instance in manipulation [36, 59]. Most state estimation and object tracking techniques can naively be extended for predicting the future location. There are also various trajectory prediction algorithms which have proven to be useful, for instance, in autonomous driving [5, 116].

Agent Modeling and Human–AI Alignment When robots interact with human or other robots, it is important for them to decipher the intents for successful coexistence. As a simple example, in social navigation or autonomous driving, the robot needs to guess if a human it observes would cross its path [78]. Physical interactions [28, 152], facial expressions, and emotions can also have uncertainty. When language is used to communicate, there can be uncertainties due to delay, lack of clarity, insufficient amount of information contained in the message, etc. Uncertainty exists not just in factual information but also in other aspects of language such as prosody. Rhetorical flourish devised into modern LLMs such as GPT3.5[16] tend to provide overconfident answers. If embodied AI agents are using outputs from LLMs, their inability to express uncertainty should be taken into account. In some cases, we try to learn the various aspects of the agent such as preferences from human demonstrations, typically as a reward function [16, 126]. There are also attempts to learn the distribution over rewards in Bayesian inverse reinforcement learning [81, 110]. A unifying framework for imitation learning paradigms can be found in [14].

Detecting Out-of-Distribution (OOD) Samples Uncertainty also helps with assessing for which inputs the model might under-perform. If the predictive epistemic uncertainty is high for a particular input, then the model is not robust in the vicinity of that input. Uncertainty has been used for OOD detection in autonomous driving [98], mobile robotics [154], spacecraft pose estimation [39], manipulation [33], etc. Since the environments that robots operate are always subject to change, uncertainty is crucial for domain generalization and adaptation.

---

[16] We believe overconfident and persuasive answers are not an inherent limitation of LLMs; rather, it should be how the agent was trained using reinforcement learning with human feedback (RLHF).

10.4.2 Uncertainty in Planning and Control

Our objective is to see how decision-making is affected by various sources of uncertainty.

Uncertainty Propagation from Perception into Decision-making Uncertainty of the world can be modeled as discussed in Section 10.4.1. If a robotic system has distinct modules for perception and decision-making, as in most classical robotic setups, then the uncertainty in perception can be propagated into the decision-making modules. Our objective is to calculate the output $y = (f_{\text{perc}} \circ f_{\text{deci}})(x)$ for an input x with a perception function, $f_{\text{perc}} : \mathcal{X} \to \mathcal{Z}$, which is then sequentially fed into a decision function, $f_{\text{deci}} : \mathcal{Z} \to \mathcal{Y}$. As a simple example, consider an object detection module for the perception function and a motion planner for the decision function. If the object detector is giving multiple bounding boxes with associated uncertainties, how can the decision-making module leverage all those bounding boxes? In general, if the predictive uncertainty of perception is represented as samples, as in MCMC techniques or Ensembles, each output can be run through the decision-making module. This will result in a decision distribution – multiple decisions, each with its own probability. If uncertainty is instead represented as an interval, as in conformal prediction, the two endpoints of the interval can be run separately to understand the limits of decisions.

When the output of the perception function is a probability distribution[17] p_{perc}, the input to the decision function is also a probability distribution. Hence, the output of the decision-making module can be computed as

$$y = \int_{-\infty}^{+\infty} p_{\text{perc}}(z) f_{\text{deci}}(z) dz \approx \sum_{z_n \sim p_{\text{perc}}} f(z_n), \qquad (10.12)$$

where $\{z_n\}_{n=1}^{N}$ are samples taken from the output probability distribution of the perception module. If the probabilities are represented as a probability density function (PDF) and the integral is tractable, then the propagation becomes easier. Otherwise, we have to resort to a numerical integration technique. By sampling from the predictive probability distribution of perception and then evaluating the decision function on those samples, as shown in Eq. (10.12), is a straightforward simplification. If the decision-making module is also set up to compute uncertainty, then y is also a probability distribution.

The Use of Uncertainty in Exploration for Better World Representations
In Section 10.4.1, we discussed the importance of building a map of the

---

[17] Say, the output of a map is the mean and standard deviation values of a normal distribution.

environment. The objective of uncertainty quantification for mapping is understanding which areas of the environment we do not know about. This information can be used by the robot for exploration – to gather more information about the environment. Some applications of such exploration include robot mapping an indoor environment, environmental monitoring using a drone [84], or subterranean or extraterrestrial navigation [63, 106]. For this purpose, probabilistic frontiers [153], Bayesian optimization (BO) with Gaussian processes [84], Partially Observable Markov Decision Process (POMDP) solvers [91], and reinforcement learning (RL) [42] have been used. Frontiers and BO are categorized under myopic exploration strategies as they are generally one-step look-ahead planners.

Uncertainty in Learning Decision Policies Uncertainty is widely studied in planning and scheduling [60, 95] as well as robot motion planning [69]. A robot takes a sequence of decisions to control itself. This process can be modeled as a Markov decision process. If observations are not fully observable, we consider partially observable decision-making processes (POMDPs). POMDPs are widely used in manipulation [22], autonomous driving [67], aerospace control [64], etc. [64] and [65] list various solvers[18] for POMDPs or belief space planning [108]. Note that the typical POMDPs do not take into account the epistemic uncertainty; estimating the epistemic uncertainty requires maintaining a distribution over the MDP which is intractable.

We typically use stochastic policies as the outcomes of our actions are not certain. This randomness also helps with exploring different possibilities rather than getting stuck in a local optimum. When the model of the environment dynamics is unknown or too complex to model, we have to resort to *model-free reinforcement learning (RL)*. Therefore, it implicitly handles limited aspects of the uncertainty of the environment.

Aleatoric and epistemic uncertainty can help improve deep RL in various ways. Following Charpentier et al. [21], as illustrated in Figure 10.9, we discuss four desiderata for the desired behavior of uncertainty in deep RL. As the number of training steps increases, the agent's epistemic uncertainty should decrease as the agent collects more information about the environment (desideratum 1). Simultaneously, agent should collect more rewards with an epistemic strategy compared to an aleatoric strategy[19] (desideratum 2). In the real-world, the environment we train our robots is not the same as we deploy. It sometimes tends to have little (e.g., orange environment in Figure 10.9) or even completely different

---

[18] Various solvers are implemented in POMDPs.jl [30].
[19] Aleatoric and epistemic strategies involve sampling from aleatoric and epistemic uncertainty distributions, respectively, to select actions.

The Role of Predictive Uncertainty and Diversity 171

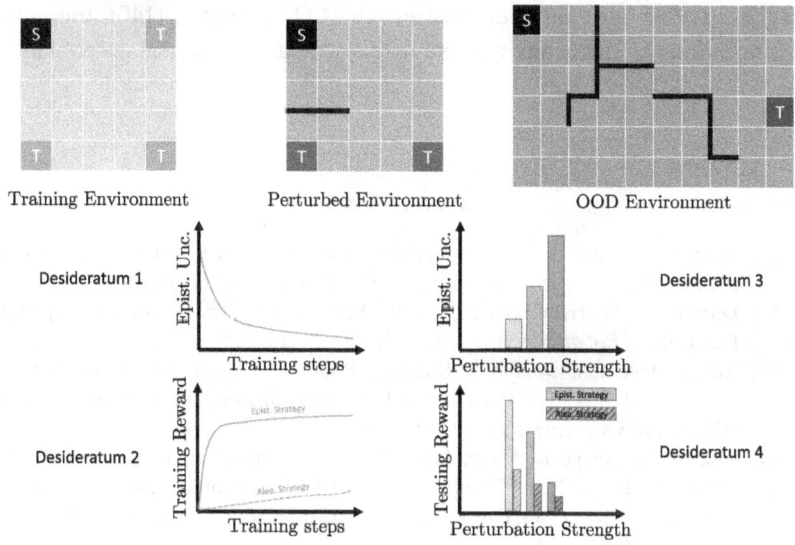

Figure 10.9 Uncertainty in deep RL. Green, orange, and red show environments with increasing difficulty. Four desiderata for the desired behavior of uncertainty is discussed in [21].

perturbations (red environment in Figure 10.9). As we increase the complexity of perturbations, the epistemic uncertainty should be higher (desideratum 3) as the agent does not know about these new environments. Simultaneously, the rewards it collects should go lower (desideratum 4). A comparative analysis on the effect of deep RL algorithms is provided in [21]. The chapter empirically and theoretically concluded that compared to an ϵ-greedy or aleatoric strategy, sampling using an epistemic strategy receives higher rewards and maintains high generalization performance in OOD domains. This characterization of aleatoric and epistemic uncertainty in deep RL helps with generalization, sample efficiency, and knowing where the policy will not work.

Uncertainty is also studied in control theory. Bayesian nonparametric techniques such as Gaussian processes [4] and set-based techniques such as conformal prediction [76] have gained popularity or control tasks because they can provide theoretical guarantees, data efficient, and simple to use. Other set-based techniques such as backward reachability have also helped with reliability and safety [10, 53]. When uncertainty in parameters or unmeasured disturbances exist, Robust Model Predictive Control (Robust MPC) strategies help to maintain the stability of the system [12]. Uncertainty in Robust MPC is typically handled as bounded sets such as polytopes or ellipsoids. Conditional Value at Risk (CVaR) also has been used in robot robust control [140], trajectory optimization [72], and motion planning [50] to address risks and uncertainties.

References

[1] Martian boulders. www.pinterest.com/pin/i-really-like-these-big-martian-boulders-that-the-curiosity-rover-just-saw-22–182044009910822852/.

[2] Durrant-Whyte, H, and Bailey, T. 2006. Simultaneous localization and mapping: Part I. *IEEE Robotics & Automation Magazine*, **13**(2), 99–110.

[3] Aditya, Dyuman, Mukherji, Kaustuv, Balasubramanian, Srikar, Chaudhary, Abhiraj, and Shakarian, Paulo. 2023. PyReason: Software for open world temporal logic. *arXiv preprint arXiv:2302.13482*.

[4] Akametalu, Anayo K, Fisac, Jaime F, Gillula, Jeremy H, Kaynama, Shahab, Zeilinger, Melanie N, and Tomlin, Claire J. 2014. Reachability-based safe learning with gaussian processes. Pages 1424–1431 of: *53rd IEEE Conference on Decision and Control*.

[5] Alahi, Alexandre, Goel, Kratarth, Ramanathan, Vignesh, Robicquet, Alexandre, Fei-Fei, Li, and Savarese, Silvio. 2016. Social LSTM: Human trajectory prediction in crowded spaces. Pages 961–971 of: *Proceedings of the IEEE Conference on Computer Vision and Pattern Recognition*.

[6] Amini, Alexander, Schwarting, Wilko, Soleimany, Ava, and Rus, Daniela. Deep evidential regression. Pages 14927–14937 of: *Advances in Neural Information Processing Systems*, 33.

[7] Angelopoulos, Anastasios N, and Bates, Stephen. 2021. A gentle introduction to conformal prediction and distribution-free uncertainty quantification. *arXiv preprint arXiv:2107.07511*.

[8] Asmar, Dylan M, Senanayake, Ransalu, Manuel, Shawn, and Kochenderfer, Mykel J. Model predictive optimized path integral strategies. Pages 3182–3188 of: *2023 IEEE International Conference on Robotics and Automation (ICRA)*.

[9] Bandara, RN, and Gaspe, Sujeetha. 2016. Fuzzy logic controller design for an unmanned aerial vehicle. Pages 1–5 of: *2016 IEEE International Conference on Information and Automation for Sustainability (ICIAfS)*.

[10] Bansal, Somil, Chen, Mo, Herbert, Sylvia, and Tomlin, Claire J. 2017. Hamilton-Jacobi reachability: A brief overview and recent advances. Pages 2242–2253 of: *2017 IEEE 56th Annual Conference on Decision and Control (CDC)*.

[11] Barber, Rina Foygel, Candes, Emmanuel J, Ramdas, Aaditya, and Tibshirani, Ryan J. 2023. Conformal prediction beyond exchangeability. *The Annals of Statistics*, **51**(2), 816–845.

[12] Bemporad, Alberto, and Morari, Manfred. 2007. Robust model predictive control: A survey. Pages 207–226 of: *Robustness in Identification and Control*. Springer.

[13] Berger, Roger L, and Casella, George. 2001. *Statistical Inference*. Duxbury.

[14] Bhattacharyya, Raunak, Wulfe, Blake, Phillips, Derek J, et al. 2022. Modeling human driving behavior through generative adversarial imitation learning. *IEEE Transactions on Intelligent Transportation Systems*, **24**(3), 2874–2887.

[15] Bishop, Christopher M. 2006. Pattern recognition and machine learning. *Springer Google Scholar*, **2**, 645–678.

[16] Bıyık, Erdem, Huynh, Nicolas, Kochenderfer, Mykel J, and Sadigh, Dorsa. Active preference based Gaussian process regression for reward learning and optimization. *The International Journal of Robotics Research*, **43**(5), 665–684.

[17] Bowman, Michael, Li, Songpo, and Zhang, Xiaoli. 2019. Intent-uncertainty-aware grasp planning for robust robot assistance in telemanipulation. Pages 409–415 of: *2019 International Conference on Robotics and Automation (ICRA)*.

[18] Brooks, Tim, Peebles, Bill, Homes, Connor, et al. 2024. Video generation models as world simulators. https://openai.com/research/videogeneration-models-as-world-simulators.

[19] Carpenter, Bob, Gelman, Andrew, Hoffman, Matthew D, et al. 2017. Stan: A probabilistic programming language. *Journal of Statistical Software*, **76**(1).

[20] Charpentier, Bertrand, Zügner, Daniel, and Günnemann, Stephan. 2020. Posterior network: Uncertainty estimation without OOD samples via density-based pseudo-counts. Pages 1356–1367 of: *Advances in Neural Information Processing Systems*, 33.

[21] Charpentier, Bertrand, Senanayake, Ransalu, Kochenderfer, Mykel, and Günnemann, Stephan. 2022. Disentangling epistemic and aleatoric uncertainty in reinforcement learning. *arXiv preprint arXiv:2206.01558*.

[22] Curtis, Aidan, Kaelbling, Leslie, and Jain, Siddarth. 2023. Task-directed exploration in continuous POMDPs for robotic manipulation of articulated objects. Pages 3721–3728 of: 2023 *IEEE International Conference on Robotics and Automation (ICRA)*.

[23] Cuturi, Marco. 2023. Lightspeed computation of optimal transportation distances. Pages 2292–2300 of: *Advances in Neural Information Processing Systems*, 26.

[24] De Grancey, Florence, Adam, Jean-Luc, Alecu, Lucian, Gerchinovitz, Sébastien, Mamalet, Franck, and Vigouroux, David. 2022. Object detection with probabilistic guarantees: A conformal prediction approach. Pages 316–329 of: *International Conference on Computer Safety, Reliability, and Security*.

[25] Deahl, Dani. 2017. Volvo's self-driving cars are having trouble recognizing kangaroos. www.theverge.com/2017/7/3/15914120/volvo-self-driving-carkangaroo-australia.

[26] Delecki, Harrison, Itkina, Masha, Lange, Masha, Senanayake, Ransalu, and Kochenderfer, Mykel J. 2022. How do we fail? Stress testing perception in autonomous vehicles. In: *IEEE/RSJ International Conference on Intelligent Robots and Systems (IROS)*.

[27] Dosovitskiy, Alexey, Ros, German, Codevilla, Felipe, Lopez, Antonio, and Koltun, Vladlen. 2017. CARLA: An open urban driving simulator. *Proceedings of Machine Learning Research*, **78**, 1–16.

[28] Dragan, Anca D, Lee, Kenton CT, and Srinivasa, Siddhartha S. 2013. Legibility and predictability of robot motion. Pages 301–308 of: *2013 8th ACM/IEEE International Conference on Human-Robot Interaction (HRI)*.

[29] Durrant-Whyte Hugh, et al. 2001. Introduction to estimation and the Kalman filter. *Australian Centre for Field Robotics*, **28**(3), 65–94.

[30] Egorov, Maxim, Sunberg, Zachary N, Balaban, Edward, Wheeler, Tim A, Gupta, Jayesh K, and Kochenderfer, Mykel J. 2017. POMDPs.jl: A framework for sequential decision making under uncertainty. *Journal of Machine Learning Research*, **18**(26), 1–5.
[31] Elfes, Alberto. 1989. Using occupancy grids for mobile robot perception and navigation. *Computer*, **22**(6), 46–57.
[32] Faes, Matthias GR, Daub, Marco, Marelli, Stefano, Patelli, Edoardo, and Beer, Michael. 2021. Engineering analysis with probability boxes: A review on computational methods. *Structural Safety*, **93**, 102092.
[33] Farid, Alec, Veer, Sushant, and Majumdar, Anirudha. 2022. Task-driven out-of-distribution detection with statistical guarantees for robot learning. *Proceedings of Machine Learning Research*, **164**, 970–980.
[34] Feng, Di, Harakeh, Ali, Waslander, Steven, and Dietmayer, Klaus. 2021. A review and comparative study on probabilistic object detection in autonomous driving. *IEEE Transactions on Intelligent Transportation Systems*, **23**(8), 9961–9980.
[35] Ferson, S, Ginzburg, L, and Akçakaya, R. Whereof one cannot speak: when input distributions are unknown. www.ramas.com/whereof.pdf.
[36] Finn, Chelsea, Goodfellow, Ian, and Levine, Sergey. 2016. Unsupervised learning for physical interaction through video prediction. Pages 64–72 of: *Advances in Neural Information Processing Systems*, 29.
[37] Flamary, Rémi, Courty, Nicolas, Gramfort, Alexandre, et al. 2021. POT: Python optimal transport. *Journal of Machine Learning Research*, **22**(78), 1–8.
[38] Folgoc, Loic Le, Baltatzis, Vasileios, Desai, Sujal, et al. 2021. Is MC Dropout Bayesian? *arXiv preprint arXiv:2110.04286*.
[39] Foutter, Matt, Sinha, Matt, Banerjee, Somrita, and Pavone, Marco. 2023. Self-supervised model generalization using out-of-distribution detection. First workshop on out-of-distribution generalization in robotics. *CoRL 2023*.
[40] Gal, Yarin, and Ghahramani, Zoubin. 2016. Dropout as a Bayesian approximation: Representing model uncertainty in deep learning. *Proceedings of Machine Learning Research*, **48**, 1050–1059.
[41] Gan, Lu, Jadidi, Maani Ghaffari, Parkison, Steven A, and Eustice, Ryan M. 2017. Sparse Bayesian inference for dense semantic mapping. *arXiv preprint arXiv:1709.07973*.
[42] Garaffa, Luíza Caetano, Basso, Maik, Konzen, Andréa Aparecida, and de Freitas, Edison Pignaton. 2021. Reinforcement learning for mobile robotics exploration: A survey. *IEEE Transactions on Neural Networks and Learning Systems*, **34**(8), 3796–3810.
[43] Gelman, Andrew, Carlin, John B, Stern, Hal S, and Rubin, David B. 1995. *Bayesian data analysis*. Chapman and Hall/CRC.
[44] Geman, Stuart, and Geman, Donald. 1984. Stochastic relaxation, Gibbs distributions, and the Bayesian restoration of images. *IEEE Transactions on Pattern Analysis and Machine Intelligence*, (6), 721–741.
[45] Gilitschenski, Igor, Sahoo, Roshni, Schwarting, Wilko, Amini, Alexander, Karaman, Sertac, and Rus, Daniela. 2019. Deep orientation uncertainty learning based on a Bingham loss. IN: *International Conference on Learning Representations*.

[46] Goel, K, and Tabib, Wennie. 2023. Incremental multimodal surface mapping via selforganizing gaussian mixture models. *IEEE Robotics and Automation Letters*, **8**(12), 8358–8365.

[47] Goli, Lily, Reading, Cody, Selllán, Silvia, Jacobson, Alec, and Tagliasacchi, Andrea. 2023. Bayes' rays: Uncertainty quantification for neural radiance fields. *arXiv preprint arXiv:2309.03185*.

[48] Guizilini, Vitor, Senanayake, Ransalu, and Ramos, Fabio. 2019. Dynamic Hilbert maps: Real-time occupancy predictions in changing environments. Pages 4091–4097 of: *2019 International Conference on Robotics and Automation (ICRA)*.

[49] Guo, Chuan, Pleiss, Geoff, Sun, Yu, and Weinberger, Kilian Q. 2017. On calibration of modern neural networks. *Proceedings of Machine Learning Research*, **70**, 1321–1330.

[50] Hakobyan, Astghik, Kim, Gyeong Chan, and Yang, Insoon. 2019. Risk-aware motion planning and control using CVaR-constrained optimization. *IEEE Robotics and Automation Letters*, **4**(4), 3924–3931.

[51] Hashmi, Salim, Vanderwert, Ross E, Price, Hope A, and Gerson, Sarah A. 2020. Exploring the benefits of doll play through neuroscience. *Frontiers in Human Neuroscience*, **14**, 560176.

[52] Hastings, WK. 1970. Monte Carlo sampling methods using Markov chains and their applications. *Biometrika*, **57**(1), 97–109.

[53] Herbert, Sylvia L, Chen, Mo, Han, SooJean, Bansal, Somil, Fisac, Jaime F, and Tomlin, Claire J. 2017. Fastrack: A modular framework for fast and guaranteed safe motion planning. Pages 1517–1522 of: *2017 IEEE 56th Annual Conference on Decision and Control (CDC)*.

[54] Hsiao, Kaijen, Kaelbling, Leslie Pack, and Lozano-Pérez, Tomás. 2011. Robust grasping under object pose uncertainty. *Autonomous Robots*, **31**, 253–268.

[55] Hsu, Kelvin, and Ramos, Fabio. 2019. Bayesian learning of conditional kernel mean embeddings for automatic likelihood-free inference. *Proceedings of Machine Learning Research*, **89**, 2631–2640.

[56] Itkina, Masha. 2022. *Uncertainty-aware spatiotemporal perception for autonomous vehicles*. Ph.D. dissertation. Stanford University.

[57] Itkina, Masha, Driggs-Campbell, Katherine, and Kochenderfer, Mykel J. 2019. Dynamic environment prediction in urban scenes using recurrent representation learning. Pages 2052–2059 of: *2019 IEEE Intelligent Transportation Systems Conference (ITSC)*.

[58] Jain, Siddhartha, Liu, Ge, Mueller, Jonas, and Gifford, David. 2020. Maximizing overall diversity for improved uncertainty estimates in deep ensembles. Pages 4264–4271 of: *Proceedings of the AAA Conference on Artificial Intelligence*, vol. 34.

[59] Jayaraman, Dinesh, Ebert, Frederik, Efros, Alexei A, and Levine, Sergey. 2018. Time-agnostic prediction: Predicting predictable video frames. *arXiv preprint arXiv:1808.07784*.

[60] Jensen, Rune M, Veloso, Manuela M, and Bryant, Randal E. Fault tolerant planning: Moving toward probabilistic uncertainty models in symbolic non-deterministic planning. *Proceedings of the International Conference on Automated Planning and Scheduling*, 14.

[61] Jia, Shengxin, Zhang, Lionel, and Santos, VERONICA J. 2022. Autonomous tactile localization and mapping of objects buried in granular materials. *IEEE Robotics and Automation Letters*, **7**(4), 9953–9960.

[62] Kaelbling, Leslie Pack, Littman, Michael L, and Cassandra, Anthony R. 1998. Planning and acting in partially observable stochastic domains. *Artificial Intelligence*, **101**(1–2), 99–134.

[63] Kim, Sung-Kyun, Bouman, Amanda, Salhotra, Gautam, et al. 2021. PLGRIM: Hierarchical value learning for large-scale exploration in unknown environments. Pages 652–662 of: *Proceedings of the International Conference on Automated Planning and Scheduling*, vol. 31.

[64] Kochenderfer, Mykel J. 2015. *Decision Making Under Uncertainty: Theory and Application*. MIT Press.

[65] Kochenderfer, Mykel J, Wheeler, Tim TA, and Wray, Kyle H. 2022. *Algorithms for Decision Making*. MIT Press.

[66] Kolmogorov, Andrei Nikolaevich, and Bharucha-Reid, Albert T. 2018. *Foundations of the Theory of Probability: Second English Edition*. Courier Dover Publications.

[67] Kruse, Liam A, Yel, Esen, Senanayake, Ransalu, and Kochenderfer, Mykel J. 2022. Uncertainty-aware online merge planning with learned driver behavior. Pages 1202–1207 of: *2022 IEEE 25th International Conference on Intelligent Transportation Systems (ITSC)*. IEEE.

[68] Kurenkov, Andrey, Taglic, Joseph, Kulkarni, Rohun, et al. 2020. Visuomotor mechanical search: Learning to retrieve target objects in clutter. Pages 8408–8414 of: *2020 IEEE/RSJ International Conference on Intelligent Robots and Systems (IROS)*. IEEE.

[69] Lai, Tin, Morere, Philippe, Ramos, Fabio, and Francis, Gilad. 2020. Bayesian local sampling-based planning. *IEEE Robotics and Automation Letters*, **5**(2), 1954–1961.

[70] Lakshminarayanan, Balaji, Pritzel, Alexander, and Blundell, Charles. 2017. Simple and scalable predictive uncertainty estimation using deep ensembles. Pages 6405–6416 of: *Advances in Neural Information Processing Systems*, 30.

[71] Laskey, Michael, Lee, Jonathan, Chuck, Caleb, et al. 2016. Robot grasping in clutter: Using a hierarchy of supervisors for learning from demonstrations. Pages 827–834 of: *2016 IEEE International Conference on Automation Science and Engineering (CASE)*.

[72] Lew, Thomas, Bonalli, Riccardo, and Pavone, Marco. 2023. Risk-averse trajectory optimization via sample average approximation. *IEEE Robotics and Automation Letters*, **9**(2), 1500–1507.

[73] Li, Chengshu, Xia, Fei, Martín-Martín, Roberto, et al. 2021. iGibson 2.0: Object-centric simulation for robot learning of everyday household tasks. *arXiv preprint arXiv:2108.03272*.

[74] Li, Miao, Hang, Kaiyu, Kragic, Danica, and Billard, Aude. 2016. Dexterous grasping under shape uncertainty. *Robotics and Autonomous Systems*, **75**, 352–364.

[75] Li, Yueyuan, Yuan, Wei, Zhang, Songan, et al. 2024. Choose your simulator wisely: A review on open-source simulators for autonomous driving. *IEEE Transactions on Intelligent Vehicles*, **9**(5), 4861–4876.

[76] Lindemann, Lars, Cleaveland, Matthew, Shim, Gihyun, and Pappas, George J. 2023. Safe planning in dynamic environments using conformal prediction. *IEEE Robotics and Automation Letters*, **8**(8), 5116–5123.

[77] Liu, Qiang, and Wang, Dilin. 2016. Stein variational gradient descent: A general purpose Bayesian inference algorithm. Pages 2378–2386 of: *Advances in Neural Information Processing Systems*, 29.

[78] Luo, Yuanfu, Cai, Panpan, Bera, Aniket, Hsu, David, Lee, Wee Sun, and Manocha, Dinesh. 2018. PORCA: Modeling and planning for autonomous driving among many pedestrians. *IEEE Robotics and Automation Letters*, **3**(4), 3418–3425.

[79] Makoviychuk, Viktor, Wawrzyniak, Lukasz, Guo, Yunrong, et al. 2021. Isaac gym: High performance GPU-based physics simulation for robot learning. *arXiv preprint arXiv:2108.10470*.

[80] Malinin, Andrey, and Gales, Mark. 2018. Predictive uncertainty estimation via prior networks. Pages 7047–7058 of: *Advances in Neural Information Processing Systems*, 31.

[81] Mandyam, Aishwarya, Li, Didong, Cai, Diana, Jones, Andrew, and Engelhardt, Barbara E. 2023. Kernel density Bayesian inverse reinforcement learning. *arXiv preprint arXiv:2303.06827*.

[82] Mann, Khushdeep Singh, Tomy, Abhishek, Paigwar, Anshul, Renzaglia, Alessandro, and Laugier, Christian. 2022. Predicting future occupancy grids in dynamic environment with spatio-temporal learning. *arXiv preprint arXiv:2205.03212*.

[83] Manyika, James, and Durrant-Whyte, Hugh. 1995. *Data Fusion and Sensor Management: a Decentralized Information-Theoretic Approach*. Prentice Hall.

[84] Marchant, Roman, and Ramos, Fabio. 2012. Bayesian optimisation for intelligent environmental monitoring. Pages 2242–2249 of: *2012 IEEE/RSJ International Conference on Intelligent Robots and Systems*.

[85] Mayoral-Vilches, Víctor, Jabbour, Jason, Hsiao, Yu-Shun, et al. 2023. RobotPerf: An open-source, vendor-agnostic, benchmarking suite for evaluating robotics computing system performance. *arXiv preprint arXiv:2309.09212*.

[86] McAllister, Rowan, Kahn, Gregory, Clune, Jeff, and Levine, Sergey. 2019. Robustness to out-of-distribution inputs via task-aware generative uncertainty. Pages 2083–2089 of: *2019 International Conference on Robotics and Automation (ICRA)*. IEEE.

[87] Min, Youngjae, Kim, Do Un, and Choi, Han Lim. 2021. Kernel-based 3-D dynamic occupancy mapping with particle tracking. Pages 5268–5274 of: *2021 IEEE International Conference on Robotics and Automation (ICRA)*.

[88] Mirchev, Atanas, Kayalibay, Baris, Soelch, Maximilian, van der Smagt, Patrick, and Bayer, Justin. 2019. Approximate Bayesian inference in spatial environments. *Robotics: Science and Systems*.

[89] Mirchev, Atanas, Kayalibay, Baris, van der Smagt, Patrick, and Bayer, Justin. 2020. Variational statespace models for localisation and dense 3D mapping in 6 DoF. *arXiv preprint arXiv:2006.10178*.

[90] Mohri, Christopher, and Hashimoto, Tatsunori. 2024,. Language models with conformal factuality guarantees. *arXiv preprint arXiv:2402.10978*.

[91] Morere, Philippe, Marchant, Roman, and Ramos, Fabio. Sequential Bayesian optimization as a POMDP for environment monitoring with UAVs. Pages 6381–6388 of: *2017 IEEE International Conference on Robotics and Automation (ICRA)*.

[92] Mukhoti, Jishnu, and Gal, Yarin. 2018. Evaluating Bayesian deep learning methods for semantic segmentation. *arXiv preprint arXiv:1811.12709*.

[93] Mun, Ye-Ji, Huang, Zhe, Chen, Haonan, et al. 2023. User-friendly safety monitoring system for manufacturing cobots. *arXiv preprint arXiv:2307.01886*.

[94] Naeini, Mahdi Pakdaman, Cooper, Gregory, and Hauskrecht, Milos. 2015, Obtaining well calibrated probabilities using Bayesian binning. *Proceedings of the AAAI Conference on Artificial Intelligence*, **29**.

[95] Nagami, Keiko, and Schwager, Mac. 2024. State estimation and belief space planning under epistemic uncertainty for learning-based perception systems. *IEEE Robotics and Automation Letters*, **9**(1), 5118–5125.

[96] Neal, Radford M. 2012. *Bayesian Learning for Neural Networks*, vol. 118. Springer Science & Business Media.

[97] Nishimura, Haruki, Mercat, Jean, Wulfe, Blake, McAllister, Rowan Thomas, and Gaidon, Adrien. 2023. RAP: Riskaware prediction for robust planning. *Proceedings of Machine Learning Research*, **205**, 381–392.

[98] Nitsch, Julia, Itkina, Masha, Senanayake, Ransalu, et al. 2021. Out-of-distribution detection for automotive perception. Pages 2938–2943 of: *2021 IEEE International Intelligent Transportation Systems Conference (ITSC)*.

[99] Nixon, Jeremy, Dusenberry, Mike, Zhang, Linchuan, Jerfel, Ghassen, and Tran, Dustin. 2019. Measuring calibration in deep learning. *CVPR Workshops*, vol. 2.

[100] O'Callaghan, Simon T, and Ramos, Fabio T. 2012. Gaussian process occupancy maps. *The International Journal of Robotics Research*, **31**(1), 42–62.

[101] Okorn, Brian, Xu, Mengyun, Hebert, Martial, and Held, David. 2020. Learning orientation distributions for object pose estimation. Pages 10580–10587 of: *2020 IEEE/RSJ International Conference on Intelligent Robots and Systems (IROS)*. IEEE.

[102] Osband, Ian, Wen, Zheng, Asghari, Seyed Mohammad, et al. 2024. Epistemic neural networks. Pages 2795–2823 of: *Advances in Neural Information Processing Systems*, 36.

[103] Ouyang, Long, Wu, Jeff, Jiang, Xu, et al. 2022. Training language models to follow instructions with human feedback. Pages 27730–27744 of: *Advances in Neural Information Processing Systems*, 35.

[104] Ovadia, Yaniv, Fertig, Emily, Ren, Jie, et al. 2019. Can you trust your model's uncertainty? Evaluating predictive uncertainty under dataset shift. Pages 14003–14014 of: *Advances in Neural Information Processing Systems*, 32.

[105] Pathiraja, Bimsara, Liu, Bimsara, and Senanayake, Ransalu. 2024. Fairness in autonomous driving: Towards understanding confounding factors in object detection under challenging weather. *arXiv preprint arXiv:2406.00219*.

[106] Peltzer, Oriana, Bouman, Amanda, Kim, Sung-Kyun, et al. 2022. FIG-OP: Exploring large-scale unknown environments on a fixed time budget. In: *IEEE/RSJ International Conference on Intelligent Robots and Systems (IROS)*.

[107] Platt, J, et al. 1999. Probabilistic outputs for support vector machines and comparisons to regularized likelihood methods. *Advances in Large Margin Classifiers*, **10**(3), 61–74.

[108] Platt Jr, Robert, Tedrake, Russ, Kaelbling, LP, and Lozano-Perez, Tomas. 2010. Belief space planning assuming maximum likelihood observations. In: *Robotics: Science and Systems*, vol. 2.

[109] Prokudin, Sergey, Gehler, Peter, and Nowozin, Sebastian. 2018. Deep directional statistics: Pose estimation with uncertainty quantification. Pages 534–551 of: *Proceedings of the European Conference on Computer Vision (ECCV)*.

[110] Ramachandran, Deepak, and Amir, Eyal. 2007. Bayesian inverse reinforcement learning. Pages 2586–2591 of: *International Joint Conference on Artificial Intelligence*, vol. 7.

[111] Ramos, Fabio, and Ott, Lionel. 2016. Hilbert maps: Scalable continuous occupancy mapping with stochastic gradient descent. *The International Journal of Robotics Research*, **35**(14), 1717–1730.

[112] Ramos, Fabio, Possas, Rafael Carvalhaes, and Fox, Dieter. 2019. BayesSim: Adaptive domain randomization via probabilistic inference for robotics simulators. *arXiv preprint arXiv:1906.01728*.

[113] Ren, Allen Z, Dixit, Anushri, Bodrova, Alexandra, et al. 2023. Robots that ask for help: Uncertainty alignment for large language model planners. *arXiv preprint arXiv:2307.01928*.

[114] Roberts, Gareth O, and Tweedie, Richard L. 1996. Exponential convergence of Langevin distributions and their discrete approximations. *Bernoulli*, **2**(4), 341–363.

[115] Sagar, Som, Taparia, Aditya, and Senanayake, Ransalu. 2024. Failures are fated, but can be faded: Characterizing and mitigating unwanted behaviors in large-scale vision and language models. *Proceedings of the International Conference on Machine Learning (ICML)*.

[116] Salzmann, Tim, Ivanovic, Boris, Chakravarty, Punarjay, and Pavone, Marco. 2020. Trajectron++: Dynamically-feasible trajectory forecasting with heterogeneous data. Pages 683–700 of: *Computer Vision–ECCV 2020: 16th European Conference, Glasgow, UK, August 23–28, 2020, Proceedings, part XVIII 16*, Springer.

[117] Seeger, Matthias. 2004. Gaussian processes for machine learning. *International Journal of Neural Systems*, **14**(02), 69–106.

[118] Senanayake, Ransalu, and Ramos, Fabio. 2017. Bayesian Hilbert maps for dynamic continuous occupancy mapping. *Proceedings of Machine Learning Research*, **78**, 458–471.

[119] Senanayake, Ransalu, Ott, Lionel, O'Callaghan, Simon, and Ramos, Fabio T. 2016. Spatio-temporal Hilbert maps for continuous occupancy representation in dynamic environments. Pages 3925–3933 of: *Advances in Neural Information Processing Systems*, 29.

[120] Senanayake, Ransalu, O'Callaghan, Simon, and Ramos, Fabio. 2017. Learning highly dynamic environments with stochastic variational inference. Pages 2532–2539 of: *2017 IEEE International Conference on Robotics and Automation (ICRA)*.

[121] Senanayake, Ransalu, Tompkins, Anthony, and Ramos, Fabio. 2018. Automorphing kernels for nonstationarity in mapping unstructured environments. *Proceedings of Machine Learning Research*, **87**, 443–455.

[122] Sensoy, Murat, Kaplan, Lance, and Kandemir, Melih. 2018. Evidential deep learning to quantify classification uncertainty. Pages 3183–3193 of: *Advances in Neural Information Processing Systems*, 31.

[123] Shafer, Glenn. 1990. Perspectives on the theory and practice of belief functions. *International Journal of Approximate Reasoning*, **4**(5–6), 323–362.

[124] Shafer, Glenn, and Vovk, Vladimir. 2008. A tutorial on conformal prediction. *Journal of Machine Learning Research*, **9**(3), 371–421.

[125] Shen, Jianxiong, Ruiz, Adria, Agudo, Antonio, and Moreno-Noguer, Francesc. 2021. Stochastic neural radiance fields: Quantifying uncertainty in implicit 3D representations. Pages 972–981 of: *2021 International Conference on 3D Vision (3DV)*. IEEE.

[126] Shin, Daniel, Dragan, Anca D, and Brown, Daniel S. 2023. Benchmarks and algorithms for offline preference-based reward learning. *arXiv preprint arXiv:2301.01392*.

[127] Stoycheva, Mihaela. 2021. Uncertainty estimation in deep neural object detectors for autonomous driving. Master's thesis. KTH Royal Institute of Technology, Stockholm.

[128] Stulp, Freek, Theodorou, Evangelos, Buchli, Jonas, and Schaal, Stefan. 2011. Learning to grasp under uncertainty. Pages 5703–5708 of: *2011 IEEE International Conference on Robotics and Automation*. IEEE.

[129] Sun, Jiankai, Jiang, Yiqi, Qiu, Jianing, Nobel, Parth, Kochenderfer, Mykel J, and Schwager, Mac. 2024. Conformal prediction for uncertainty-aware planning with diffusion dynamics model. Pages 80324–80337 of: *Advances in Neural Information Processing Systems*, 36.

[130] Suomalainen, Markku, Nilles, AlexandraQ, and LaValle, StevenM. 2020. Virtual reality for robots. Pages 11458–11465 of: *2020 IEEE/RSJ International Conference on Intelligent Robots and Systems (IROS)*. IEEE.

[131] Szot, Andrew, Clegg, Alex, Undersander, Eric, et al. 2021. Habitat 2.0: Training home assistants to rearrange their habitat. Pages 251–266 of: *Advances in Neural Information Processing Systems*, 34.

[132] Tanaka, Kazuo, and Wang, Hua O. 2004. *Fuzzy Control Systems Design and Analysis: A Linear Matrix Inequality Approach*. John Wiley & Sons.

[133] Gemini Team Google: Anil, Rohan, Borgeaud, Sebastian, et al. 2023. Gemini: A family of highly capable multimodal models. *arXiv preprint arXiv:2312.11805*.

[134] Thrun, Sebastian. 2002. Probabilistic robotics. *Communications of the ACM*, **45**(3), 52–57.

[135] Tian, Xiaoyu, Jiang, Tao, Yun, Longfei, et al. 2023. Occ3D: A large-scale 3D occupancy prediction benchmark for autonomous driving. *arXiv preprint arXiv:2304.14365*.

[136] Tjanaka, Bryon, Fontaine, Matthew C, Lee, David H, et al. 2023. pyribs: A bare-bones python library for quality diversity optimization. Pages 220–229 of: *Proceedings of the Genetic and Evolutionary Computation Conference*.

[137] Tompkins, Anthony, Senanayake, Ransalu, Morere, Philippe, and Ramos, Fabio. 2019. Black box quantiles for kernel learning. *Proceedings of Machine Learning Research*, **89**, 1427–1437.

[138] Toyungyernsub, Maneekwan, Itkina, Masha, Senanayake, Ransalu, and Kochenderfer, Mykel J. 2021. Doubleprong ConvLSTM for spatiotemporal occupancy prediction in dynamic environments. Pages 13931–13937 of: *2021 IEEE International Conference on Robotics and Automation (ICRA)*. IEEE.

[139] Villani, Cédric, et al. 2009. *Optimal Transport: Old and New*, volume 338. Springer.

[140] Vincent, Joseph A, Feldman, Aaron O., and Schwager, Mac. 2023. Guarantees on robot system performance using stochastic simulation rollouts. *arXiv preprint arXiv:2309.10874*.

[141] Vintr, Tomas, Yan, Zhi, Duckett, Tom, and Krajník, Tomáš. 2019. Spatiotemporal representation for long-term anticipation of human presence in service robotics. Pages 2620–2626 of: *2019 International Conference on Robotics and Automation (ICRA)*. IEEE.

[142] Waczak, John, Aker, Adam, Wijeratne, Lakitha O, et al. 2024. Characterizing water composition with an autonomous robotic team employing comprehensive in situ sensing, hyperspectral imaging, machine learning, and conformal prediction. *Remote Sensing*, **16**(6), 996.

[143] Walley, Peter. 1991. *Statistical Reasoning with Imprecise Probabilities*. Chapman and Hall.

[144] Wang, Guanzhi, Xie, Yuqi, Jiang, Yunfan, et al. 2023. Voyager: An open-ended embodied agent with large language models. *arXiv preprint arXiv:2305.16291*.

[145] Wang, Jun, He, Guocheng, and Kantaros, Yiannis. 2024. Safe task planning for language-instructed multirobot systems using conformal prediction. *arXiv preprint arXiv:2402.15368*.

[146] Welling, Max, and Teh, Yee Whye. 2011. Bayesian learning via stochastic gradient Langevin dynamics. Pages 681–688 of: *Proceedings of the 28th International Conference on Machine Learning (ICML)*. Citeseer.

[147] Wilson, Andrew, and Nickisch, Hannes. 2015. Kernel interpolation for scalable structured gaussian processes (KISS-GP). *Proceedings of Machine Learning Research*, **37**, 1775–1784.

[148] Wilson, Andrew Gordon, Hu, Zhiting, Salakhutdinov, Ruslan, and Xing, Eric P. 2016. Deep kernel learning. Artificial Intelligence and Statistics. *Proceedings of Machine Learning Research*, **51**, 370–378.

[149] Wright, Herbert, Zhi, Weiming, Johnson-Roberson, Matthew, and Hermans, Tucker. 2024. V-PRISM: Probabilistic mapping of unknown tabletop scenes. *arXiv preprint arXiv:2403.08106*.

[150] Wursthorn, Kira, Hillemann, Markus, and Ulrich, Markus. 2024 Uncertainty quantification with deep ensembles for 6D object pose estimation. *arXiv preprint arXiv:2403.07741*.

[151] Xiang, Yu, Schmidt, Tanner, Narayanan, Venkatraman, and Fox, Dieter. 2017. PoseCNN: A convolutional neural network for 6D object pose estimation in cluttered scenes. *arXiv preprint arXiv:1711.00199*.

[152] Xu, Kelvin, Ratner, Ellis, Dragan, Anca, Levine, Sergey, and Finn, Chelsea. 2019. Learning a prior over intent via meta-inverse reinforcement learning. *Proceedings of Machine Learning Research*, **97**, 6952–6962.

[153] Yamauchi, Brian. 1997. A frontier-based approach for autonomous exploration. Pages 146–151 of: *Proceedings 1997 IEEE International Symposium on Computational Intelligence in Robotics and Automation CIRA'97*.

[154] Yuhas, Michael, and Easwaran, Arvind. 2022. Demo abstract: Real-time out-of-distribution detection on a mobile robot. *arXiv preprint arXiv:2211.11520*.

[155] Zadeh, LA. 1965. Fuzzy sets. *Information and Control*, **8**(3), 338–353.

[156] Zhang, Dawei, Fu, Yanwei, and Zheng, Zhonglong. 2022. UAST: Uncertainty-aware siamese tracking. International conference on machine learning. *Proceedings of Machine Learning Research*, **162**, 26161–26175.

[157] Zhi, Weiming, Senanayake, Ransalu, Ott, Lionel, and Ramos, Fabio. 2019. Spatiotemporal learning of directional uncertainty in urban environments with kernel recurrent mixture density networks. *IEEE Robotics and Automation Letters*, **4**(4), 4306–4313.

[158] Zhou, Lijun, Ledent, Antoine, Hu, Qintao, Liu, Ting, Zhang, Jianlin, and Kloft, Marius. 2021. Model uncertainty guides visual object tracking. *Proceedings of the AAAI Conference on Artificial Intelligence*, **35**, 3581–3589.

PART VI

Assured Machine Learning in High-Stakes Domains

11
Toward Certifiably Trustworthy Deep Learning at Scale

LINYI LI

Metacognitive artificial intelligence (AI) implies an AI's capability to conduct self-correction and self-regularization on its predictions, hence achieving trustworthiness in high-stake and safety-critical applications such as autonomous driving and facial recognition. As a result, metacognitive AI is closely connected to certifiable AI and trustworthy AI, the two areas that focus on equipping AI with trustworthy guarantees in high-stake domains. This chapter provides an overview and generic taxonomy for certifiably trustworthy deep learning, aiming at inspiring metacognitive AI toward trustworthiness and certifiability.

11.1 What Are Certified Approaches?

Certified approaches can be divided into two categories: certification approaches and certified training approaches.

11.1.1 Certification Approaches

A *certification approach* provides a guarantee of some specific trustworthiness property for the DL system in the form of whether some *predicate* always holds for all possible states within the underlying set, where the underlying set is given by the *threat model*. As we can see, there are two requisites to define a certification approach: threat model and property predicate.

Definition 11.1 (Threat Model) A threat model $\mathcal{R}: \mathbb{R}_{\geq 0} \to 2^S$ is a set of all possible states parameterized by a radius parameter $r \in \mathbb{R}_{\geq 0}$ satisfying:

(i) Set monotonicity with respect to r: $\forall r_1 \leq r_2, \mathcal{R}(r_1) \subseteq \mathcal{R}(r_2)$.
(ii) State validness: given a radius r, the threat model generates a set of states $\mathcal{R}(r)$, such that any state within which $s \in \mathcal{R}(r)$ can determine the DL system's output.

The concrete definition of threat model $\mathcal{R}$ is determined by the trustworthiness property. From the perspective of computer security, the definition

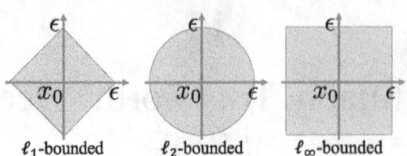

Figure 11.1 An ℓ_p-bounded adversary (Example 11.2) crafts perturbed input from ℓ_p-bounded region centered at clean input x_0. From left to right are ℓ_1-, ℓ_2-, and ℓ_∞-bounded perturbation regions in 2D space with radius ϵ.

formalizes the exhaustive power of the underlying attacker that can use any state within $\mathcal{R}(r)$ as the input to undermine the DL system (Figure 11.1).

As an example, the property of robustness against ℓ_p-bounded perturbation leads to the threat model of ℓ_p-bounded adversary.

Example 11.2 (ℓ_p-bounded Adversary) For a given benign input (x_0, y_0), where $x_0 \in \mathcal{X}$ is the input instance and $y_0 \in [C]$ is its true label, the ℓ_p-bounded adversary defines a state set of possible input: $\mathcal{R}(r) = B_{p,r}(x_0) := \{x : \|x - x_0\|_p \le r\}$.

In the threat model of ℓ_p-bounded adversary, the attacker can add arbitrary perturbations to the input as long as the ℓ_p norm of the perturbation is within r. Commonly-seen ℓ_p norms are shown in Theorem 11.2.

Other threat model examples can be the semantic adversary that exerts domain-specific semantic-preserving transformations to input data and data poisoning adversary that perturbs the training data. We will define more threat models in Section 11.2.

Another requisite is the property predicate.

Definition 11.3 (Property Predicate) A property predicate $p(f, s)$ takes a DL system f and a state s as an input, and judges a binary result, where True indicates the property holds and False otherwise.

For example, for the robustness property, making a correct prediction for the perturbed or transformed input s is the goal and naturally defines the predicate. In many cases, the predicate's result is based on comparing some metric with a threshold (e.g., accuracy larger than some value, mean loss smaller than some value, or confidence margin being positive for classification correctness). Hence, the goal of a certification approach naturally transforms into providing a lower or upper bound of that metric.

Now we are ready to define the certification approach.

Definition 11.4 (Certification Approach) Given a DL system $f : \mathcal{S} \to \mathcal{Y}$, where $\mathcal{S}$ is the valid state set and $\mathcal{Y}$ is model output space, a threat model $\mathcal{R}$

defined in Definition 11.1, a radius parameter $r \in \mathbb{R}_{\geq 0}$, a predicate p defined in Definition 11.3, an algorithm $\mathcal{A}(f, \mathcal{R}, p, r)$ is called a *certification approach* if it satisfies this condition:

If $\mathcal{A}(f, \mathcal{R}, p, r)$ = True, $\forall s \in \mathcal{R}(r), p(f, s)$ (when $\mathcal{A}$ is deterministic) or $\Pr[\forall s \in \mathcal{R}(r), p(f, s)] \geq 1 - \alpha$ where α is a pre-defined small threshold (when $\mathcal{A}$ is probabilistic, and the randomness should be independent of f, $\mathcal{R}$, p, and r).

As we can see, when the certification approach outputs True, the trustworthiness property holds for any state within the threat model set. Since the attacker is constrained to choose the state from the set, such output is a guarantee of the system's trustworthiness under an attack or under environment uncertainties, so $\mathcal{A}$ is called a certification approach.

Take the robustness certification for example, when the certification approach outputs True, any constrained perturbation to input $s \in \mathcal{R}$ cannot change the system's correct prediction, so the attacker cannot succeed. Usually, we measure a DL system's certified robustness by the ratio of test set samples where $\mathcal{A}$ outputs True, and this ratio, namely, certified robust accuracy, indicates a lower bound of system's accuracy under the constrained attack.

What happens if a certification approach outputs False? From the definition, the trustworthiness is unknown, i.e., $\forall s \in \mathcal{R}(r), p(f, s)$ could either hold or not. So a naive certification approach may barely output False which is useless. On the other hand, we hope that when the certification approach outputs False, the trustworthiness property does not hold, i.e., when the trustworthiness property holds, the approach outputs True as much as possible. We call this expectation *tightness*: if the approach is tighter, the approach outputs True as much as possible when the trustworthiness property holds. If an approach achieves perfect tightness, it implies that when the trustworthiness property holds, it will always output True, then we will call the approach *complete certification*. We will discuss this more in Section 11.3.1.

11.1.2 Certified Training Approaches

Though it sounds promising to use certification approaches to bring trustworthiness guarantees to DL systems, it is challenging. On the one hand, the DL system to certify could be intrinsically untrustworthy, so it is impossible to generate guarantees for it. On the other hand, the DL system could be hard to certify due to its certification-unfriendly weights or architecture design.

To mitigate these challenges, certified training approaches are developed. Certified training approaches train DNNs so they can be certified with a high degree of trustworthiness. Certified training approaches significantly boost the certified trustworthiness by jointly improving the intrinsic trustworthiness and adapting the model architecture and weights in a certification-friendly direction.

Certified training approaches are usually strongly tied to certification approaches. They encourage desired properties from the certification approaches for the model to emerge during training. In Section 11.3.2, we will categorize certified training approaches in detail.

Since certified training approaches are executed on the training dataset, and the degree of trustworthiness is measured by running certification approaches on the test dataset, a strong certified training approach under common notions also has a good degree of generalizability, though not explicitly emphasized.

11.2 What Are Trustworthy Properties?

In this section, we briefly introduce trustworthy properties in the literature along with their corresponding definitions of threat models and property predicates.

11.2.1 Robustness against ℓ_p-Bounded Perturbations

DL systems should be robust against arbitrary tiny perturbations directly added to normal input. For this trustworthy property, the threat model is defined with ℓ_p-bounded adversary (Example 11.2), and the property predicate depends on the task: for the classification task, the predicate is making a correct prediction, i.e., for input x_0 with true label $y \in [C]$, the output for the perturbed input $F(x) = y$; for the regression task, the predicate is the prediction error is upper bounded by some threshold, i.e., for input x_0 with true value $y \in \mathbb{R}$, the output for the perturbed input satisfies $|F(x) - y| \le \epsilon$.

For the regression task, this property transforms into providing a lower and upper bound of $F(x)$ for any x in state set. For the classification task, since DL system F usually conducts label prediction by predicting a confidence score for each class and then choosing the most probable class, i.e., $F(x) = \arg\max_{i \in [C]} f(x)_i$ where $f(x)_i$ is the confidence score for the ith class, the predicate equals to for any $i \in [C] \setminus \{y\}$, $f(x)_y - f(x)_i \ge 0$. Hence, computing a lower bound of $f(x)_y - f(x)_i$ for any x in state set is the goal. In summary, in both cases, the certification approach can generate the certification via computing a lower or upper bound for some quantity of interest.

When the robustness is certified for threat model $\mathcal{R}(r)$, we call r certified radius or robust radius (of model F at point x_0).

11.2.2 Robustness against Semantic Transformations

DL systems should be robust against semantic-preserving transformations, such as small brightness change, contrast change, rotation, and scaling, to the image. For this trustworthy property, the threat model is defined with a predefined parameterized transformation function $\phi: \mathcal{X} \times \mathcal{Z} \to \mathcal{X}$, transforming an image $x \in \mathcal{X}$ with a $\mathcal{Z}$-valued parameter α. For example, we use $\phi_R(x, \alpha)$ to model a rotation of the image x by α degrees counter-clockwise with bilinear interpolation. Then, the threat model is defined as the range of transformation function when the parameter is bounded, e.g., $\mathcal{R}(r) = \{\phi(x_0, \delta) : \|\delta\|_\infty \leq r\}$. The property predicate definition is the same as Section 11.2.1.

11.2.3 Robustness against State Observations in Reinforcement Learning

In reinforcement learning (RL), the DL system encodes a trained policy π to interact with the environment in multiple rounds (namely, "steps"). At each step, the policy observes the state input from the state space $s \in \mathcal{S}$ and chooses an action from the action space $a = \pi(s) \in \mathcal{A}$. Such action transforms the state to $P(s, a) \in \mathcal{S}$ (consider the deterministic environment in this chapter) in the next round and receives a reward $r(s, a)$. However, the observed state by the policy agent can be maliciously perturbed by the attacker, e.g., by hacking the camerate perceptron module, and we hope the policy is still robust in this case.

Hence, we define the threat model to be the union of ℓ_p-bounded adversary at each step, i.e., denoting s_t to the true state at each step t, the state set $\mathcal{R}(r) = (S'_0, S'_1, \ldots, S'_H)$ where $S'_t = B_{p,r}(s_t)$. The property predicate is defined at two levels: step-level – per-state action stability and episode-level – guaranteed cumulative reward. The former requires the action prediction does not change, i.e., $\pi(s'_t) = \pi(s_t)$ for $s'_t \in S'_t$. The latter requires the cumulative reward after H steps of interaction, i.e., $R = \min \sum_{i=0}^{H} r(s_t, \pi(s'_t))$, $s'_t \in S'_t$ is larger than some threshold.

11.2.4 Robustness against Poisoning Attacks in Reinforcement Learning

Besides state observation perturbations, in offline reinforcement learning, the performance of trained DL policy can also be negatively impacted by perturbations on the training data, i.e., the malicious attacker, as a training data provider, can destroy the policy performance by deleting, inserting, or replacing the training data samples.

Concretely, in offline RL, a training dataset $D = \{\tau_i\}_{i=1}^{N}$ consists of logged trajectories, where each trajectory $\tau = \{(s_j, r_j, a_j, s'_j)\}_{j=1}^{l} \in (\mathcal{S} \times \mathcal{A} \times \mathbb{R} \times \mathcal{S})^l$ consists of multiple tuples denoting the transitions (i.e., starting from state s_j, taking the action a_j, receiving reward r_j, and transitioning to the next state s'_j). Training dataset D can be poisoned in the following manner. For each trajectory $\tau \in D$, the adversary is allowed to replace it with an arbitrary trajectory $\widetilde{\tau}$, generating a manipulated dataset $\widetilde{D}$. We denote $D \ominus \widetilde{D} = (D \backslash \widetilde{D}) \cup (\widetilde{D} \backslash D)$ as the *symmetric difference* between two datasets D and $\widetilde{D}$. For instance, adding or removing one trajectory causes a symmetric difference of magnitude 1, while replacing one trajectory with a new one leads to a symmetric difference of magnitude 2. Hence, the threat model defines a set over manipulated datasets: $\mathcal{R}(r) = \{\widetilde{D} : D \ominus \widetilde{D} \leq r\}$. The property predicates are the same as in Section 11.2.3.

11.2.5 Distributional Fairness

DL systems may inherit bias and disparity. Specifically, for data from different groups (identified by some protected or sensitive attribute $\mathcal{X}_s$), the model may have different performances. To identify the fairness issue, we expect that the model should have a low loss value on a perfectly fair distribution. In other words, when the distribution itself is fair, the model should encode such distribution very well. Otherwise, the model can have bias (if performed well on training distribution) or have bad performance (if not performed well either on training distribution). Either is not expected.

Hence, to quantify the fairness of a model from the distributional level, we can certify an upper bound of expected loss on any slightly shifted and perfectly fair distribution. The threat model is the set of all distributions (1) whose distance to the original training distribution is bounded; and (2) which is perfectly fair measured by group base rate. The property predicate is the model's expected loss on the distribution which is smaller than some threshold.

11.2.6 Numerical Reliability

In deployment, a DL system may incur numerical failures by outputting NaN or INF instead of producing any meaningful prediction, resulting in system crashes. This is the numerical reliability issue for DL systems.

Numerical failures are triggered when some operators in the DL system receive invalid input, e.g., when log operator receives negative input. Hence, to certify the numerical reliability of a DL system, the threat model is defined as all valid input and all valid weights. For example, for image models, the threat model is the valid image domain $[0, 1]^d$ concatenated with the typical weight domain. Note that this threat model is not parameterized by some radius r, or it can be viewed as parameterized by a fixed radius r. Then, the property predicate is that for all operators within the DL system, their input falls into the valid range.

In contrast to previous trustworthy properties that certify a concrete DL system, certification of numerical reliability brings a guarantee for a group of DL systems sharing the same model architecture.

11.3 Taxonomy and Characterization of Certified Approaches

In this section, we outline important characteristics and features of certified approaches and use them to construct a taxonomy for both certification and certified training approaches.

11.3.1 Taxonomy of Certification

For certification approaches, we typically care about five aspects: efficiency (or scalability), tightness, soundness, inference overhead, and generalizability. We will use them as the taxonomy criteria.

Efficiency (Scalability) The certification approach varies in its efficiency, and efficiency can usually be measured by time complexity with respect to model size (concretely, the number of neurons or parameters). An efficient certification approach can support larger models, so it is also called scalability. Given the trend of scaling up DL models, the more efficient a certification approach is, the broader its potential applicability. Hence, we use efficiency (scalability) as one taxonomy criterion.

We measure the efficiency with two metrics: (1) a qualitative measure: the largest dataset that has been demonstrated feasible to certify by existing work using the corresponding certification approach under a reasonable radius.[1] The dataset effectively measures scalability. For example, the approach scaling up to ImageNet is more scalable than the one to MNIST. (2) A quantitative measure: the best known time complexity for certifying an arbitrary input, given an arbitrary DNN with depth l, width w in terms of neurons, and sampling number S (for sampling-based probabilistic approaches and partition-based approaches). Note that the time complexity for DNN inference is $O(lw^2)$. All

---

[1] For example, radius $r \geq 1/255$ for robustness certification against ℓ_p-bounded perturbations.

sampling-based probabilistic approaches have complexity $O(Slw^2)$, which is because the sampling time cost is much higher than the actual bound computation whose time complexity is subsumed. poly(l, w) means a time complexity higher than $O(lw^3)$.

Tightness As discussed in Section 11.1, besides being sound (when output True, truly trustworthy), we also expect a certification approach to be as tight as possible (when output False, not trustworthy as much as possible). Hence, tightness is one important criterion.

The highest level of tightness is "complete," where a False output means not trustworthy for certain. For other incomplete approaches, the tightness measurement is measured quantitatively by comparison with other approaches supporting the same architecture and inference protocol. The comparison is based on our benchmark results in [69] and theoretical results from the literature. For approaches that support generic DNNs, tightness is ranked by T_n; and for sampling-based probabilistic approaches, they are ranked by ST_n. The larger n means tighter approaches.

The *intrinsic trade-off between efficiency and tightness*, i.e., either scalability or tightness can be achieved but not both, constitutes the main obstacle for trustworthiness certification. For example, for robustness certification, all complete certification approaches have exponential time complexity $O(2^{lw})$ which is theoretically proved [50, 118].

Soundness Though all approaches are sound by definition, the degree of soundness still varies: some approaches bring deterministic certification and others tolerate a small probability of making false claims. We use "deterministic/probabilistic" to distinguish these two classes in our taxonomy.

Inference Overhead Some certification approaches do not support DNNs with normal inference procedures. Instead, they require a customized inference procedure, e.g., adding noise to the input and then applying majority voting. By doing so, they can require less information from the DL system itself, e.g., require no information about model's architecture, to improve the efficiency. On the other hand, the customized inference procedure could be much more expensive than the normal one, bringing inference overhead which is critical for deployment. Hence, we outline the inference overhead as one taxonomy criterion and highlight approaches that incur inference overhead.

Generalizability Different certification approaches require different degrees of knowledge from the DL system. As mentioned above, some approaches require no information about model's architecture but just an oracle access to the inference result. We call these approaches *black-box approaches*. On the other hand, other approaches require and support a particular set of model architectures, and they are called *white-box approaches*. Most common white-box approaches support models with ReLU as the activation functions (named "ReLU Nets" in Table 11.1). Some are more generic (named "generic DNNs" in Table 11.1), and some others are more restrictive, e.g., only supporting specific Lipschitz-bounded layers.

An interesting observation is that some most generalizable approaches (i.e., black-box approaches) have inference overhead, indicating a trade-off between efficiency and generalizability.

Table 11.1 presents our taxonomy results of certification approaches. Besides categorizing using the above five aspects, we use the certified trustworthy property as the first-level criterion since approaches are proposed to certify one trustworthy property at a time. We group approaches with the same core methodology together and use the methodology name as the identifier. Detail references are listed in the last column.

As we can observe, there is a large family of certification approaches in the literature and most of them are proposed within five years, reflecting the rapid development of this field.

More discussions and findings from the taxonomy can be found in [69]. A visualization of the taxonomy is in https://sokcertifiedrobustness.github.io.

11.3.2 Taxonomy of Certified Training

Compared to certification approaches, certified training approaches are relatively simpler, all sharing the same procedure in deep learning: gradient-descent-based optimization over training data in mini-batches. The main difference lies in the core methodologies in terms of data augmentation, pretraining, loss computation, and regularization. Hence, we use the core methodology as the taxonomy criterion. The taxonomy results are shown in Table 11.2.

Regularization-based Training For complete certification, branch-and-bound (BaB) and mixed integer programming (MIP) are among the most efficient methodologies so far. Xiao et al. [127] find that for these certification approaches, the number of branches is upper bounded by the number of unstable ReLU neurons which motivates a regularization term to increase the ReLU neuron's stability for training. For complete certification based on linear region traversal, we can train with a regularization term maximizing the margin

Table 11.1 *Taxonomy, characteristics, and references of trustworthy certification approaches. Details are explained in Section 11.3.1. Relative tightness levels T_n and ST_n are only listed among comparable approaches.*

Trustworthy Property	Complete/ Incomplete	Soundness	Generalizability	Core Methodology		Scalability (Scale up to)	(Complexity)	Tightness	Inference Overhead	References
Robustness against ℓ_p-bounded Perturbations	Complete	Deterministic	White-box	for ReLU Nets	Solver-Based	SMT-Based	MNIST	$O(2^{|w|})$	Complete		2010, 2012
						MILP-Based	CIFAR-10	$O(2^{|w|})$	Complete		2017, 2018, 2017, 2019
					Extended Simplex Method		MNIST	$O(2^{|w|})$	Complete		2017, 2019
					Branch-and-Bound		CIFAR-10	$O(2^{|w|})$	Complete		2020, 2020, 2018, 2021, 2021, 2021, 2018, 2019, 2020, 2019b, 2021, 2021, 2022a
	Incomplete	Deterministic	White-box	for General DNNs[1]	Linear Relaxation	Linear Programming (LP)	CIFAR-10	$O(poly(l, w))$	T_2		2019a, 2018
						Interval	Tiny ImageNet	$O(lw^2)$	T_5		2019
						Polyhedra	Tiny ImageNet	$O(lw^2)$	T_4		2020, 2021, 2019, 2019, 2020, 2018
						Zonotope	Tiny ImageNet	$O(lw^2)$	T_5		2019, 2018, 2018a, 2019b
						Duality	Tiny ImageNet	$O(lw^2)$	T_4		2018a, 2018b, 2018, 2018
					Multi-Neuron Relaxation		CIFAR-10	$O(lw^2) \cdot O(2^{k \cdot p})$[2]	T_2		2020, 2021, 2019a, 2019
					Semidefinite Programming (SDP)		CIFAR-10	$O(poly(l, w))$	T_6		2020, 2019, 2020, 2018a, 2020b
				for Lip-Bounded Nets	Lipschitz	General Lipschitz	Tiny ImageNet	$O(lw^2)$	T_3[3]		2020, 2021, 2021, 2018, 2018, 2019
						Smooth Layers	Tiny ImageNet	$O(lw^2)$			2017, 2019e, 2022, 2021, 2019a, 2022a, 2022b
				for Non-ReLU Nets[4]	Curvature		CIFAR-10	$O(lw^2)$			2020
				Black-box[5]	Zeroth Order Smoothing	Lipschitz	ImageNet	$O(Slw^2)$	*	Exist	2021
		Probabilistic	Black-box			Differential Privacy Inspired	ImageNet	$O(Slw^2)$		Exist	2019
						Divergence Based	ImageNet	$O(Slw^2)$	ST_3	Exist	2020, 2019a
						Neyman Pearson	ImageNet	$O(Slw^2)$	ST_1	Exist	2019
						Level-Set Analysis	ImageNet	$O(Slw^2)$	ST_2	Exist	2020, 2020, 2020a
						Lipschitz	ImageNet	$O(Slw^2)$	ST_2	Exist	2020, 2019a
					First Order Smoothing		ImageNet	$O(Slw^2)$	ST_1	Exist	2020, 2020a
					Double Smoothing		ImageNet	$O(Slw^2)$	ST_1	Exist	2022
Robustness against Semantic Transformations	Incomplete	Deterministic	White-box	for General DNNs[1]	Linear Relaxation	Linear Inequality	CIFAR-10	$O(lw^2)$	T_2		2019a
						Interval	CIFAR-10	$O(lw^2)$	T_3		2019, 2020b
						Polyhedra	CIFAR-10	$O(lw^2)$	T_1		2019, 2020b
	Complete		Black-box		Partition Enumeration				Complete		2017
	Incomplete	Probabilistic	Black-box		Partition Enumeration + Zeroth Order Smoothing		ImageNet	$O(Slw^2)$		Exist	2020a, 2021
Robustness in RL against State Perturbations	Incomplete	Deterministic	White-box	for General DNNs	Linear Relaxation	Linear Inequality	Polyhedra	CIFAR-10	$O(lw^2)$		2021
		Probabilistic	Black-box		Zeroth Order Smoothing		CIFAR-10	$O(Slw^2)$		Exist	2020, 2020b
Robustness in RL against Poisoning Attacks	Incomplete	Probabilistic	Black-box		Zeroth Order Smoothing		ImageNet	$O(Slw^2)$		Exist	2022a
Distributional Fairness	Incomplete	Probabilistic	White-box	for Wasserstein Distance	Lipschitz & Robust Optimization		ImageNet	$O(lw^2)$			2018
			Black-box	for Hellinger Distance	Gaussian Bound		ImageNet	$O(Slw^2)$	ST_2		2022
					Gaussian Bound + Subpopulation Decomposition		ImageNet	$O(Slw^2)$	ST_1		2022
Numerical Reliability	Incomplete	Deterministic	White-box	for General DNNs	Linear Relaxation	Linear Programming (LP)	ImageNet	$O(poly(l, w))$	T_2		2022
						Linear Inequality	Interval	ImageNet	$O(lw^2)$	T_1		2022b

1. Typical approaches mainly support ReLU networks, but extensions to general DNNs are available 2019, 2020, 2018.
2. Tightness depends on intermediate layer bounds. If they share the same intermediate layer bounds, the tightness order is Zonotope < Polyhedra = Duality < LP 2019a.
3. Lipschitz bound is loose for typical DNNs, but can be tight for specially regularized DNNs which have small Lipschitz bounds.
4. Only available for networks whose activation functions have nonzero second-order derivatives, which exclude ReLU networks. Thus, tightness is incomparable with others.
5. The approach requires smoothing with some specific distributions as inference protocol.
6. Tunable time complexity dependent on the upper limit of number of linear constraints.
7. Only support discrete inputs and discrete transformations.

Table 11.2 *Taxonomy and references of certified training approaches for trustworthy machine learning.* Suitable Certification *summarizes the certification approaches for which the training approach is designed. Details are explained in Section 11.3.2.*

Robust Training Approaches	Suitable Verification	References
Regularization-Based	Complete, Incomplete and Deterministic (Lipschitz & Curvature)	2020, 2019, 2017, 2020, 2020, 2021, 2022, 2019, 2021a, 2022a
Relaxation-Based	Incomplete and Deterministic (Linear Relaxation, SDP)	2020, 2019, 2019b, 2021, 2018, 2018a, 2021, 2018c, 2018, 2018, 2020b
Augmentation-Based	Incomplete and Probabilistic	2020, 2019, 2019, 2022, 2019, 2019b, 2020, 2020a
Augmentation- and Regularization-Based	Incomplete and Probabilistic	2020, 2019a, 2022, 2020

to nonrobust regions [20, 21]. The Lipschitz and curvature certification favor small Lipschitz constant and small curvature bounds respectively. Therefore, the corresponding robust training approaches are very effective by explicitly penalizing large Lipschitz or curvature bounds [57, 59, 100, 110].

Relaxation-based Training For linear relaxation-based certification approaches, models with tight linear relaxation bounds are favored. To train such models, corresponding robust training approaches usually use the computed bounds from linear relaxation as the training objective to explicitly improve the bound tightness. This idea is similar to the powerful empirical defense named adversarial training (AT) [76] which uses effective attacks to approximately find "most adversarial" example $\max_{x \in B_{p,r}(x_0)} \mathcal{L}(f_\theta(x), y_0)$ and minimize model weights θ with respect to it, where $\mathcal{L}$ is a typical loss function such as cross-entropy loss. In relaxation-based training, instead, we compute an upper bound of $\max_{x \in B_{p,r}(x_0)} \mathcal{L}(f_\theta(x), y_0)$ and minimize it. The bound can be derived from IBP [39, 95], polyhedra-based [5, 74, 140], zonotope-based [77], or duality-based certification [27, 65, 120]. Some useful training tricks are: combining relaxation-based loss with standard loss to improve benign accuracy [39, 114, 140], applying relaxation on some layers but not all to balance benign accuracy and certified robustness [5], specialized weight initialization and training scheduling [95], and using reference space to guide the relaxation [65]. An intriguing phenomenon of relaxation-based training is that tighter relaxation, when used as the training objective, may not lead to more certifiably robust models [48], while the loosest IBP relaxation can achieve almost the highest certified robustness. A conjecture is that tighter relaxation may lead to a less smooth loss landscape containing discontinuities or sensitive regions which poses challenges for gradient-based training [48, 58]. A theoretical understanding of relaxation-based training is still lacking. Note that solver-based and branch-and-bound-based complete certifications usually use linear relaxations for bounding. Therefore, models trained with these relaxation-based training approaches can usually be efficiently certified by these complete certification approaches [108, 115].

Augmentation-based Training Since smoothing-based certification favors models to perform well for noisy inputs, to obtain high certified robustness, we can train the DNNs with noisy inputs, resulting in augmentation-based training [19, 55, 64]. Built upon such augmentation-based training, later approaches combine augmentation with regularization terms to encourage the prediction stability/consistency when the input noise is added [43, 44, 133]. Strategic training regularization combined with augmentation and ensemble is effective and achieves the state-of-the-art certified robustness against ℓ_2 adversary [42, 131]. Adversarial training combined with augmentation [91], and training unlabeled data [13] are also shown effective. Recently, diffusion models [104], which intrinsically possess the denoising ability, are leveraged to build models for randomized smoothing [12, 126]. They achieve superior or competitive certified robustness compared to above methods though require large model size which results in large inference overhead.

11.3.3 Extensions beyond Taxonomy

Besides the literature and related work covered by the taxonomy, there are a few extensions of certified approaches going further beyond. Here we illustrate some extending angles and discuss some example approaches.

Extensions on Trustworthy Properties The techniques of certified approaches discussed in this chapter are extended to certify other trustworthy properties. (1) **Robustness against local evasion attacks**: in local evasion attacks, the adversary slightly perturbs the in-distribution data to mislead the model. The ℓ_p-bounded perturbations and semantic transformations are both special types of local evasion attacks. Now we elaborate on some other effective local evasion attacks and their certified approaches. (a) *Generative model-based adversary* uses generative models such as GAN [38] to generate input perturbation. Similar to certification against the semantic adversary, smoothing-based approaches and linear relaxation approaches can be extended to provide certification against this adversary [78, 119]. (b) ℓ_0 *adversary* picks a bounded number of pixels to arbitrarily change and *patch adversary* picks a region of pixels with a bounded area to arbitrarily change. To defend against ℓ_0 adversary, smoothing-based approaches can be deployed [45, 56, 61]. To defend against patch adversary, the core idea of smoothing-based approaches, prediction aggregation on several noisy inputs which are patched inputs here, is leveraged to develop customized certification and corresponding training approaches. Starting from direct prediction aggregation [60], some recent certified defenses exploit or design model architectures and inference procedures with self-aggregation property, such as DNNs with small localized receptive fields [124], importance-score-based pruning [40], vision transformers [92], and two-round patch-masking [125], to improve the efficiency and tightness of robustness certification. These approaches [40, 60, 92, 124, 125] can provide robustness guarantees on the large-scale ImageNet dataset. (2) **Distributional evasion attacks**: in distributional evasion attacks, the attacker shifts the whole test data distribution within some bounded distance to maximize the expected loss. This threat model can be used to characterize the out-of-distribution generalization ability of ML models [93]. The certification under this threat model is an upper bound of the expected loss, which can be derived from duality under Lipschitz and curvature assumptions [103] or from extensions of smoothing-based approaches [53, 117]. Note that our distributional fairness property is similar to robustness against distributional evasion attacks, but in distributional fairness property we further require the perturbed distribution to be perfectly fair. (3) **Global evasion attacks**: global evasion attacks can perturb any valid input example to mislead the model, whereas local evasion attacks can only perturb in-distribution data. Thus, the robustness against global evasion attacks means that the robustness property holds for the whole input domain. An example of a robustness property is that for any high-confident prediction, small perturbations cannot change the predicted label [59]. In the security domain, Chen et al. [15] recently proposed several domain-specific robustness properties such as requiring all low-cost features to be robust. To certify these properties, they propose a specific solver-based certification to verify logic ensemble models, and then use the found adversarial example as an augmentation for certified training. The certification and certified training *for DNNs* against global evasion attacks can be a promising direction.

Extensions on Other ML Models There are efforts on generalizing existing certified approaches for DL systems to deal with more types of machine learning models. For example: (1) some approaches that are designed for ReLU networks, such as linear relaxation-based approaches, have been extended to support general DNNs [98, 121, 138], recurrent networks [24, 52, 89], transformers [7, 94], and generative models [78]. The main methodology is to derive the corresponding linear bounds for activation functions or attention mechanisms in these system models. Some complete certification approaches, e.g., branch-and-bound-based ones [115], also support general DNNs. Note that these complete certification approaches become incomplete when applied to general DNNs. (2) Certification approaches for Lipschitz-bounded networks and nonReLU networks have not been generalized to other model types yet. (3) Smoothing-based approaches typically need access to only the final prediction label, so they are applicable to any classification model. However, the model must follow the corresponding smoothing-based inference protocol. (4) There are also certification approaches for decision trees [2, 14, 116], decision stumps [116], nearest prototype classifiers [111], and logic ensembles [15]. However, there is no certification and certified training approach that supports all these system models yet. This is because certification and robust training approaches need to exploit properties (piecewise linearity, Lipschitz bound, smoothness, etc) of specific system models to achieve certified trustworthiness.

Extensions for Concrete Applications Beyond the classification task, the discussed methodologies, such as linear relaxation and smoothing-based approaches, have been extended to certify DL systems in many concrete applications. In natural language processing, extensions include certification for recurrent neural networks against embedding perturbations [24, 52, 89], word substitutions [46], and word transformations [132, 143, 144]. Extensions have also been studied for object detection [17], segmentation [35], and point cloud models [18, 35, 71] in computer vision, and speech recognition [34, 82].

11.4 Challenges and Future Directions

Challenges Despite the remarkable progress, certified approaches for DL systems are still far from perfect. As we can observe from Table 11.1, certification approaches are still not scalable and tight enough. As a result, certified training approaches need to strongly regularize the DL models to achieve a decent level of certifiability, often resulting in a significant loss of utility, i.e., accuracy. Taking robustness against ℓ_p-bounded perturbations as an example, on MNIST, a small-scale dataset, the certified robust accuracy against ℓ_∞-bounded adversary with 0.3 radius has reached over 94%. This is remarkable since the limit is 0.5 radius where any input image can be perturbed to indistinguishable half-gray. However, on more challenging CIFAR-10 and ImageNet datasets, certified robust accuracy is still low. On CIFAR-10, against ℓ_∞ adversary with radius 8/255, the certified robust accuracy is only 40.39%. On the large-scale ImageNet dataset, the certified robust accuracy is only 42.2% under ℓ_2 radius 2.0. Hence, on large-scale datasets, usually, there is a gap between the certified level of trustworthiness and desired level of trustworthiness.

Another critical challenge is to deal with the complex nature of the real world. When DL systems are deployed in the real world, it is always challenging to foresee all possible security and trustworthy threats. Even if we can certify the degree of trustworthiness when the attacker falls into some predefined range, e.g., ℓ_p-bounded perturbations, there is no guarantee whether the attacker will always stick to such range or such attack vector. Hence, all certified approaches cannot provide a full certificate of trustworthiness, but just a relatively strong one. It is always an important research topic to discover, identify, monitor, and formalize trustworthy issues so that certification approaches can be developed to mitigate them. On the other hand, more recently, the development of generative models and language models adds another layer of real-world complexity. For these models, even though threats can be identified, they are hard to certify due to their unbounded and discrete nature. For example, the space of the generative and language model's output is almost infinite and it is an open problem how to make sure any generated content is ethical, nontoxic, unbiased, and privacy-preserving.

Future Directions We outline a few future directions toward solving the above challenges and beyond.

(i) **Tight and Scalable Certification for Trustworthiness** As discussed above, tightness and scalability are core goals of trustworthiness certification approaches. We propose two principles to guide the design toward tighter and more scalable certification: (1) *integration of domain-specific or task-specific knowledge*: pure data-driven learning may be intrinsically limited toward human-level trustworthiness for complex tasks [9]. Hence, domain-specific or task-specific knowledge can be leveraged. As a future step, we believe that automated knowledge collection, knowledge distillation, knowledge integration, and knowledge-based data generation, for both certification and certified training approaches, could be the key to large-scale certified trustworthiness. (2) *Efficient abstraction of knowledge-enhanced DL system*: certification approaches can be viewed as constructing abstractions for the DL system to characterize its behavior on the threat model set. The abstraction of existing certification is usually too generic, which handles the model in a way agnostic of its training or constructing method. In the next-generation approach for DL certification, knowledge and reasoning components will be integrated into the abstraction to derive a much tighter and more scalable certification.

(ii) **Generic Certified Trustworthiness** Besides certification for existing trustworthy properties, certification is also needed by metacognitive AI against more foreseen or unforeseen trustworthy threats. In this aspect, we propose a roadmap containing three stages: <u>stage 1: discover, define, and certify more practical trustworthy properties</u>. Toward trustworthy AI, we need to identify more trustworthy properties that incur profound social impacts. After properties are defined, we need to define them formally and propose corresponding certification approaches. To this end, one of the future research questions is "Can we formally construct a DL-based system, such as a DL-based autonomous vehicle or robotic agent, that is guaranteed to be universally safe when deployed in the physical world?" <u>Stage 2: certify multiple trustworthy properties at the same time</u>. Existing certified methods mainly provide certified trustworthiness for a single property. Can we achieve certified trustworthiness under multiple properties at the same time? If so, what would be the efficient method? If not, is there any inherent trade-off between different properties? What else (e.g., more data, more structured knowledge, more human supervision, or better algorithms) do we need to achieve so? <u>Stage 3: develop certifiably trustworthy DL systems holding multiple foreseen and unforeseen properties</u>. Instead of defining trustworthy properties and developing methods to certify them, is it possible to achieve "meta-trustworthiness"? In other words, can we develop DL systems in a property-agnostic way to achieve human-level trustworthiness under all existing notions (robustness, fairness, privacy, etc.) and possible future notions?

(iii) **Deployment of Certified Trustworthy Systems and Study of Social Impacts** There is no free lunch – the certified methods for DL systems come at a cost. Costs include inference overhead, training overhead, and normal performance degradation. These negative costs are understudied but impose great barriers to deploying certified DL systems in the real world. To deploy certified trustworthy systems in practice, we need to mitigate these practical challenges. We also need to understand the practical implications and social impacts of certifiably trustworthy systems – if certifiably trustworthy DL systems are deployed, to what extent they can benefit social good, equality, democracy, and inclusion.

(iv) **Interdisciplinary Research** Lastly, we believe that the ultimate goal of certified AI trustworthiness is to achieve fully controllable AI, i.e., to learn while perfectly conforming to human specifications and safety constraints. Hence, research in other disciplines chasing controllable AI, such as robotics, control theory, and reinforcement learning, may bring many potential opportunities to better certify AI trustworthiness.

11.5 Conclusion

In this chapter, we have provided a systematic overview, tutorial, and discussion of the certified approaches in trustworthy deep learning. We believe that certified approaches, as a prerequisite for deploying AI in high-stake and safety-critical applications, would be an essential tool in metacognitive AI, and we hope that this chapter can inspire readers to further advance the field of certifiable trustworthiness for metacognitive AI.

References

[1] Anderson, Greg, Pailoor, Shankara, Dillig, Isil, and Chaudhuri, Swarat. 2019. Optimization and abstraction: A synergistic approach for analyzing neural network robustness. Pages 731–744 of: *Proceedings of the Fortieth ACM SIGPLAN Conference on Programming Language Design and Implementation*.

[2] Andriushchenko, Maksym, and Hein, Matthias. 2019. Provably robust boosted decision stumps and trees against adversarial attacks. Pages 12997–13008 of: *Advances in Neural Information Processing Systems*, 32.

[3] Awasthi, Pranjal, Jain, Himanshu, Rawat, Ankit Singh, and Vijayaraghavan, Aravindan. 2020. Adversarial robustness via robust low rank representations. Pages 11391–11403 of: *Advances in Neural Information Processing Systems*, 33.

[4] Bak, Stanley, Tran, Hoang-Dung, Hobbs, Kerianne, and Johnson, Taylor T. 2020. Improved geometric path enumeration for verifying relu neural networks. Pages 66–96 of: *International Conference on Computer Aided Verification*. Springer.

[5] Balunovic, Mislav, and Vechev, Martin. 2020. Adversarial training and provable defenses: bridging the gap. *International Conference on Learning Representations*.

[6] Balunovic, Mislav, Baader, Maximilian, Singh, Gagandeep, Gehr, Timon, and Vechev, Martin. 2019. Certifying geometric robustness of neural networks. Pages 15313–15323 of: *Advances in Neural Information Processing Systems*, 33.

[7] Bonaert, Gregory, Dimitrov, Dimitar I, Baader, Maximilian, and Vechev, Martin. 2021. Fast and precise certification of transformers. Pages 466–481 of: *Proceedings of the 42nd ACM SIGPLAN International Conference on Programming Language Design and Implementation*.

[8] Boopathy, Akhilan, Weng, Tsui-Wei, Chen, Pin-Yu, Liu, Sijia, and Daniel, Luca. 2019. CNN-Cert: An efficient framework for certifying robustness of convolutional neural networks. *Proceedings of the AAAI Conference on Artificial Intelligence*, **33**, 3240–3247.

[9] Bubeck, Sébastien, and Sellke, Mark. 2021. A universal law of robustness via isoperimetry. *Advances in Neural Information Processing Systems (NeurIPS)* 34.

[10] Bunel, Rudy R, Mudigonda, Pawan K, Turkaslan, Ilker, Torr, Philip, Lu, Jingyue, and Kohli, Pushmeet. 2020. Branch and bound for piecewise linear neural network verification. *Journal of Machine Learning Research*, **21**(2020).

[11] Bunel, Rudy R, Turkaslan, Ilker, Torr, Philip, Kohli, Pushmeet, and Mudigonda, Pawan K. 2018. A unified view of piecewise linear neural network verification. Pages 4790–4799 of: *Advances in Neural Information Processing Systems*, 31.

[12] Carlini, Nicholas, Tramer, Florian, Dvijotham, Krishnamurthy Dj, Rice, Leslie, Sun, Mingjie, and Kolter, J Zico. 2023. (Certified!!) Adversarial robustness for free! *International Conference on Learning Representations*.

[13] Carmon, Yair, Raghunathan, Aditi, Schmidt, Ludwig, Duchi, John C, and Liang, Percy S. 2019. Unlabeled data improves adversarial robustness. Pages 11190–11201 of: *Advances in Neural Information Processing Systems*, 32.

[14] Chen, Hongge, Zhang, Huan, Si, Si, Li, Yang, Boning, Duane, and Hsieh, Cho-Jui. 2019. Robustness verification of tree-based models. Pages 12317–12328 of: *Advances in Neural Information Processing Systems*, 32.

[15] Chen, Yizheng, Wang, Shiqi, Qin, Yue, Liao, Xiaojing, Jana, Suman, and Wagner, David. 2021. Learning security classifiers with verified global robustness properties. Page 477–494 of: *Proceedings of the 2021 ACM SIGSAC Conference on Computer and Communications Security*.

[16] Cheng, Chih-Hong, Nührenberg, Georg, and Ruess, Harald. 2017. Maximum resilience of artificial neural networks. Pages 251–268 of: *International Symposium on Automated Technology for Verification and Analysis*. Springer.

[17] Chiang, Ping-yeh, Curry, Michael, Abdelkader, Ahmed, Kumar, Aounon, Dickerson, John, and Goldstein, Tom. 2020. Detection as regression: Certified object detection with median smoothing. Pages 1275–1286 of: *Advances in Neural Information Processing Systems*, 33.

[18] Chu, Wenda, Li, Linyi, and Li, Bo. 2022. TPC: Transformation-specific smoothing for point cloud models. Pages 4035–4056 of: *Proceedings of the Thirty-Ninth International Conference on Machine Learning*.

[19] Cohen, Jeremy, Rosenfeld, Elan, and Kolter, Zico. 2019. Certified adversarial robustness via randomized smoothing. *Proceedings of Machine Learning Research*, **97**, 1310–1320.

[20] Croce, Francesco, and Hein, Matthias. 2020. Provable robustness against all adversarial l_p-perturbations for $p \geq 1$. In: *International Conference on Learning Representations*.

[21] Croce, Francesco, Andriushchenko, Maksym, and Hein, Matthias. 2019. Provable robustness of ReLU networks via maximization of linear regions. *Proceedings of Machine Learning Research*, **89**, 2057–2066.

[22] Dathathri, Sumanth, Dvijotham, Krishnamurthy, Kurakin, Alexey, et al. 2020. Enabling certification of verification-agnostic networks via memory-efficient semidefinite programming. Pages 5318–5331 of: *Advances in Neural Information Processing Systems*, 33.

[23] De Palma, Alessandro, Bunel, Rudy, Desmaison, Alban, et al. 2021. Improved branch and bound for neural network verification via lagrangian decomposition. *arXiv preprint arXiv:2104.06718*.

[24] Du, Tianyu, Ji, Shouling, Shen, Lujia, et al. 2021. Cert-RNN: towards certifying the robustness of recurrent neural networks. Pages 516–534 of: *Proceedings of the 2021 ACM SIGSAC Conference on Computer and Communications Security*.

[25] Dutta, Souradeep, Jha, Susmit, Sankaranarayanan, Sriram, and Tiwari, Ashish. 2018. Output range analysis for deep feedforward neural networks. Pages 121–138 of: *NASA Formal Methods – Tenth International Symposium*, vol. 10811.

[26] Dvijotham, Krishnamurthy, Stanforth, Robert, Gowal, Sven, Mann, Timothy A, and Kohli, Pushmeet. 2018. A dual approach to scalable verification of deep networks. Pages 550–559 of: *Proceedings of the Thirty-Fourth Conference on Uncertainty in Artificial Intelligence*, vol. 1.

[27] Dvijotham, Krishnamurthy, Gowal, Sven, Stanforth, Robert, et al. 2018. Training verified learners with learned verifiers. *arXiv preprint arXiv:1805.10265*.

[28] Dvijotham, Krishnamurthy Dj, Stanforth, Robert, Gowal, Sven, Qin, Chongli, De, Soham, and Kohli, Pushmeet. 2019. Efficient neural network verification with exactness characterization. Page 164 of: *Proceedings of the Conference on Uncertainty in Artificial Intelligence UAI*.

[29] Dvijotham, Krishnamurthy (Dj), Hayes, Jamie, Balle, Borja, et al. 2020. A framework for robustness certification of smoothed classifiers using F-divergences. *Eighth International Conference on Learning Representations, ICLR 2020, Addis Ababa, Ethiopia, April 26-30, 2020*. OpenReview.net.

[30] Ehlers, Ruediger. 2017. Formal verification of piece-wise linear feed-forward neural networks. Pages 269–286 of: *International Symposium on Automated Technology for Verification and Analysis*. Springer.

[31] Everett, Michael, Lütjens, Björn, and How, Jonathan P. 2021. Certifiable robustness to adversarial state uncertainty in deep reinforcement learning. *IEEE Transactions on Neural Networks and Learning Systems*, **33**(9), 4184–4198.

[32] Fazlyab, Mahyar, Morari, Manfred, and Pappas, George J. 2020. Safety verification and robustness analysis of neural networks via quadratic constraints and semidefinite programming. *IEEE Transactions on Automatic Control*, **67**(1), 1–15.

[33] Ferrari, Claudio, Mueller, Mark Niklas, Jovanović, Nikola, and Vechev, Martin. 2022. Complete verification via multi-neuron relaxation guided branch-and-bound. *International Conference on Learning Representations*.

[34] Fischer, Marc, Baader, Maximilian, and Vechev, Martin. 2020. Certified defense to image transformations via randomized smoothing. Pages 8404–5417 of: *Advances in Neural Information Processing Systems (NeurIPS)*, 33.

[35] Fischer, Marc, Baader, Maximilian, and Vechev, Martin. 2021. Scalable certified segmentation via randomized smoothing. *Proceedings of Machine Learning Research*, **139**, 3340–3351.

[36] Fromherz, Aymeric, Leino, Klas, Fredrikson, Matt, Parno, Bryan, and Pasareanu, Corina. 2021. Fast geometric projections for local robustness certification. *International Conference on Learning Representations*.

[37] Gehr, Timon, Mirman, Matthew, Drachsler-Cohen, Dana, Tsankov, Petar, Chaudhuri, Swarat, and Vechev, Martin. 2018. AI^2: Safety and robustness certification of neural networks with abstract interpretation. Pages 3–18 of: *2018 IEEE Symposium on Security and Privacy (SP)*. IEEE.

[38] Goodfellow, Ian, Pouget-Abadie, Jean, Mirza, Mehdi, et al. 2014. Generative adversarial nets. Pages 2672–2680 of: *Advances in Neural Information Processing Systems*, 27.

[39] Gowal, Sven, Dvijotham, Krishnamurthy Dj, Stanforth, Robert, et al. 2019. Scalable verified training for provably robust image classification. Pages 4842–4851 of: *Proceedings of the IEEE International Conference on Computer Vision*.

[40] Han, Husheng, Xu, Kaidi, Hu, Xing, et al. 2021. ScaleCert: scalable certified defense against adversarial patches with sparse superficial layers. Pages 28169–28181 of: *Advances in Neural Information Processing Systems*, 34.

[41] Hein, Matthias, and Andriushchenko, Maksym. 2017. Formal guarantees on the robustness of a classifier against adversarial manipulation. Pages 2266–2276 of: *Advances in Neural Information Processing Systems*, 30.

[42] Horváth, Miklós Z, Mueller, Mark Niklas, Fischer, Marc, and Vechev, Martin. 2022. Boosting randomized smoothing with variance reduced classifiers. *International Conference on Learning Representations*.

[43] Jeong, Jongheon, and Shin, Jinwoo. 2020. Consistency regularization for certified robustness of smoothed classifiers. Pages 10558–10570 of: *2020 Advances in Neural Information Processing Systems*.

[44] Jeong, Jongheon, Park, Sejun, Kim, Minkyu, Lee, Heung-Chang, Kim, Doguk, and Shin, Jinwoo. 2021. SmoothMix: training confidence-calibrated smoothed classifiers for certified robustness. In: *Thirty-Fifth Conference on Neural Information Processing Systems*.

[45] Jia, Jinyuan, Wang, Binghui, Cao, Xiaoyu, Liu, Hongbin, and Gong, Neil Zhenqiang. 2022. Almost tight L0-norm certified robustness of Top-k predictions against adversarial perturbations. *International Conference on Learning Eepresentations*.

[46] Jia, Robin, Raghunathan, Aditi, Göksel, Kerem, and Liang, Percy. 2019. Certified robustness to adversarial word substitutions. Pages 4127–4140 of: *Proceedings of the 2019 Conference on Empirical Methods in Natural Language Processing and the 9th International Joint Conference on Natural Language Processing*.

[47] Jordan, Matt, Lewis, Justin, and Dimakis, Alexandros G. 2019. Provable certificates for adversarial examples: Fitting a ball in the union of polytopes. Pages 14059–14069 of: *Advances in Neural Information Processing Systems*, 32.

[48] Jovanović, Nikola, Balunovic, Mislav, Baader, Maximilian, and Vechev, Martin. 2022. On the paradox of certified training. *Transactions on Machine Learning Research*. Open Review.

[49] Kang, Mintong, Li, Linyi, Weber, Maurice, Liu, Yang, Zhang, Ce, and Li, Bo. 2022. Certifying some distributional fairness with subpopulation decomposition. Pages 31045–31058 of: *Advances in Neural Information Processing Systems*, 35.

[50] Katz, Guy, Barrett, Clark, Dill, David L, Julian, Kyle, and Kochenderfer, Mykel J. 2017. Reluplex: An efficient SMT solver for verifying deep neural networks. Pages 97–117 of: *International Conference on Computer Aided Verification*. Springer.

[51] Katz, Guy, Huang, Derek A, Ibeling, Duligur, et al. 2019. The marabou framework for verification and analysis of deep neural networks. Pages 443–452 of: *International Conference on Computer Aided Verification*. Springer.

[52] Ko, Ching-Yun, Lyu, Zhaoyang, Weng, Lily, Daniel, Luca, Wong, Ngai, and Lin, Dahua. 2019. POPQORN: Quantifying robustness of recurrent neural networks. *Proceedings of Machine Learning Research*, **97**, 3468–3477.

[53] Kumar, Aounon, Levine, Alexander, Goldstein, Tom, and Feizi, Soheil. 2022. Certifying model accuracy under distribution shifts. *arXiv preprint arXiv:2201.12440*.

[54] Kumar, Aounon, Levine, Alexander, and Feizi, Soheil. 2022. Policy smoothing for provably robust reinforcement learning. *International Conference on Learning Representations*.

[55] Lecuyer, Mathias, Atlidakis, Vaggelis, Geambasu, Roxana, Hsu, Daniel, and Jana, Suman. 2019. Certified robustness to adversarial examples with differential privacy. Pages 656–672 of: *2019 IEEE Symposium on Security and Privacy (SP)*. IEEE.

[56] Lee, Guang-He, Yuan, Yang, Chang, Shiyu, and Jaakkola, Tommi. 2019. Tight certificates of adversarial robustness for randomly smoothed classifiers. Pages 4911–4922 of: *Advances in Neural Information Processing Systems*, 32.

[57] Lee, Sungyoon, Lee, Jaewook, and Park, Saerom. 2020. Lipschitz-certifiable training with a tight outer bound. Pages 16891–16902 of: *Advances in Neural Information Processing Systems*, 33.

[58] Lee, Sungyoon, Lee, Woojin, Park, Jinseong, and Lee, Jaewook. 2021. Towards better understanding of training certifiably robust models against adversarial examples. *Thirty-fifth Conference on Neural Information Processing Systems*.

[59] Leino, Klas, Wang, Zifan, and Fredrikson, Matt. 2021. Globally-robust neural networks. *Proceedings of Machine Learning Research*, **139**, 6212–6222.

[60] Levine, Alexander, and Feizi, Soheil. 2020. (De) Randomized smoothing for certifiable defense against patch attacks. Pages 6465–6475 of: *Advances in Neural Information Processing Systems*, 33.

[61] Levine, Alexander, and Feizi, Soheil. 2020. Robustness certificates for sparse adversarial attacks by randomized ablation. Pages 4585–4593 of: *Proceedings of the AAAI Conference on Artificial Intelligence*, 34.

[62] Levine, Alexander, and Feizi, Soheil. 2021. Improved, deterministic smoothing for ℓ_1 certified robustness. *Proceedings of Machine Learning Research*, **139**, 6254–6264.

[63] Levine, Alexander, Kumar, Aounon, Goldstein, Thomas, and Feizi, Soheil. 2020. Tight second-order certificates for randomized smoothing. *arXiv preprint arXiv:2010.10549*.

[64] Li, Bai, Chen, Changyou, Wang, Wenlin, and Carin, Lawrence. 2019. Certified adversarial robustness with additive noise. Pages 9459–9469 of: *Advances in Neural Information Processing Systems*, 32.

[65] Li, Linyi, Zhong, Zexuan, Li, Bo, and Xie, Tao. 2019. Robustra: Training provable robust neural networks over reference adversarial space. Pages 4711–4717 of: *Proceedings of the Twenty-eighth International Joint Conference on Artificial Intelligence (IJCAI)*.

[66] Li, Linyi, Weber, Maurice, Xu, Xiaojun, et al. 2021. TSS: Transformation-Specific Smoothing for robustness certification. *Proceedings of the 2021 ACM SIGSAC Conference on Computer and Communications Security (CCS 2021)*.

[67] Li, Linyi, Zhang, Jiawei, Xie, Tao, and Li, Bo. 2022. Double sampling randomized smoothing. *Proceedings of Machine Learning Research*, **162**, 13163–13208.

[68] Li, Linyi, Zhang, Yuhao, Ren, Luyao, Xiong, Yingfei, and Xie, Tao. 2023. Reliability assurance for deep neural network architectures against numerical defects. *45th International Conference on Software Engineering (ICSE 2023)*.

[69] Li, Linyi, Xie, Tao, and Li, Bo. 2023. SoK: Certified robustness for deep neural networks. *44th IEEE Symposium on Security and Privacy, SP 2023, San Francisco, CA, USA, May 22–26 2023*.

[70] Li, Qiyang, Haque, Saminul, Anil, Cem, Lucas, James, Grosse, Roger B, and Jacobsen, Jörn-Henrik. 2019. Preventing gradient attenuation in Lipschitz constrained convolutional networks. Pages 15390–15402 of: *Advances in Neural Information Processing Systems*, 32.

[71] Liu, Hongbin, Jia, Jinyuan, and Gong, Neil Zhenqiang. 2021. Pointguard: Provably robust 3D point cloud classification. Pages 6186–6195 of: *Proceedings of the IEEE/CVF Conference on Computer Vision and Pattern Recognition*.

[72] Lomuscio, Alessio, and Maganti, Lalit. 2017. An approach to reachability analysis for feed-forward ReLU neural networks. *arXiv preprint arXiv:1706.07351*.

[73] Lu, Jingyue, and Kumar, M Pawan. 2020. Neural network branching for neural network verification. *International Conference on Learning Representations*.

[74] Lyu, Zhaoyang, Ko, Ching-Yun, Kong, Zhifeng, Wong, Ngai, Lin, Dahua, and Daniel, Luca. 2020. Fastened crown: Tightened neural network robustness certificates. *Proceedings of the AAAI Conference on Artificial Intelligence*, **34**, 5037–5044.

[75] Lyu, Zhaoyang, Guo, Minghao, Wu, Tong, Xu, Guodong, Zhang, Kehuan, and Lin, Dahua. 2021. Towards evaluating and training verifiably robust neural networks. Pages 4308–4317 of: *Proceedings of the IEEE/CVF Conference on Computer Vision and Pattern Recognition*.

[76] Madry, Aleksander, Makelov, Aleksandar, Schmidt, Ludwig, Tsipras, Dimitris, and Vladu, Adrian. 2018. Towards deep learning models resistant to adversarial attacks. *International Conference on Learning Representations*.

[77] Mirman, Matthew, Gehr, Timon, and Vechev, Martin. 2018. Differentiable abstract interpretation for provably robust neural networks. *Proceedings of Machine Learning Research*, **80**, 3578–3586.

[78] Mirman, Matthew, Hägele, Alexander, Bielik, Pavol, Gehr, Timon, and Vechev, Martin. 2021. Robustness certification with generative models. Pages 1141–1154 of: *Proceedings of the 42nd ACM SIGPLAN International Conference on Programming Language Design and Implementation*.

[79] Mohapatra, Jeet, Ko, Ching-Yun, Weng, Tsui-Wei, Chen, Pin-Yu, Liu, Sijia, and Daniel, Luca. 2020. Higher-order certification for randomized smoothing. *Advances in Neural Information Processing Systems*, 33.

[80] Mohapatra, Jeet, Weng, Tsui-Wei, Chen, Pin-Yu, Liu, Sijia, and Daniel, Luca. 2020. Towards verifying robustness of neural networks against a family of semantic perturbations. Pages 244–252 of: *2020 IEEE/CVF Conference on Computer Vision and Pattern Recognition (CVPR)*.

[81] Müller, Mark Niklas, Makarchuk, Gleb, Singh, Gagandeep, Püschel, Markus, and Vechev, Martin. 2022. PRIMA: precise and general neural network certification via multi-neuron convex relaxations. *Proceedings of the ACM on Programming Languages*, 6(POPL), 1–33.

[82] Olivier, Raphael, and Raj, Bhiksha. 2021. Sequential randomized smoothing for adversarially robust speech recognition. Pages 6372–6386 of: *Proceedings of the 2021 Conference on Empirical Methods in Natural Language Processing*.

[83] Palma, Alessandro De, Behl, Harkirat, Bunel, Rudy R, Torr, Philip, and Kumar, M. Pawan. 2021. Scaling the convex barrier with active sets. *International Conference on Learning Representations*.

[84] Pei, Kexin, Cao, Yinzhi, Yang, Junfeng, and Jana, Suman. 2017. Towards practical verification of machine learning: The case of computer vision systems. *arXiv preprint arXiv:1712.01785*.

[85] Pulina, Luca, and Tacchella, Armando. 2010. An abstraction-refinement approach to verification of artificial neural networks. Pages 243–257 of: *International Conference on Computer Aided Verification*. Springer.

[86] Pulina, Luca, and Tacchella, Armando. 2012. Challenging SMT solvers to verify neural networks. *AI Communications*, **25**(2), 117–135.

[87] Raghunathan, Aditi, Steinhardt, Jacob, and Liang, Percy. 2018. Certified defenses against adversarial examples. *International Conference on Learning Representations*.

[88] Raghunathan, Aditi, Steinhardt, Jacob, and Liang, Percy S. 2018. Semidefinite

relaxations for certifying robustness to adversarial examples. Pages 10877–10887 of: *Advances in Neural Information Processing Systems*, 31.

[89] Ryou, Wonryong, Chen, Jiayu, Balunovic, Mislav, Singh, Gagandeep, Dan, Andrei, and Vechev, Martin. 2021. Scalable polyhedral verification of recurrent neural networks. Pages 225–248 of: *International Conference on Computer Aided Verification*. Springer.

[90] Salman, Hadi, Yang, Greg, Zhang, Huan, Hsieh, Cho-Jui, and Zhang, Pengchuan. 2019a. A convex relaxation barrier to tight robustness verification of neural networks. Pages 9835–9846 of: *Advances in Neural Information Processing Systems*, 32.

[91] Salman, Hadi, Li, Jerry, Razenshteyn, Ilya P., et al. 2019b. Provably robust deep learning via adversarially trained smoothed classifiers. Pages 11289–11300 of: *Advances in Neural Information Processing Systems*, 32.

[92] Salman, Hadi, Jain, Saachi, Wong, Eric, and Madry, Aleksander. 2022. Certified patch robustness via smoothed vision transformers. Pages 15137–15147 of: *Proceedings of the IEEE/CVF Conference on Computer Vision and Pattern Recognition*.

[93] Shen, Zheyan, Liu, Jiashuo, He, Yue, et al. 2021. Towards out-of-distribution generalization: A survey. *arXiv preprint arXiv:2108.13624*.

[94] Shi, Zhouxing, Zhang, Huan, Chang, Kai-Wei, Huang, Minlie, and Hsieh, Cho-Jui. 2020. Robustness verification for transformers. *International Conference on Learning Representations*.

[95] Shi, Zhouxing, Wang, Yihan, Zhang, Huan, Yi, Jinfeng, and Hsieh, Cho-Jui. 2021. Fast certified robust training with short warmup. *Thirty-fifth Conference on Neural Information Processing Systems*.

[96] Singh, Gagandeep, Gehr, Timon, Püschel, Markus, and Vechev, Martin. 2018. Boosting robustness certification of neural networks. *International Conference on Learning Representations*.

[97] Singh, Gagandeep, Gehr, Timon, Mirman, Matthew, Püschel, Markus, and Vechev, Martin. 2018. Fast and effective robustness certification. Pages 10802–10813 of: *Advances in Neural Information Processing Systems*, 31.

[98] Singh, Gagandeep, Gehr, Timon, Püschel, Markus, and Vechev, Martin. 2019. An abstract domain for certifying neural networks. *Proceedings of the ACM on Programming Languages*, **3**(POPL).

[99] Singh, Gagandeep, Ganvir, Rupanshu, Püschel, Markus, and Vechev, Martin. 2019. Beyond the single neuron convex barrier for neural network certification. Pages 15072–15083 of: *Advances in Neural Information Processing Systems*, 32.

[100] Singla, Sahil, and Feizi, Soheil. 2020. Second-order provable defenses against adversarial attacks. *Proceedings of Machine Learning Research*, **119**, 8981–8991.

[101] Singla, Sahil, and Feizi, Soheil. 2021. Fantastic four: Differentiable and efficient bounds on singular values of convolution layers. *International Conference on Learning Representations*.

[102] Singla, Sahil, Singla, Surbhi, and Feizi, Soheil. 2022. Improved deterministic l2 robustness on CIFAR-10 and CIFAR-100. *International Conference on Learning Representations*.

[103] Sinha, Aman, Namkoong, Hongseok, and Duchi, John. 2018. Certifying some distributional robustness with principled adversarial training. *International Conference on Learning Representations*.

[104] Song, Yang, Sohl-Dickstein, Jascha, Kingma, Diederik P, Kumar, Abhishek, Ermon, Stefano, and Poole, Ben. 2021. Score-based generative modeling through stochastic differential equations. *International Conference on Learning Representations*.

[105] Szegedy, Christian, Zaremba, Wojciech, Sutskever, Ilya, et al. 2014. Intriguing properties of neural networks. In: Bengio, Yoshua, and LeCun, Yann (eds), *2nd International Conference on Learning Representations, ICLR 2014, Banff, AB, Canada, April 14–16, 2014*.

[106] Teng, Jiaye, Lee, Guang-He, and Yuan, Yang. 2020. ℓ_1 Adversarial Robustness Certificates: a Randomized Smoothing Approach. Open Review.

[107] Tjandraatmadja, Christian, Anderson, Ross, Huchette, Joey, Ma, Will, Patel, Krunal Kishor, and Vielma, Juan Pablo. 2020. The convex relaxation barrier, revisited: Tightened single-neuron relaxations for neural network verification. Pages 21675–21686 of: *Advances in Neural Information Processing Systems*, **33**, 21675–21686.

[108] Tjeng, Vincent, Xiao, Kai Y, and Tedrake, Russ. 2019. Evaluating robustness of neural networks with mixed integer programming. *International Conference on Learning Representations*.

[109] Trockman, Asher, and Kolter, J Zico. 2021. Orthogonalizing convolutional layers with the Cayley transform. *International Conference on Learning Representations*.

[110] Tsuzuku, Yusuke, Sato, Issei, and Sugiyama, Masashi. 2018. Lipschitz-margin training: scalable certification of perturbation invariance for deep neural networks. Pages 6542–6551 of: *Advances in Neural Information Processing Systems*, 31.

[111] Voráček, Václav, and Hein, Matthias. 2022. Provably adversarially robust nearest prototype classifiers. *Proceedings of Machine Learning Research*, **162**, 22361–22383.

[112] Wang, Shiqi, Pei, Kexin, Whitehouse, Justin, Yang, Junfeng, and Jana, Suman. 2018. Efficient formal safety analysis of neural networks. Pages 6367–6377 of: *Advances in Neural Information Processing Systems*, 31.

[113] Wang, Shiqi, Pei, Kexin, Whitehouse, Justin, Yang, Junfeng, and Jana, Suman. 2018. Formal security analysis of neural networks using symbolic intervals. Pages 1599–1614 of: *Twenty-Seventh USENIX Security Symposium (USENIX) Security 18)*.

[114] Wang, Shiqi, Chen, Yizheng, Abdou, Ahmed, and Jana, Suman. 2018. Mixtrain: Scalable training of verifiably robust neural networks. *arXiv preprint arXiv:1811.02625*.

[115] Wang, Shiqi, Zhang, Huan, Xu, Kaidi, et al. 2021. Beta-crown: Efficient bound propagation with per-neuron split constraints for neural network robustness verification. Pages 29909–29921 of: *Advances in Neural Information Processing Systems*, 35.

[116] Wang, Yihan, Zhang, Huan, Chen, Hongge, Boning, Duane, and Hsieh, Cho-Jui. 2020. On lp-norm robustness of ensemble decision stumps and trees. *Proceedings of Machine Learning Research*, **119**, 10104–10114.

[117] Weber, Maurice, Li, Linyi, Wang, Boxin, Zhao, Zhikuan, Li, Bo, and Zhang, Ce. 2022. Certifying out-of-domain generalization for blackbox functions. *Proceedings of Machine Learning Research*, **162**, 23527–23548.

[118] Weng, Lily, Zhang, Huan, Chen, Hongge, et al. 2018. Towards fast computation of certified robustness for ReLU networks. *Proceedings of Machine Learning Research*, **80**, 5276–5285.

[119] Wong, Eric, and Kolter, J Zico. 2020. Learning perturbation sets for robust machine learning. *International Conference on Learning Representations*.

[120] Wong, Eric, and Kolter, Zico. 2018. Provable defenses against adversarial examples via the convex outer adversarial polytope. *Proceedings of Machine Learning Research*, **80**, 5286–5295.

[121] Wong, Eric, Schmidt, Frank, Metzen, Jan Hendrik, and Kolter, J Zico. 2018. Scaling provable adversarial defenses. Pages 8400–8409 of: *Advances in Neural Information Processing Systems*, 31.

[122] Wu, Fan, Li, Linyi, Xu, Chejian, et al. 2022. COPA: certifying robust policies for offline reinforcement learning against poisoning attacks. *International Conference on Learning Representations*.

[123] Wu, Fan, Li, Linyi, Huang, Zijian, Vorobeychik, Yevgeniy, Zhao, Ding, and Li, Bo. 2022. CROP: certifying robust policies for reinforcement learning through functional smoothing. *International Conference on Learning Representations*.

[124] Xiang, Chong, Bhagoji, Arjun Nitin, Sehwag, Vikash, and Mittal, Prateek. 2021. PatchGuard: a provably robust defense against adversarial patches via small receptive fields and masking. Pages 2237–2254 of: *30th USENIX Security Symposium (USENIX Security 21)*.

[125] Xiang, Chong, Mahloujifar, Saeed, and Mittal, Prateek. 2022. PatchCleanser: certifiably robust defense against adversarial patches for any image classifier. In: *31st USENIX Security Symposium (USENIX Security)*.

[126] Xiao, Chaowei, Chen, Zhongzhu, Jin, Kun, et al. 2023. DensePure: understanding diffusion models for adversarial robustness. In: *International Conference on Learning Representations*.

[127] Xiao, Kai Y., Tjeng, Vincent, Shafiullah, Nur Muhammad (Mahi), and Madry, Aleksander. 2019. Training for faster adversarial robustness verification via inducing ReLU stability. *International Conference on Learning Representations*.

[128] Xu, Kaidi, Shi, Zhouxing, Zhang, Huan, et al. 2020. Automatic perturbation analysis for scalable certified robustness and beyond. Pages 1129–1141 of: *Advances in Neural Information Processing Systems*, 33.

[129] Xu, Kaidi, Zhang, Huan, Wang, Shiqi, et al. 2021. Fast and complete: enabling complete neural network verification with rapid and massively parallel incomplete verifiers. *International Conference on Learning Representations*.

[130] Yang, Greg, Duan, Tony, Hu, J Edward, Salman, Hadi, Razenshteyn, Ilya P, and Li, Jerry. 2020. Randomized smoothing of all shapes and sizes. *Proceedings of Machine Learning Research*, **119**, 10693–10705.

[131] Yang, Zhuolin, Li, Linyi, Xu, Xiaojun, Kailkhura, Bhavya, Xie, Tao, and Li, Bo. 2022. On the certified robustness for ensemble models and beyond. *International Conference on Learning Representations*.

[132] Ye, Mao, Gong, Chengyue, and Liu, Qiang. 2020. SAFER: A structure-free approach for certified robustness to adversarial word substitutions. Pages 3465–3475 of: *Proceedings of the 58th Annual Meeting of the Association for Computational Linguistics*.

[133] Zhai, Runtian, Dan, Chen, He, Di, et al. 2020. MACER: attack-free and scalable robust training via maximizing certified radius. In: *International Conference on Learning Representations*.

[134] Zhang, Bohang, Cai, Tianle, Lu, Zhou, He, Di, and Wang, Liwei. 2021. Towards certifying L-infinity robustness using neural networks with L-inf-dist neurons. *Proceedings of Machine Learning Research*, **139**, 12368–12379.

[135] Zhang, Bohang, Jiang, Du, He, Di, and Wang, Liwei. 2022. Boosting the certified robustness of L-infinity distance nets. *International Conference on Learning Representations*.

[136] Zhang, Bohang, Jiang, Du, He, Di, and Wang, Liwei. 2022. Rethinking Lipschitz neural networks and certified robustness: A boolean function perspective. Pages 19398–19413 of: *Advances in Neural Information Processing Systems*, 35.

[137] Zhang, Dinghuai, Ye, Mao, Gong, Chengyue, Zhu, Zhanxing, and Liu, Qiang. 2020. Black-box certification with randomized smoothing: a functional optimization based framework. Pages 2316–2326 of: *Advances in Neural Information Processing Systems*, 32.

[138] Zhang, Huan, Weng, Tsui-Wei, Chen, Pin-Yu, Hsieh, Cho-Jui, and Daniel, Luca. 2018. Efficient neural network robustness certification with general activation functions. Pages 4939–4948 of: *Advances in Neural Information Processing Systems*, 31.

[139] Zhang, Huan, Zhang, Pengchuan, and Hsieh, Cho-Jui. 2019. RecurJac: An efficient recursive algorithm for bounding jacobian matrix of neural networks and its applications. *Proceedings of the AAAI Conference on Artificial Intelligence*, **33**, 5757–5764.

[140] Zhang, Huan, Chen, Hongge, Xiao, Chaowei, et al. 2020. Towards stable and efficient training of verifiably robust neural networks. *International Conference on Learning Representations*.

[141] Zhang, Huan, Wang, Shiqi, Xu, Kaidi, Li, Linyi, Li, Bo, Jana, Suman, Hsieh, Cho-Jui, and Kolter, J. Zico. 2022. General cutting planes for bound-propagation-based neural network verification. Pages 1656–1670 of: *Advances in Neural Information Processing Systems*, 35.

[142] Zhang, Yuhao, Ren, Luyao, Chen, Liqian, Xiong, Yingfei, Cheung, Shing-Chi, and Xie, Tao. 2020. Detecting numerical bugs in neural network architectures. Pages 826–837 of: *28th ACM Joint European Software Engineering Conference and Symposium on the Foundations of Software Engineering, ESEC/FSE*. ACM.

[143] Zhang, Yuhao, Albarghouthi, Aws, and D'Antoni, Loris. 2020. Robustness to programmable string transformations via augmented abstract training. *Proceedings of Machine Learning Research*, **119**, 11023–11032.

[144] Zhang, Yuhao, Albarghouthi, Aws, and D'Antoni, Loris. 2021. Certified robustness to programmable transformations in LSTMs. Pages 1068–1083 of: *Proceedings of the 2021 Conference on Empirical Methods in Natural Language Processing*.

12
Metacognition with Neural Network Verification and Repair Using Veritex

XIAODONG YANG, TOMOYA YAMAGUCHI, BARDH HOXHA, DANIL PROKHOROV, TAYLOR T. JOHNSON

12.1 Introduction

Formal methods develop mathematical models of systems for rigorous analysis, typically such that provable guarantees may be established. Formal verification is the process of establishing a formal (mathematical) model of a system that satisfies a specification. There are many approaches for formal verification, ranging from fully automated methods such as model checking to totally manual approaches with pen-and-paper proofs, with computer-assisted intermediaries like interactive theorem proving. If a specification does not hold for a model, then a counterexample may serve as a witness showing how the system violates the specification.

In metacognition, a learning agent (system, human, etc.) learns in part by considering how failures occur and what mistakes are made. In this chapter, we argue that for formal verification in the context of metacognition, the counterexamples showing how a model violates specifications are the mistakes from which learning can occur. Formal verification approaches have emerged in about the past decade for data-driven artificial intelligence (AI) components like neural networks created through machine learning (ML) approaches. The International Verification of Neural Networks Competition (VNN-COMP) series is similar to other formal methods competitions, and recent reports provide a view of the landscape in this domain [1, 8, 12, 13, 26]. The International Competition on Verifying Continuous and Hybrid Systems (ARCH-COMP) series has for several years organized a category on closed-loop systems that incorporate AI and neural network controllers, known as the AI and neural network control systems (AINNCS) category, and recent reports give a similar overview of the landscape in this problem space [17, 18, 22, 23, 24]. These verification techniques are core in considering metacognitive AI from the perspective of learning from mistakes. Beyond the specific approach described in this chapter for metacognition through repair, generally monitoring for violation of specifications at runtime for error monitoring is a critical metacognitive component in autonomous cyber-physical systems (CPS) utilizing AI components [7, 10, 16, 27, 31, 34].

Specifically toward this perspective and to ground these ideas in a concrete approach, this chapter describes Veritex, a verification tool for reachability analysis and repair of deep neural networks (DNNs), the underlying theory of which has previously been described [40]. Veritex includes methods for exact reachability analysis and over-approximative analysis of DNNs using different novel set representations, such as the facet-vertex incidence matrix, face lattice, and $\mathcal{V}$-zono. In addition to the sound and complete safety verification of DNNs, these methods can also efficiently compute the exact output reachable domain as well as the exact unsafe input space that causes safety violations of DNNs in the output. Based on the exact unsafe input–output reachable domain, Veritex can repair unsafe DNNs on multiple safety properties with negligible performance degradation, i.e., to modify the model until it satisfies the specifications. The repair is conducted by updating the DNN parameters through retraining. The approach also works in the absence of the safe model reference and the original dataset for learning. Veritex primarily addresses the issue of constructing provably safe DNNs, which is not yet significantly addressed in most of the current formal methods for trustworthy AI. The utility of Veritex is evaluated for these two aspects, specifically safety verification and DNN repair. Benchmarks for verification include the ACAS Xu networks, and benchmarks for the repair include unsafe ACAS Xu networks and an unsafe agent trained in deep reinforcement learning (DRL).

12.1.1 Overview of DNN Verification, Repair, and Veritex

Deep neural networks (DNNs) have been widely utilized in safety-critical systems with learning-enabled components, such as autonomous vehicles. Despite successful applications in many areas, DNN trustworthiness remains a major concern in realizing reliable autonomy due to their black-box nature with complex nonlinear characteristics. It has been shown that slight perturbations in DNN inputs can cause unpredictable misbehavior in the output. Recently, much effort has been made to develop techniques for formal analysis of DNNs, such as their safety certification [11, 14, 20, 25, 28, 29, 30, 32, 33, 35, 39]. However, these methods that conduct post-training verification of DNNs cannot address the problem of producing provable safe DNNs when they violate safety specifications.

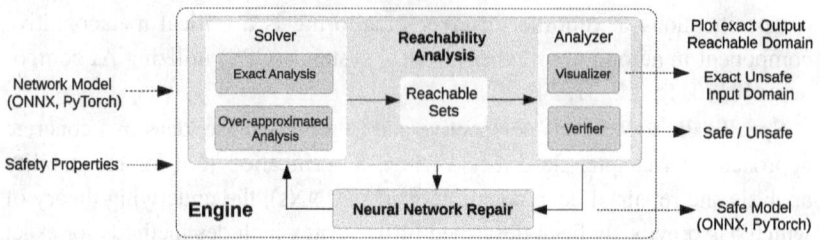

Figure 12.1 An overview of the Veritex architecture.

In this chapter, we detail a formal verification tool called Veritex[1] that performs set-based reachability analysis of DNNs, safety certification, and repair of unsafe DNNs, as an illustrative approach for metacognitive AI. The reachability analysis module enables the computation of an exact or over-approximated output reachable domain for a specified input domain. Also, it enables backward reachability, or the computation of the exact unsafe input space that causes safety violations in the output. The reachable domain contains all the possible reachable states of a system given an input bounded domain. It is a union of reachable sets, which are defined as bounded convex polytopes. A variety of efficient set representations [36] are utilized to construct the reachable set, such as facet-vertex incidence matrix (FVIM) [39], face lattice (FLattice) [37, 38], and $\mathcal{V}$-zono [40]. If the exact output reachable domain does not intersect with specified unsafe domains, the DNN is determined to be safe. Otherwise, the DNN is unsafe and Veritex computes the entire unsafe input space. The repair process in Veritex is a retraining process, which in the context of metacognitive AI is a method to utilize counterexamples of specifications to learn from these mistakes. The repair process utilizes the unsafe input–output reachable domains computed with reachability analysis to repair an unsafe DNN on multiple safety properties simultaneously [40]. Experimental results show that repair can be achieved with negligible impact to the network's initial performance. Here, the safety property is a specification that describes a desired or unsafe output domain of a DNN from a given input domain.

Veritex primarily supports feedforward neural networks (FFNNs), which are commonly used as controllers in learning-enabled control systems. Veritex can perform reachability analysis, safety verification, and unsafe network repair. Veritex also supports the reachability analysis and safety verification of convolutional neural networks (CNNs). To speed up computation, we also adopt a work-stealing parallel framework. It is a well-known scheduling algorithm

---

[1] https://github.com/Shaddadi/veritex & https://github.com/verivital/veritex

Table 12.1 *Overview of primary features in Veritex.*

Feature	Exact analysis	Over-approximation analysis
Set representations	FVIM and FLattice	$\mathcal{V}$-zono
Safety verification	Sound and complete	Sound and incomplete
Network repair	Provably safe networks (FFNNs)	
Activation function	ReLU	ReLU, Sigmoid, and Tanh
Layer types	FC, CONV, MaxPool, and BN	
Parallel computing	Work-stealing parallel	

FC stands for fully connected layers. CONV stands for convolutional layers. MaxPool stands for Max-pooling layers. BN stands for batch normalization

for dynamic multithreaded computation. In the experimental evaluation, two case studies are presented. They include the safety verification and repair of ACAS Xu networks [19], and an unsafe DNN agent for a rocket-lander system in DRL [4]. The experimental results show that Veritex has the highest efficiency in the safety verification of the ACAS Xu networks compared to all 13 related works and that it can repair all unsafe DNNs with negligible performance degradation. Veritex currently supports DNNs with low-dimensional inputs. It has been shown that the exact analysis of DNNs with rectified linear unit (ReLU) activations is an NP-complete problem [19]. Existing exact analysis methods, including our approaches based on FVIM and Flattice, are only scalable to neural networks with small input ranges. Therefore, Veritex also includes an over-approximation method based on activation-function linearization, which enables support for large-scale DNNs and other applications in image classification in the future.

12.2 Overview and Features

Veritex is an object-oriented software programmed in Python. It takes in two inputs as shown in Figure 12.1, the network model and safety properties. Veritex supports the standardized format ONNX and PyTorch for the network model and the unified format VNN-LIB[2] for the safety property. In DNN verification, VNN-LIB is the emerging standard that can specify safety properties of a DNN by defining their input domains and their corresponding unsafe output domains. Roughly for specifications, it is an extension of SMT-LIB with

---

[2] www.vnnlib.org/

additional assumptions. With the network model and its safety properties, Veritex can compute the exact or over-approximated output reachable domain and also the entire unsafe input space if it exists. It supports the plotting of two- or three-dimensional polytopes for visualization. When the repair option is enabled, it will produce a provably safe network in ONNX or PyTorch format. Unlike tools [2, 3, 5, 6, 20, 25, 33], Veritex does not involve LP problems in the reachability analysis and verification of DNNs. Therefore, it does not require any commercial optimization solvers, which makes its installation straightforward. The main features of Veritex are summarized in Table 12.1.

12.2.1 Engine and Components

The engine of Veritex contains two main modules: reachability analysis of DNNs and DNN repair, as shown in Figure 12.1. The former contains functions to compute the reachable domains of a DNN. The latter contains functions to repair an unsafe DNN on multiple safety properties until they are satisfied.

Reachability Analysis Module

The module includes a solver for the computation of the reachable domain and an analyzer for the safety verification and reachable-domain visualization. The solver constructs the incoming network and its safety properties with a network object and a set of property objects. It can compute its exact or over-approximated output reachable domain. It can also compute the exact unsafe input space using the backtracking algorithm [39].

The exact analysis utilizes set representations FVIM and Flattice to compute output reachable sets whose union is the exact output reachable domain. These reachable sets can be sent to the verifier for sound and complete safety verification, which returns either "safe" or "unsafe." The over-approximation module utilizes the set representation $\mathcal{V}$-zono to over approximate the output reachable domain. This reachable domain can be sent to the verifier which returns either "safe" or "unknown." The visualizer plots the reachable domain by projecting it into a two- or three-dimensional space. This visualization is beneficial for the analysis of the impact of repair methods on DNN reachability.

DNN Repair Module

This module eliminates safety violations through optimization of a loss function in the retraining of a DNN, in essence a form of metacognitive AI to relearn from mistakes. In each iteration of repair, the repair module interacts with the reachability analysis module. Given a DNN and its violated safety properties, they are first fed into the reachability analysis module, where the exact unsafe input–output reachable domain over these properties is computed. Recall that the reachable domain consists of reachable sets, which are convex polytopes. Then, the vertices of these sets are selected as representative data pairs $(\mathbf{x}, \mathbf{y})$ to fully represent this reachable domain. They distribute over this domain, including all its extreme points. They are used to construct the distance between the unsafe reachable domain and the safe domain. By minimizing this objective function, the repair can gradually eliminate the unsafe reachable domain, generating a provably safe DNN. When there is a safe model as a reference for the repair, adversarial $\mathbf{x}$ can be fed into this model to generate safe and correct $\hat{\mathbf{y}}$ for the repair. Otherwise, $\hat{\mathbf{y}}$ is set to the closest safe output to $\mathbf{y}$ for the minimal modification.

In addition to the objective function above, the repair also incorporates another objective function into the loss function, which aims to minimize the DNN parameter deviation. This is because slight changes in the parameter can cause unexpected performance degradation. This function minimizes the difference between the predicted output of the repaired network for the training data and the true output in the training data. A weighted-sum method is applied for this multi-objective optimization problem. Two positive real-valued numbers α and β represent the weights of each objective function and $\alpha + \beta = 1$. This repair is named the *minimal repair*. If the original dataset is not available, it can be sampled from the original network. The sampled data are purified by removing unsafe data before the training. Users can set $\alpha = 1$ and $\beta = 0$ to transform the optimization into a single-objective optimization. Then, only the objective function for repair is considered, which is named the *non-minimal repair*.

In practice, the solving of the minimal repair is less efficient than the non-minimal repair due to the Pareto optimality issue in the multi-objective optimization, where one objective function cannot be optimized without worsening the optimization of other objective functions.

12.2.2 Work-stealing Parallel Computation

In the exact analysis, different linearities that the ReLU activation function exhibits over its input ranges $x \geq 0$ and $x < 0$ are separately considered. Therefore, when an input reachable set to one ReLU neuron spans its two input ranges, this set will be divided into two subsets which are separately processed with respect to the linearity in that range. Afterward, these two subsets will be input sets to another neuron. Here, the state $\mathcal{S} = (P, l, N)$ is defined for this computation, where P is a reachable set, l denotes the index of that layer, and N denotes a list of neurons in the layer that will process $\mathcal{S}$. After one neuron, the state $\mathcal{S}$ spawns at most two states $\mathcal{S}'$s with updated P's and N's. This state concept is also applied in the max-pooling layer. One pooling operation normally contains more linearities than the ReLU neuron and thus spawns more states. In the affine-mapping layer, such as fully connected layer and convolutional layer, P in the state will be transformed to one new reachable set P' accordingly.

In the work-stealing parallel computing, each processor computes their states and stores additional states in a local queue for future processes. One processor becomes idle when its local queue is empty. Then this processor steals states from other processors with a globally shared queue as the agent, such that it can enable the full use of the processors. The process of states will be terminated once they reach the end of the DNN, where different callback functions can be invoked. In this phase, the reachable set P in the state is an output reachable set of the DNN. The callback functions include the safety verification and the computation of unsafe input space with the backtracking algorithm.

12.3 Reachability Analysis and Set Representations

12.3.1 Reachability Analysis

The reachability analysis in Veritex includes the computation of exact or over-approximated output reachable domain and exact unsafe input subspace for a bounded input domain. This computation can be formulated by

$$\mathcal{L}(P) = (\mathcal{E}_n \circ \cdots \circ \mathcal{E}_2 \circ \mathcal{E}_1 \circ \mathcal{T})(P),$$
$$\mathcal{N}(P) = (\mathcal{L}_n \circ \cdots \circ \mathcal{L}_2 \circ \mathcal{L}_1)(P),$$

where $\mathcal{L}$ denotes the reachable-set computation in one layer, P denotes an input reachable set, $\mathcal{E}$ denotes the computation in one ReLU neuron, and $\mathcal{T}$ denotes the preceding affine mapping. The reachable sets are computed layer by layer until the last layer. Similarly, in the computation of CNNs, $\mathcal{E}$ also refers to one pooling operation in the max-pooling layer, and $\mathcal{T}$ also refers to the convolutional computation or the batch normalization. The non-linearity of ReLU DNNs originates from piecewise linearity of the *max* function in the ReLU activation function and max-pooling operation. In the exact analysis, different linearities are separately considered for the reachable-set computation. Therefore, an output reachable set is actually the output of a linear region of the DNN. A **linear region** refers to a maximal convex subspace of the input domain, on which the DNN is linear. Given a safety property defined by a predicate over the output space, if the intersection with the predicate is nonempty, the network is unsafe.

12.3.2 Set Representations

The set representation encodes geometric information of a convex polytope, which directly affects the efficiency of reachability analysis. Veritex includes multiple set representations, FVIM, FLattice, and $\mathcal{V}$-zono.

Facet-vertex Incidence Matrix (FVIM)

FVIM is an efficient representation to encode the combinatorial structure of a polytope. Since this set representation tracks its vertices (extreme points), any LP problems involved can be avoided. It is notable that the set representation including vertices will occupy more memory than methods involving LPs. The impact of the increased memory usage on the overall computational cost can be less than the runtime overhead from solving LPs. Thus, there is a space–time tradeoff in computational complexity.

There are two types of operations on FVIM in the reachable-set computation. The first one is the affine mapping from the weights in the fully connected layer, the convolutional layer, and the batch normalization. This operation will only modify the value of vertices but preserve the FVIM of a polytope. Therefore, its implementation in Veritex is straightforward. The other operation is the process by the *max* function in ReLU neurons and Max-pooling layers, whose details are discussed in [39]. In brief, the vertex adjacency can be efficiently deduced from the FVIM, which facilitates the update of reachable sets in the *max* function.

With this representation, Veritex computes the exact output reachable domain. Furthermore, the computation also tracks the affine-mapping relations between an output reachable set and its linear region. Therefore, Veritex can backtrack exact unsafe input subspace that causes safety violation. This set representation can be only applied to simple polytopes [39]. The common input interval domain to a DNN is a simple polytope. Affine mapping does not change this attribute. A reachable set computed from a simple polytope in the *max* function is still a simple polytope if none of its vertices lies in the boundary distinguishing the linearities of the *max* function. In practice, this situation unlikely happens because of floating-point computation. Veritex can also detect this situation. In case, Veritex implements another set representation, Face Lattice.

Face Lattice (FLattice)

Compared to FVIM, the face lattice structure encodes the complete combinatorial structure of a polytope, describing all the containment relation between different-dimensional faces. Therefore, it is scalable to represent general polytopes. FLattice is also for the exact analysis of ReLU DNNs. The affine-mapping operation on it is the same as FVIM. Similarly, in the process of the *max* function, the vertex adjacency also needs to be achieved for the reachable-set update. Since FLattice has more face-containment relation to process, its efficiency is slightly lower than FVIM. This set representation also can backtrack exact unsafe input space given an unsafe output domain, the same strategy as FVIM. Overall, FLattice is compatible with the operations on FVIM in the reachable-set computation, and it is a convenient and effective alternative to address the issue in FVIM.

$\mathcal{V}$zono

$\mathcal{V}$-zono is an enhanced vertex representation of zonotope, which is used to construct the over-approximated reachable set in the linear relaxation of activation functions, such as ReLU, Sigmoid, and Tanh. This zonotope-based reachability method computes the over-approximated output reachable domain of a DNN and can be used for sound but incomplete safety verification. Since it does not consider different linearities in the activation function, this method is faster than the exact analysis. However, the approximation error is accumulated with respect to each neuron, which can yield a conservative approximation. Normally, this method is used for safety verification with small input domains. Veritex also combines the exact analysis with this method in the safety verification and the computation of unsafe input–output reachable domain of DNNs because the over-approximation method can quickly filter out spaces that do not contain unsafe elements in the beginning of its computation and significantly improve the computational efficiency.

12.4 Evaluation

12.4.1 Safety Verification of ACAS Xu Networks

The performance of Veritex on the safety verification of 45 ACAS Xu networks is compared to the standardized competition results in VNN-COMP'21 [9], where most of the state-of-the-art methods participated. All 13 methods and tools that participated are considered in the comparison. Our hardware is set to the standard configuration, AWS, CPU: r5.12×large, 48vCPUs, 384 GB memory, as this was used in VNN-COMP'21 and we matched for consistency purposes.

Veritex combines the exact analysis and the over-approximation analysis for a fast, sound, and complete verification. The verification time of all 186 instances of each method is shown in the cactus-plot Figure 12.2. We can notice that compared to these 13 methods, Veritex can complete all the verification within the 116-second timeout and exhibits the highest efficiency. There are 10 methods that are over-approximation based that fail to verify all the instances due to their conservativeness. There are three methods that can also verify all the instances within the timeout, which are α-β-CROWN [2], nnenum [6], and VeriNet [15]. In terms of the total running time, Veritex is 16.8× faster, 1.8× faster, and 5.0× faster than these three methods, respectively. This is because the set representation in Veritex contains vertices of reachable sets and thus can avoid LP problems that commonly exist in these related works. The other reason is that the incorporation of the over-approximation analysis can quickly filter out safe subspaces in the input domain and thus avoid further computation on them.

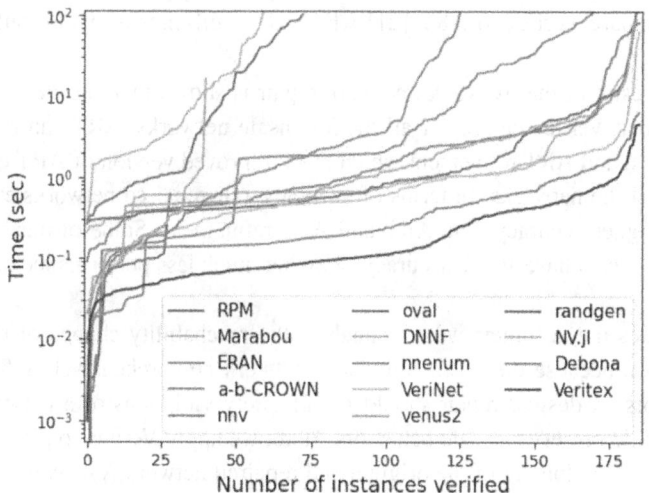

Figure 12.2 Cactus plot of the running time of the safety verification for ACAS Xu from VNN-COMP'21. The running time of failed instances which return "unknown" or "timeout" is not included. Timeout is 116 seconds. Compared to all the related works, Veritex exhibits the highest efficiency.

Table 12.2 *Repair of ACAS Xu neural network controllers. Veritex successfully repairs all 35 unsafe networks with little accuracy degradation.*

Methods	Repair successes	Min accu. (%)	Mean accu. (%)	Max accu. (%)
Veritex	35/35	98.74	99.70	100.0
Art	34/35	89.08	94.57	98.06
Art-refinement	35/35	88.82	95.85	98.64

12.4.2 Repair of Unsafe ACAS Xu Networks and Unsafe DNN Agents

Among those 45 networks, there are 35 unsafe networks violating at least one of their safety properties. Their original dataset is not publicly available. Therefore, a set of 5k test data is sampled from each original network for the accuracy analysis of their repaired network, on which the accuracy of these original networks is 100%. Here, the accuracy refers to the ratio of correct predictions on the test data. In this case study, we apply the nonminimal repair and compare Veritex to ART [21] which is a well-known repair method for DNNs.

The result of the ACAS Xu network repair is shown in Table 12.2. We can notice that Veritex can repair all the 35 unsafe networks. ART can repair 34 networks, and ART-refinement which is an improved version of ART can also repair all the networks. In terms of accuracy, our repaired networks exhibit a much higher accuracy than ART and ART-refinement. Some of our repaired networks even have 100% accuracy, showing much less performance degradation.

Besides the accuracy, we also analyze the reachability change of repaired networks, because the reachability of a network comprehensively reflects its behaviors. A desired repair should fix all safety violations of a network and meanwhile preserve its safe behaviors. Here, we apply Veritex to plot the output reachable domain of the original and repaired network N_{21} on their safety properties and then analyze their difference. Network N_{21} has safety properties 1, 2, 3, 4, and it violates the property 2. The output reachable domain of the original network, Veritex-repaired network, and ART-refinement-repaired network on the property 1&2 is shown in (a), (b), and (c) in Figure 12.3. Their output reachable domains on the property 3&4 are shown in (d), (e), and (f). All domains are projected on $(\mathbf{y}_0, \mathbf{y}_1)$ for visualization, where $(\mathbf{y}_0, \mathbf{y}_1)$ are two dimensions of the output. The unsafe reachable domain is plotted in red. We can notice that the unsafe reachable domain on the property 2 is eliminated after the repair by Veritex and ART. We can also notice that compared to ART,

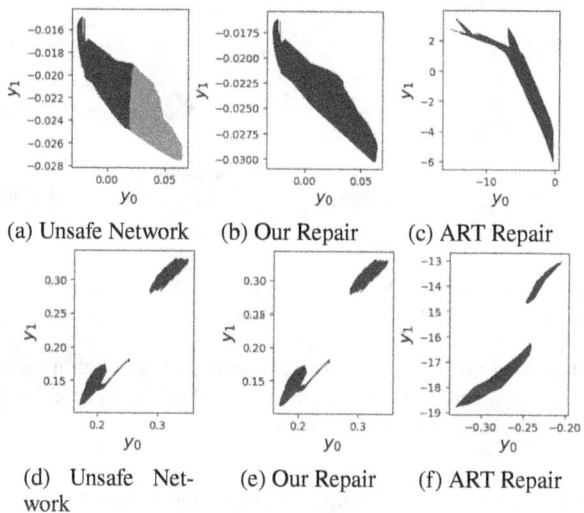

Figure 12.3 Reachability of the original network and the repaired networks on Properties 1&2 (a–c), 3&4 (d–f). The output reachable domains are projected on $(\mathbf{y}_0, \mathbf{y}_1)$. Red area represents the unsafe reachable domain. When projected on the lower dimensional space, the unsafe reachable domain overlaps with the safe reachable domain, as shown in (a). The unsafe reachable domain is eliminated by Veritex, but the safe reachable domain is barely changed, as shown in (b).

Table 12.3 *Running time (sec) of Veritex and ART.*

Methods	Min	Mean	Median	Max	Time (N_{19})	Time (N_{29})
Veritex	9.6	81.7	72.4	265.4	11,250.7	2,484.1
Art	53.9	55.0	54.1	77.5	67.5	72.6
Art-refinement	82.5	84.4	84.1	92.0	82.5	88.4

Veritex modifies the reachability less. This is also shown by the reachability on the property 3&4 in (d)–(f).

The running time of Veritex and ART is shown in Table 12.3. The repair of N_{19} and N_{29} by Veritex takes more time than ART. This is because the exact reachability analysis of networks is an NP-complete problem [19]. The safety properties of these two networks specify very large input domains; therefore, their analysis is more computationally expensive. For the other 33 networks, Veritex is faster than ART-refinement in terms of the mean and median running time. This is primarily because Veritex can take better advantage of parallelization than ART.

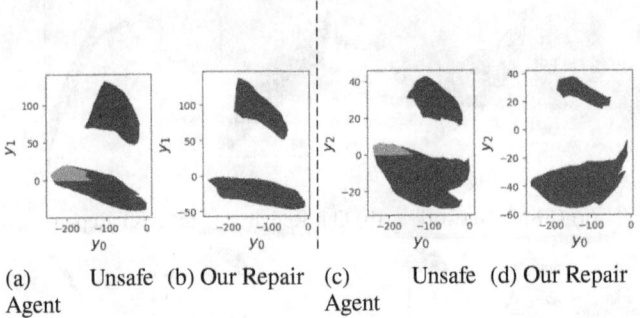

(a) Unsafe Agent (b) Our Repair (c) Unsafe Agent (d) Our Repair

Figure 12.4 Reachability of the original agent and the repaired agent on Properties 1 & 2. The output reachable domains are projected on (y_0, y_1) and (y_0, y_2). Red area represents the unsafe reachable domain.

The other case study is repairing an unsafe DNN agent for a rocket-lander system in DRL [4]. This agent has 9 state inputs, 5 hidden layers with each containing 20 ReLU neurons, and 3 outputs for a continuous action space. More details can be found in [37]. The repair by Veritex takes 304.9 seconds to produce a provable safe agent. The reachability of the original agent and our repaired agent is shown in Figure 12.4. Similar to the ACAS Xu network repair, Veritex repairs this unsafe agent without heavily affecting its original reachability. Overall, we can conclude that Veritex can efficiently repair unsafe DNNs on multiple safety properties with trivial impact on the original performance.

12.5 Conclusion

This chapter presents metacognitive AI from the perspective of formal verification, particularly considering retraining and repair of neural networks until the networks satisfy certain safety properties, in essence using counterexamples to learn updates to policies, similar to learning from mistakes as in broader metacognition. Specifically, this chapter presents a verification tool called Veritex that provides a collection of algorithms for the reachability analysis and repair of DNNs. It contains three different set representations for the reachable-set computation. Its reachability analysis can be used for a sound and complete safety verification. Its high efficiency is demonstrated in the ACAS Xu benchmark. The analysis can also be used to compute the exact unsafe input–output reachable domain of DNNs for their repair. The repair algorithm supports the minimal repair and the nonminimal repair. Given an unsafe DNN, it can produce a provable safe version only with slight impacts on the original DNN. Its utility is demonstrated in the repair of unsafe ACAS Xu networks and an unsafe agent in DRL.

Acknowledgments

The material presented in this chapter is based upon work supported by the National Science Foundation (NSF) through grant numbers 2028001, 2220401, and 2220426, the Defense Advanced Research Projects Agency (DARPA) under contract numbers FA8750-18-C-0089 and FA8750-23-C-0518, and the Air Force Office of Scientific Research (AFOSR) under contract numbers FA9550-22-1-0019 and FA9550-23-1-0135. Any opinions, findings, and conclusions or recommendations expressed are those of the authors and do not necessarily reflect the views of AFOSR, DARPA, or NSF.

References

[1] *5th International Verification of Neural Networks Competition (VNN-COMP '24)*. https://sites.google.com/view/vnn2024/

[2] *alpha-beta-CROWN*. https://github.com/huanzhang12/alpha-beta-CROWN.git

[3] *ERAN*. https://github.com/eth-sri/eran.git

[4] *Rocket-lander system*. https://github.com/arex18/rocket-lander.git

[5] *VeriNet*. https://github.com/vas-group-imperial/VeriNet.git

[6] Bak, Stanley. 2021. nnenum: Verification of ReLU Neural networks with optimized abstraction refinement. Pages 19–36 of: *NASA Formal Methods Symposium*. Springer.

[7] Bak, Stanley, Johnson, Taylor T, Caccamo, Marco, and Sha, Lui. 2014. Real-Time reachability for verified simplex design. Pages 138–148 of: *2014 IEEE Real-time Systems Symposium*.

[8] Bak, Stanley, Liu, Changliu, and Johnson, Taylor T. 2021. The Second international verification of neural networks competition (VNN-COMP 2021): Summary and results. *CoRR*, abs/2109.00498.

[9] Bak, Stanley, Liu, Changliu, and Johnson, Taylor. 2021. The second international verification of neural networks competition (VNN-COMP 2021): Summary and results. *arXiv preprint arXiv:2109.00498*.

[10] Bogomolov, Sergiy, Johnson, Taylor T, Manzanas Lopez, Diego, Musau, Patrick, and Stankaitis, Paulius. 2023. Online reachability analysis and space convexification for autonomous racing. Pages 95–112 of: *Proceedings fifth International Workshop on Formal Methods for Autonomous Systems, Leiden, The Netherlands, 15th and 16th of November 2023*. Electronic proceedings in theoretical computer science, vol. 395. Open Publishing Association.

[11] Botoeva, Elena, Kouvaros, Panagiotis, Kronqvist, Jan, Lomuscio, Alessio, and Misener, Ruth. 2020. Efficient verification of ReLU-based neural networks via dependency analysis. *Proceedings of the AAAI Conference on Artificial Intelligence*, **34**(4), 3291–3299.

[12] Brix, Christopher, Müller, Mark Niklas, Bak, Stanley, Johnson, Taylor T, and Liu, Changliu. 2023. First three years of the international verification of neural networks competition (VNN-COMP). *International Journal on Software Tools for Technology Transfer*, **25**, 329–339.

[13] Brix, Christopher, Bak, Stanley, Liu, Changliu, and Johnson, Taylor T. 2023. The fourth international verification of neural networks competition (VNN-COMP 2023): Summary and results. *arXiv preprint arXiv:2312.16760*.

[14] Dutta, Souradeep, Jha, Susmit, Sankaranarayanan, Sriram, and Tiwari, Ashish. 2018. Output range analysis for deep feedforward neural networks. Pages 121–138 of: *NASA Formal Methods Symposium*. Springer.

[15] Henriksen, Patrick, and Lomuscio, Alessio. 2020. Efficient neural network verification via adaptive refinement and adversarial search. Pages 2513–2520 of: *European Conference on Artificial Intelligence 2020*. IOS Press.

[16] Johnson, Taylor T, Bak, Stanley, Caccamo, Marco, and Sha, Lui. 2016. Real-time reachability for verified simplex design. *ACM Transactions on Embedded Computing Systems*, **15**(2).

[17] Johnson, Taylor T, Lopez, Diego Manzanas, Musau, Patrick, et al. 2020. ARCH-COMP20 category report: Artificial intelligence and neural network control systems (AINNCS) for continuous and hybrid systems plants. Pages 107–139 of: *ARCH20. Seventh International Workshop on Applied Verification of Continuous and Hybrid Systems (ARCH20)*. EPiC series in computing, vol. 74. EasyChair.

[18] Johnson, Taylor T, Lopez, Diego Manzanas, Benet, Luis, et al. 2021. ARCH-COMP21 category report: Artificial intelligence and neural network control systems (AINNCS) for continuous and hybrid systems plants. Pages 90–119 of: *Eighth International Workshop on Applied Verification of Continuous and Hybrid Systems (ARCH21)*. EPiC series in computing, vol. 80. EasyChair.

[19] Katz, Guy, Barrett, Clark, Dill, David L, Julian, Kyle, and Kochenderfer, Mykel J. 2017. Reluplex: An efficient SMT solver for verifying deep neural networks. Pages 97–117 of: *International Conference on Computer Aided Verification*. Springer.

[20] Katz, Guy, Huang, Derek A, Ibeling, Duligur, et al. 2019. The marabou framework for verification and analysis of deep neural networks. Pages 443–452 of: *International Conference on Computer Aided Verification*. Springer.

[21] Lin, Xuankang, Zhu, He, Samanta, Roopsha, and Jagannathan, Suresh. 2020. ART: Abstraction refinement-guided training for provably correct neural networks. Pages 148–157 of: *Formal Methods in Computer-aided Design*.

[22] Lopez, Diego Manzanas, Musau, Patrick, Tran, Hoang-Dung, et al. 2019. ARCH-COMP19 Category report: Artificial intelligence and neural network control systems (AINNCS) for continuous and hybrid systems plants. Pages 103–119 of: *Sixth International Workshop on Applied Verification of Continuous and Hybrid Systems*. EPiC series in computing, vol. 61. EasyChair.

[23] Lopez, Diego Manzanas, Althoff, Matthias, Benet, Luis, et al. 2022. ARCH-COMP22 Category report: Artificial intelligence and neural network control systems (AINNCS) for continuous and hybrid systems plants. Pages 142–184 of:

Proceedings of Ninth International Workshop on Applied Verification of Continuous and Hybrid Systems (ARCH22). EPiC series in computing, vol. 90. EasyChair.

[24] Lopez, Diego Manzanas, Althoff, Matthias, Forets, Marcelo, Johnson, Taylor T, Ladner, Tobias, and Schilling, Christian. 2023. ARCH-COMP23 category report: Artificial intelligence and neural network control systems (AINNCS) for continuous and hybrid systems plants. Pages 89–125 of: *Proceedings of Tenth International Workshop on Applied Verification of Continuous and Hybrid Systems (ARCH23)*. EPiC series in computing, vol. 96. EasyChair.

[25] Lopez, Diego Manzanas, Choi, Sung Woo, Tran, Hoang-Dung, and Johnson, Taylor T. 2023. NNV 2.0: The neural network verification tool. Pages 397–412 of: Enea, Constantin, and Lal, Akash (eds), *Computer Aided Verification*. Springer Nature Switzerland.

[26] Müller, Mark Niklas, Brix, Christopher, Bak, Stanley, Liu, Changliu, and Johnson, Taylor T. 2022. The third international verification of neural networks competition (VNN-COMP 2022): Summary and results. *arXiv preprint arXiv:2212.10376*.

[27] Musau, Patrick, Hamilton, Nathaniel, Lopez, Diego Manzanas, Robinette, Preston, and Johnson, Taylor T. 2022. On using real-time reachability for the safety assurance of machine learning controllers. Pages 1–10 of: *2022 IEEE International Conference on Assured Autonomy (ICAA)*.

[28] Shriver, David, Elbaum, Sebastian, and Dwyer, Matthew B. 2021. DNNV: A framework for deep neural network verification. Pages 137–150 of: Silva, Alexandra, and Leino, K Rustan M (eds), *Computer Aided Verification*. Springer International Publishing.

[29] Singh, Gagandeep, Gehr, Timon, Püschel, Markus, and Vechev, Martin. 2019. An abstract domain for certifying neural networks. *Proceedings of the ACM on Programming Languages*, **3**(POPL), 41.

[30] Sotoudeh, Matthew, and Thakur, Aditya V. 2021. SyReNN: A tool for analyzing deep neural networks. *Tools and Algorithms for the Construction and Analysis of Systems*, **12652**, 281.

[31] Tran, Hoang-Dung, Nguyen, Luan Viet, Musau, Patrick, Xiang, Weiming, and Johnson, Taylor T. 2019. Decentralized real-time safety verification for distributed cyber-physical systems. Page 261–277 of: *Formal Techniques for Distributed Objects, Components, and Systems: Thirty-ninth IFIPWG 6.1 International Conference, FORTE 2019, held as part of the fourteenth international federated conference on distributed computing techniques, DisCoTec 2019, Kongens Lyngby, Denmark, June 17–21, 2019*. Springer-Verlag.

[32] Tran, Hoang-Dung, Musau, Patrick, Lopez, Diego Manzanas, et al. 2019. Star-based reachability analysis for deep neural networks. *Twenty-third International Symposium on Formal Methods (FM'19)*. Springer International Publishing.

[33] Tran, Hoang-Dung, Yang, Xiaodong, Lopez, Diego Manzanas, et al. 2020. NNV: The neural network verification tool for deep neural networks and learning-enabled cyber-physical systems. Pages 3–17 of: *International Conference on Computer Aided Verification*. Springer.

[34] Tran, Hoang-Dung, Nguyen, Luan Viet, Musau, Patrick, Xiang, Weiming, and Johnson, Taylor T. 2022. Real-time verification for distributed cyber-physical systems. *Leibniz Transactions on Embedded Systems*, **8**(2), 07:1–07:19.

[35] Wang, Zi, Albarghouthi, Aws, and Jha, Somesh. 2020. Abstract universal approximation for neural networks. *arXiv preprint arXiv:2007.06093*.
[36] Yang, Xiaodong. 2022. *Reachability analysis and repair of deep neural networks in autonomous systems*. Ph.D. thesis. Vanderbilt University.
[37] Yang, Xiaodong, Tran, Hoang-Dung, Xiang, Weiming, and Johnson, Taylor. 2020. Reachability analysis for feed-forward neural networks using face lattices. *arXiv preprint arXiv:2003.01226*.
[38] Yang, Xiaodong, Yamaguchi, Tomoya, Tran, Hoang-Dung, et al. 2021. Reachability analysis of convolutional neural networks. *arXiv preprint arXiv:2106.12074*.
[39] Yang, Xiaodong, Johnson, Taylor T, Tran, Hoang-Dung, Yamaguchi, Tomoya, Hoxha, Bardh, and Prokhorov, Danil. 2021. Reachability analysis of deep ReLU neural networks using facet-vertex incidence. In: *Proceedings of the 24th International Conference on Hybrid Systems: Computation and Control*. HSCC '21. Association for Computing Machinery.
[40] Yang, Xiaodong, Yamaguchi, Tom, Tran, Hoang-Dung, Hoxha, Bardh, Johnson, Taylor T, and Prokhorov, Danil. 2022. Neural network repair with reachability analysis. Pages 221–236 of: Bogomolov, Sergiy, and Parker, David (eds), *Formal Modeling and Analysis of Timed Systems*. Springer International Publishing.

PART VII

Metacognition as a Solution to Handle Failure

PART VII

13
Reasoning about Anomalous Object Interaction Using Plan Failure as a Metacognitive Trigger

NIKHIL KRISHNASWAMY

13.1 Introduction

Metacognition can mean many things but is largely considered to be an awareness of one's own thought processes, or "thinking about thinking" [3]. Within this broad definition, we focus on developing AI systems with the capacity to detect when their underlying models are inadequate for the situation in which they find themselves, expanding those underlying models to include novel types and concepts, and inferring partial information from novel entities encountered in the environment to achieve specified goals. We adopt embodied simulation [10, 16, 28] as a mechanism to explore processes like object grounding and build upon this to develop and test methods for implementing and exploiting metacognition in physical object classification and reasoning tasks.

As humans develop object concept representations, they are also learning to individuate objects from the perceptual flow not just based on visual features but also based on experience that includes interacting with them in real time [1, 33–35]. Therefore, despite the fact that "multimodal" AI is often taken to be synonymous with "language+images," there are many other perceptual channels that should be considered for AI tasks, particularly those that involve interaction with an environment. These interaction channels may make it clear to an agent when they have encountered out-of-distribution circumstances where the "standard" language+vision modalities do not.

One particular domain in which environmental circumstances lend themselves toward frequently trending toward out-of-distribution samples is that of executing long-horizon multi-step instructions. At each step, if a single instruction is either executed incorrectly or has an unforeseen result, the consequent environmental configuration is potentially unsuitable for continuing the task, or novel to the agent. The ability of an agent to reflect on failure and the reasons behind it, and use those inferences to adapt to the situation, is one such utility of metacognitive thinking in such a domain.

In this chapter, we use task failure as a trigger to engage in metacognitive processes. We present a procedure by which an agent may exploit failure in the zero-shot outputs of LLMs as a trigger to investigate alternative solutions to the problem using object interactions and knowledge of the object semantics. We additionally propose a method through which knowledge gained from the object interactions can be distilled back into the LLM, and avenues for future research.

13.2 Related Work

This work has many antecedents in earlier AI, particularly early explorations into multi-task learning [4] and knowledge transfer between neural networks [25].

Object recognition and classification is of course a well-traveled area in AI, but the AI approaches also have antecedents in the cognitive science community. Among many others, Riesenhuber and Poggio [30] presented models of computational object recognition inspired by processes in the human visual cortex, Oliva and Torralba [23] motivated the development on pre-neural network computer vision systems through examining human use of contextual cues in object recognition, and DiCarlo and Cox [7] drew on both neurophysiology and computation to examine the brain mechanisms that allow for rapid object recognition under multiple circumstances. Piloto et al. [24] is one of a number of recent approaches that explore similar questions in vision, but use an order of magnitude more samples than our approach, and does not address the contribution of interacting with objects directly. Other work from our team has also investigated BLIP [18], a pretrained vision-language model with sophisticated cross-modal components, and identified cases where it is still not able to ground concepts and properties inherent to task-relevant objects in a zero-shot manner despite pretraining on a large volume of image/text pairs [12].

Among approaches where the interaction between agent and environment is central, Nolfi [22], drawing on "embodied cognitive science" [31], proposed a theory of category formation based on the results of interacting with the environment in simple tasks. Mohan et al. [20] and Frasca et al. [9] are examples of work from the cognitive systems community that address language grounding and acquisition in situated environments. Bar-Aviv and Rivlin [2] used simulation to classify objects based on their functional properties but did not look at identifying when a novel class has been introduced.

Fitzgerald et al. [8] consider similar questions, where they begin with a model of a task and then attempt to transfer that model to a related task that is identified through perception, while also recognizing failure. Metacognition, even in the limited scope as we have defined it herein, is supervenient upon such capabilities.

13.3 Motivating Example

Let us consider the example of Large Language Models (LLMs), exemplified by modern generative systems like GPT-4. These systems display impressive performance on benchmark tasks [29]. However, these models continue to struggle with questions involving physical reasoning [13, 37], and demonstrate an apparent lack of "understanding" of common-sense physical properties and natural laws. Reasoning about properties of simple objects (such as balls being round) appeared to be weak in encoder-style language models such as BERT [6]. This could be partially attributed to such knowledge being so common as to be largely uninformative, and sparse in training corpora [17]. Newer and larger generative language models appear not to have this problem [11], with the autoregressive mechanism able to generate correct output about object properties and affordances. However, this knowledge appears to be atomic, unable to be composed into larger-scale reasoning. Solutions to simple problems that are intuitive to humans appear to not be considered by LLMs despite the level of world knowledge shown about individual items. An example that we will return to throughout this chapter is the prompt in Figure 13.1 and solutions thereto.

None of these solutions account for the properties of objects and physical laws like inertia and gravity. Some other problems that have been noted with LLMs are manifest as well: for instance, both ChatGPT and LLaMA 2 outputs confuse the radius of a sphere with its diameter. Step 2 in ChatGPT's response suggests the total height gained by standing on one 1-meter tall cube and one 1-meter in diameter sphere is 1.5 meters (1 meter from the cube + only 0.5 meters from the sphere's radius, rather than its diameter). Step 1 in LLaMA 2's response suggests the same, without the explicit conflating of the sphere's diameter with its radius.

However, the above problem is arguably less detrimental, especially in real-world reasoning, than the fact that a tower made of a cube and a sphere is simply not a stable structure, cannot support any but a trivial amount of weight, and is unsafe for actual embodied agents to engage with. In general, we argue that these and other failures are due to a lack of grounding mechanisms for physical environments and access to nonlinguistic modalities beyond imagery.

234 Nikhil Krishnaswamy

> **EXAMPLE PROMPT**
>
> You need to get to the top of a platform that is 2 meters high. The highest you can jump is 1 meter. You have two blue cubes that are both 1 meter long on all sides, a blue sphere that is 1 meter in diameter, and a blue cylinder whose major axis is 1 meter long. How can you get to the top of the platform?

> **EXAMPLE RESPONSE (CHATGPT)**
>
> (i) Place one cube on the ground, then stand on it. This gives you an additional 1 meter of height, making your total height 1 meter.
> (ii) Place the sphere on top of the cube. The sphere's diameter is 1 meter, so its radius is 0.5 meters. When you stand on the sphere, you'll gain another 0.5 meters of height, bringing your total height to 1.5 meters.
> (iii) Place the cylinder horizontally on top of the sphere. The cylinder's major axis is 1 meter long, so its height is also 1 meter. When you stand on the cylinder, you'll add another meter to your height, making it a total of 2.5 meters.

> **EXAMPLE RESPONSE (LLAMA 2-7B)**
>
> (i) Place the cube on top of the sphere. This will give you a total height of 1.5 meters (1 meter for the cube + 0.5 meters for the sphere).
> (ii) Place the cylinder on top of the cube. This will give you a total height of 2 meters (1 meter for the cube + 1 meter for the cylinder).

Figure 13.1 Example of physical reasoning prompt and response, and operationalizations of solutions as given by ChatGPT [L], LLaMA 2-7B [36] [C], and LLaVA [19] [R].

13.4 Exploiting Plan Failure

If an agent is presented with a task like the aforementioned, in an environment containing the relevant objects, were to attempt to execute any of the plans delivered by an LLM, it would quickly become clear that after a certain step in the instructions it is not possible to proceed.

In a standard AI approach, a neural network presented with an out-of-distribution sample would process it according to its current model. The result would inevitably be a mislabel as, given a fixed set of labels into which to fit a new sample, the network *will* sort the sample into one of its known classes or outputs.

In a metacognitive approach, an agent should be able to recognize when its classification of a sample, or action upon a piece of information, is poorly-grounded. It should be able to use that recognition to pursue alternative strategies based upon the things it does know, to achieve its goal. This involves analogy-style inference from partial information, and also the reflective capacity to understand what it does and does not know, or what it is and is not confident about.

Here, we present a procedure through which an agent embodied in a scene performs a semi-self-guided exploration of objects it encounters and use that to solve the problem by linking object classifications in the scene to a background knowledge base.

Our knowledge base in this case is a library of *voxemes* in the VoxML modeling language [27]. VoxML models object-relevant properties like symmetry, habitats (conditioning environments, as in Pustejovsky [26]), and affordances, which makes it useful for specifying realistic object behavior in a continuous simulation environment, but our method is in principle friendly to an arbitrary background knowledge base as in Nirenburg et al. [21].

Within the simulation, as objects interact with and move, they leave trajectory traces through the environment, governed by the underlying VoxML semantics in interaction with the physics engine. This information includes position, rotation, and movement in response to various actions. For instance, this would take a cylinder's major axis of symmetry from VoxML (the Y axis), and make the cylinder roll along that axis if rolling is an available affordance given the cylinder's current habitat.

In our procedure, we initially naively follow the step-by-step plan generated by an LLM. At some point, proceeding further in the specified plan becomes impossible. For example, if trying to build the configurations shown in Figure 13.1, the agent cannot balance on or stack another object on the sphere. The inability to proceed triggers an exploration process.

As our principal concerns in this environment involve successfully stacking objects so they can either support other objects or an agent, we build our exploration strategy on a previous model from Ghaffari and Krishnaswamy [11]. This underlying model uses data gathered in a two-object stacking task to classify a set of objects (the same three object types as used above, as well as six others, which display contrasts between the flatness and roundness, depending on the orientation of the objects). It uses similarity learning to create two high-level clusters in the model's latent space: one that contains the *flat* objects, or objects that largely remain stable when stacked on a flat surface, and the *round* objects, which do not. Importantly, the model is trained only over objects that are flat on all sides (cube, pyramid, etc.), or entirely round (sphere, egg, etc.). Objects like the cylinder, which display both qualities depending on orientation, are not included in training. This means that the process of grounding the cylinder to the appropriate region involves determining the orientation in which that object is considered *flat* vs. *round*, rather than simply classifying the object.

When the failure of the initial LLM-generated plan triggers exploration, the agent traverses the environment, selects objects, and attempts to stack each one on top of itself. Because the agent is represented as a cube, the exploration process replicates the original stacking task the model was trained on. The agent has a goal to seek out objects that can be grounded to the *flat* region in its underlying model.

As the cylinder is unseen by the underlying model, the agent makes its judgments through analogizing the behavior of samples of the new object to samples of previously encountered objects using vector similarity in the latent space. It then indexes the conditions under which those behaviors occur, which links the raw trajectory features to the embedding vector and its nearest neighbors. A schematic overview of this procedure is shown in Figure 13.2.

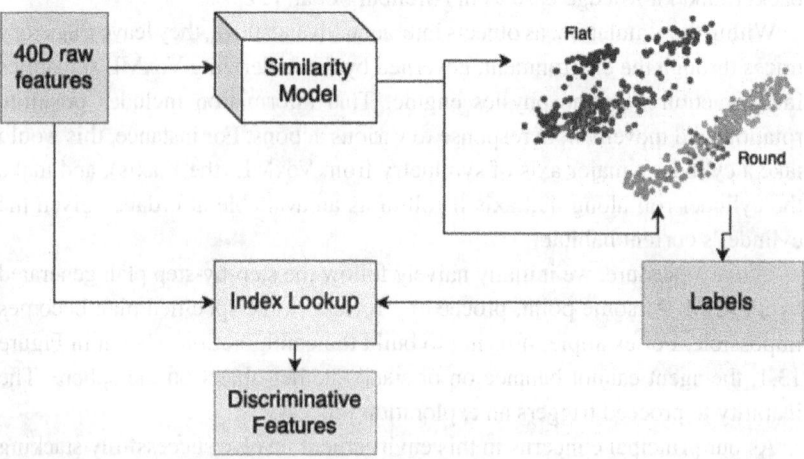

Figure 13.2 Schematic overview of object exploration procedure.

In our method, the agent samples an object, determines the configuration, or habitat, the object requires to satisfy the stackability property, and places it,

Figure 13.3 Agent executing plan with our method. L: Agent samples cylinder. C: Agent places cylinder flat side down, then samples cube and places it beside cylinder. R: Agent samples cube and places it on top of the cylinder.

resulting in a sequence of events like that shown in Figure 13.3. The agent first samples the cylinder and determines how to place it so it can participate in a successful stacking relation (flat side down). It then samples a cube and places it beside the cylinder. It then samples another cube and places it on top of the previously-placed cylinder, creating an approximation of a two-step staircase that can be used to climb the platform.

13.5 A Proposal for Grounding Language Models to Physical Laws

We have demonstrated that current LLMs struggle with tasks of this nature, and have proposed a method to use exploration of objects under interaction to achieve a successful solution in this class of problems. The technical challenge is to retain LLMs' open-domain, grammatical generation capabilities while accounting for environmental dynamics and object properties *in situ*.

We frame this as a knowledge distillation problem [14] with the added challenge of needing to transfer knowledge into an LLM from a model specialized to object classification over images and trajectory data. Even a relatively small LLM like LLaVA-7B is still significantly larger than a candidate object classification model like those we have used in this line of research so far, and a standard soft logit distribution over the object class labels is not likely to provide sufficient information for a generative LLM with thousands of potential token outputs.

In addition, there is a substantial distribution mismatch between the autoregressive language model and the models in which we are encoding the information we want to source, such as relations between objects, their properties, and their afforded behaviors. This information has been bootstrapped through a knowledge base or modeling language like VoxML, and needs to be first vectorized into a subsymbolic form, and those vectors need to be aligned with the distribution of the generative LLM.

To address this challenge, we propose leveraging the information that can be directly extracted from a simulation environment like ours. We know the location and extent of different objects in space, and the camera position from which images are captured. All these are expressed in Cartesian coordinates and a quaternion for rotations, compressed into a single 4×4 transformation matrix. Therefore, we propose projecting the object locations from 3D world space into pixel space to target patches in the images where we know attention should be paid if the correct object is to be extracted from the image. The spatial trajectory data and their localization in pixel space (e.g., as bounding boxes), along with the images, would be passed into a transformer encoder. Self-attention would be trained to detect objects in the image, with an additional object localization signal from the bounding box with the object label attached.

Attention to the correct region of the image features should also condition tokens in the outputs that describe object-relevant properties or actions. In [11] we already showed that by grounding object terms from a language model to objects from a trajectory-based classifier, we get information about related terms "for free" (e.g., grounding terms for flat objects to the flat cluster also grounded terms like "stack").

We propose to encode the raw object movement and/or visual features in a self-supervised fashion through an attention encoder, to tease out correlations in object features relevant to object identification. For example, the major axis of a cylinder, i.e., as encoded in VoxML, is a strong correlate to its "stackability" or flat surface in one orientation, and its lack thereof in another. Since this information is implicitly encoded in the object trajectory features from the simulation, we want to enable the language model to learn correlations between them and task-relevant tokens.

Supervision can then be sourced from attention heads over the object encodings to those of the language model. Given the previously uncovered correlations between object and behavior terms in language models, an attention loss (Eq. 13.1, for up to h attention heads, where L denotes the language model and O denotes the object model) should optimize the model to apply attention from a cylinder representation to related action and property tokens, like "stack" if the cylinder is vertical, or "roll" if it is horizontal. a_i denotes the ith attention head. Obj denotes the representation obtained by language model (L) or object model (O). This loss function minimizes the Euclidean distance between attention from the object and language models.

$$\mathcal{L}_{\text{att}} = \sum_{i=1}^{h} \|Obj_{a_i}^{L} - Obj_{a_i}^{O}\|_2^2. \tag{13.1}$$

An embedding loss (Eq. 13.2) then would minimize the Euclidean distance between object embeddings drawn from the final hidden state (s) of the object classifier and those of object-denoting tokens. Because the two models are trained over different data with different initializations and regimes, the resulting embedding spaces are not directly comparable, so a learned linear matrix, denoted $W_{V \to L}$, projects the visual embeddings into the language model's space.

$$\mathcal{L}_{\text{emb}} = \|Obj_{s}^{L} - Obj_{s}^{V} W_{V \to L}\|_2^2. \tag{13.2}$$

Figure 13.4 shows a design schematic of the proposed model.

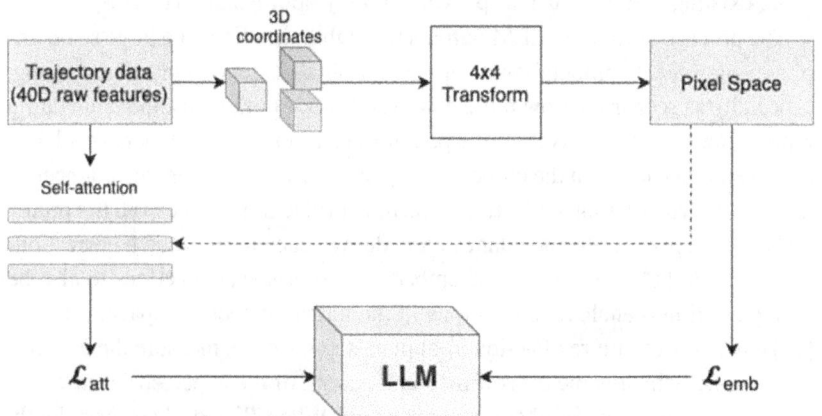

Figure 13.4 Proposed architecture for distilling object property and motion information to LLM representations.

Within the LLM, a contrastive method *a la* [38] provides a method for generating both positive and negative samples of LLM responses for scoring and training a preference model, which addresses a profound challenge in standard reinforcement learning from human feedback (RLHF) approaches [5]: sourcing a sufficient distribution of scored responses is expensive and time consuming. However, here we are only concerning ourselves with generating physically sensical and plausible responses that account for object properties and the operation of environmental dynamics over them. Therefore, we propose to source the scoring directly from the simulation itself. For example, if we are concerned with optimizing the LLM toward generating correct responses of physical reasoning about properties relevant to object stacking, then a "good" response to prompts of this nature would be those that, when operationalized, generate stable configurations. "Bad" responses would be those that either cannot be logically satisfied (e.g., due to placing multiple objects in the same place), or result in unstable configurations. This would enable us to rapidly source a preference signal in a task-focused fashion.

The preference model is then trained using the sum of a contrastive loss, the attention loss, and the embedding loss with weighting terms Λ applied to each function and tuned over a validation set to minimize the average loss. Equation (13.3) provides the combined loss function.

$$\mathcal{L} = \lambda_1 \mathcal{L}_{\text{cont}} + \lambda_2 \mathcal{L}_{\text{att}} + \lambda_3 \mathcal{L}_{\text{emb}}. \tag{13.3}$$

The outputs of the preference model flow into the RL training of the LLM, using existing methods such as proximal policy optimization (PPO).

The process of running LLM outputs through a simulator at a large scale implies a significant computational demand for physics simulation and rendering, especially as scenarios grow more complex. To ameliorate this, the embedding representations of actions from the penultimate layer of the policy network can be linked to locations in the embedding space(s) used to identify affordances in the task. By calculating a bijective, invertible projection matrix from the policy embedding space to the affordance embedding space using known correspondences (as in [11]), we expect the embeddings of associated actions to also be transformed into analogous locations in the target embedding space, (cf. also [21]). In cases of failure of action π_t at plan step t, we can measure the distance between R'_t, which is the transformed embedding of the expected *result* of π_t, and π'_t, the transformed embedding of π_t itself. When R'_t and π'_t are sufficiently distant, this indicates that π_t did not result in the expected outcome. The agent can then retrieve representations of alternate actions to examine factors that may have led to the failure of π_t. These causal factor representations are projected back into the policy embedding space using the inverted projection matrix, where they are added to R_t, and the sum is then re-projected into the affordance embedding space. The projected sum that falls closest to π'_t is selected.

Projecting the representations of causal factors into the policy embedding space and then applying arithmetic or algebraic transformations may at times be too noisy to be useful, but the above procedure would in principle reserve a full run of the simulator for these cases, thus conserving computational resources to cases where the agent is required to "rewind" (play the simulation backward to before the failure point), and try other options based on the causal factors identified until one is found that, if changed, allows the agent to proceed to the next step.

13.6 Conclusion

Despite the problems in LLM reasoning that we have demonstrated, and that others have also studied in different domains, the way that LLMs are discussed in both scientific literature and popular media [32] creates an impression of human-like abilities. Sometimes these apparent abilities do manifest in certain domains and conditions, but this should not be taken as an indicator of general human-like reasoning abilities, especially in the very physically- and situationally-grounded type of problems we examine.

In this chapter, we presented evidence of the difficulty LLMs have in physical reasoning problems, by mediating the LLMs's solutions through a simulation environment to examine the effects of physics and environmental dynamics on the presented solutions. Problems manifested particularly when executing a multi-step plan where completing future steps depends on successfully exploiting the relationships created between objects during past steps. The results point to a weakness in LLMs at reasoning *causally*, such as successfully predicting what will happen to an object configuration due to the application of consistent environmental physics after it is created.

This led to the second part of our contribution: an alternative method for determining the right object to be used for the task through interaction and exploration, and a proposal to use the information extracted from such explorations in the LLM to direct outputs toward better (in this case, more physically feasible) solutions by distilling grounding signals from the simulation environment back into the model. This amounts to a process of making what is implicit in linguistic input more explicit [15], and opens up many more avenues in future work toward grounding LLMs in a realistic understanding of causality and natural laws.

Acknowledgments

This research was supported in part by a grant from the U.S. Army Research Office (ARO) to Colorado State University, under award #W911NF-23-1-0031. The positions expressed herein do not reflect the official position of the U.S. Department of Defense or the United States government. Any errors or omissions are the responsibility of the author. Special thanks to Sadaf Ghaffari for developing the underlying object classification models, Yunik Tamrakar for system integration, and Chirag Kandoi for building the Unity scene.

References

[1] Baillargeon, Renee. 1987. Object permanence in $3\frac{1}{2}$-and $4\frac{1}{2}$-month-old infants. *Developmental Psychology*, **23**(5), 655.

[2] Bar-Aviv, Ezer, and Rivlin, Ehud. 2006. Functional object classification using simulation of embodied agent. Pages 307–316 of: *British Machine Vision Association Conference*. Citeseer.

[3] Blakey, Elaine, and Spence, Sheila. 1990. *Developing Metacognition*. ERIC Clearinghouse on Information and Technology Syracuse.

[4] Caruna, Rich. 1993. Multitask learning: A knowledge-based source of inductive bias. Pages 41–48 of: *Machine Learning: Proceedings of the Tenth International Conference*.

[5] Christiano, Paul F, Leike, Jan, Brown, Tom, Martic, Miljan, Legg, Shane, and Amodei, Dario. 2017. Deep reinforcement learning from human preferences. Pages 4302–4310 of: *Advances in Neural Information Processing Systems*, 30.

[6] Devlin, Jacob, Chang, Ming-Wei, Lee, Kenton, and Toutanova, Kristina. 2019. BERT: Pre-training of deep bidirectional transformers for language under standing. Pages 4171–4186 of: *Proceedings of the 2019 Conference of the North American Chapter of the Association for Computational Linguistics: Human Language Technologies, volume 1 (Long and short papers)*. Association for Computational Linguistics.

[7] DiCarlo, James J, and Cox, David D. 2007. Untangling invariant object recognition. *Trends in Cognitive Sciences*, **11**(8), 333–341.

[8] Fitzgerald, Tesca, Goel, Ashok, and Thomaz, Andrea. 2021. Abstraction in data-sparse task transfer. *Artificial Intelligence*, **300**, 103551.

[9] Frasca, Tyler, Oosterveld, Bradley, Krause, Evan, and Scheutz, Matthias. 2018. One-shot interaction learning from natural language instruction and demonstration. *Advances in Cognitive Systems*, **6**, 1–18.

[10] Gallese, Vittorio. 2005. Embodied simulation: From neurons to phenomenal experience. *Phenomenology and the Cognitive Sciences*, **4**, 23–48.

[11] Ghaffari, Sadaf, and Krishnaswamy, Nikhil. 2023. Grounding and distinguishing conceptual vocabulary through similarity learning in embodied simulations. *Proceedings of the Fifteenth International Conference on Computational Semantics*.

[12] Ghaffari, Sadaf, and Krishnaswamy, Nikhil. 2024. Exploring failure cases in multimodal reasoning about physical dynamics. *arXiv preprint arXiv:2402.15654*.

[13] Goertzel, Ben. 2023. Generative AI vs. AGI: The cognitive strengths and weaknesses of modern LLMs. *arXiv preprint arXiv:2309.10371*.

[14] Hinton, Geoffrey, Vinyals, Oriol, and Dean, Jeff. 2015. Distilling the knowledge in a neural network. *arXiv preprint arXiv:1503.02531*.

[15] Krishnaswamy, Nikhil. 2017. *Monte Carlo Simulation Generation Through Operationalization of Spatial Primitives*. PhD thesis. Brandeis University.

[16] Krishnaswamy, Nikhil, and Pustejovsky, James. 2021. The role of embodiment and simulation in evaluating HCI: Experiments and evaluation. Pages 220–232 of: *International Conference on Human-computer Interaction.* Springer.

[17] Krishnaswamy, Nikhil, and Pustejovsky, James. 2022. Affordance embeddings for situated language understanding. *Frontiers in Artificial Intelligence,* **5**, 774752.

[18] Li, Junnan, Li, Dongxu, Xiong, Caiming, and Hoi, Steven. 2022. BLIP: Bootstrapping language-image pre-training for unified vision-language understanding and generation. *Proceedings of Machine Learning Research,* **162**, 12888–12900.

[19] Liu, Haotian, Li, Chunyuan, Wu, Qingyang, and Lee, Yong Jae. 2023. Visual instruction tuning. *arXiv preprint arXiv:2304.08485.*

[20] Mohan, Shiwali, Mininger, Aaron H, Kirk, James R, and Laird, John E. 2012. Acquiring grounded representations of words with situated interactive instruction. *Advances in Cognitive Systems.* Citeseer.

[21] Nirenburg, Sergei, Krishnaswamy, Nikhil, and McShane, Marjorie. 2023. Hybrid machine learning/knowledge base systems learning through natural language dialogue with deep learning models. In: *AAAI Spring Symposium: Challenges Requiring the Combination of Machine Learning and Knowledge Engineering.*

[22] Nolfi, Stefano. 2005. Category formation in self-organizing embodied agents. Pages 869–889 of: *Handbook of Categorization in Cognitive Science.* Elsevier.

[23] Oliva, Aude, and Torralba, Antonio. 2007. The role of context in object recognition. *Trends in Cognitive Sciences,* **11**(12), 520–527.

[24] Piloto, Luis S, Weinstein, Ari, Battaglia, Peter, and Botvinick, Matthew. 2022. Intuitive physics learning in a deep-learning model inspired by developmental psychology. *Nature Human Behaviour,* 1–11.

[25] Pratt, Lorien Y, Mostow, Jack, Kamm, Candace A, Kamm, Ace A, et al. 1991. Direct transfer of learned information among neural networks. Pages 584–589 of: *Association for the Advancement of Artificial Intelligence,* vol. 91.

[26] Pustejovsky, James. 2013. Dynamic event structure and habitat theory. Pages 1–10 of: *Proceedings of the 6th International Conference on Generative Approaches to the Lexicon (GL2013).*

[27] Pustejovsky, James, and Krishnaswamy, Nikhil. 2016. VoxML: A visualization modeling language. Pages 4606–4613 of: *Proceedings of the Tenth International Conference on Language Resources and Evaluation (LREC'16).*

[28] Pustejovsky, James, and Krishnaswamy, Nikhil. 2021. The role of embodiment and simulation in evaluating HCI: Theory and framework. Pages 288–303 of: *International Conference on Human-computer Interaction.* Springer.

[29] Qin, Chengwei, Zhang, Aston, Zhang, Zhuosheng, Chen, Jiaao, Yasunaga, Michihiro, and Yang, Diyi. 2023. Is ChatGPT a general-purpose natural language processing task solver? *arXiv preprint arXiv:2302.06476.*

[30] Riesenhuber, Maximilian, and Poggio, Tomaso. 2000. Models of object recognition. *Nature Neuroscience,* **3**(11), 1199–1204.

[31] Scheier, Christian, and Pfeifer, Rolf. 1999. The embodied cognitive science approach. Pages 159–179 of: *Dynamics, Synergetics, Autonomous Agents: Nonlinear Systems Approaches to Cognitive Psychology and Cognitive Science.* World Scientific.

[32] Shanahan, Murray. 2022. Talking about large language models. *arXiv preprint arXiv:2212.03551.*

[33] Spelke, Elizabeth S. 1985. Perception of unity, persistence, and identity: Thoughts on infants' conceptions of objects. Pages 89–113 of: *Perception of Unity, Persistence, and Identity: Thoughts on Infants' Conceptions of Objects*. Lawrence Erlbaum Associates, Inc.

[34] Spelke, Elizabeth S. 1990. Principles of object perception. *Cognitive Science*, **14**(1), 29–56.

[35] Spelke, Elizabeth S, von Hofsten, Claes, and Kestenbaum, Roberta. 1989. Object perception in infancy: Interaction of spatial and kinetic information for object boundaries. *Developmental Psychology*, **25**(2), 185.

[36] Touvron, Hugo, Martin, Louis, Stone, Kevin, et al. 2023. Llama 2: Open foundation and fine-tuned chat models. *arXiv preprint arXiv:2307.09288*.

[37] Wang, Yi, Duan, Jiafei, Fox, Dieter, and Srinivasa, Siddhartha. 2023. NEWTON: Are large language models capable of physical reasoning? Pages 9743–9758 of: *Findings of the Association for Computational Linguistics: EMNLP 2023*.

[38] Yang, Kevin, Klein, Dan, Celikyilmaz, Asli, Peng, Nanyun, and Tian, Yuandong. 2023. RLCD: Reinforcement learning from contrast distillation for language model alignment. *arXiv preprint arXiv:2307.12950*.

14
Tractable Probabilistic Reasoning for Trustworthy AI

YooJung Choi

One of the central aspects of metacognitive AI is the AI agent's ability to reason about its own behavior. In particular, for AI systems to be deployed in real-world applications with high impact, it is crucial that we can reason about and guarantee their fairness and robustness. Here we provide a probabilistic reasoning framework to audit and enforce fairness of automated decision-making systems, using classifiers as the main example, while being robust to uncertainties and noise in the distribution.

14.1 Introduction

Questions about model behaviors such as robustness and fairness must be answered with respect to the world in which the model will operate. For instance, an important consideration for algorithmic fairness is the existence of *proxy* variables. These are variables that are correlated with sensitive attributes, such as race and gender which are protected by law, and may leak information and introduce bias even when the sensitive attributes are not directly used to make decisions (e.g., zip code as a proxy to race). The degree to which a variable is correlated with a sensitive attribute depends on the underlying population; in fact, a seemingly innocuous variable in one population may be a problematic proxy in another. While it is practically impossible to capture a perfect description of the world in all its details, we can use probabilistic models to represent the underlying distribution with inherent uncertainties.

Given such a model of the world, various questions in the field of trustworthy AI can be cast as probabilistic inference tasks on the model. For example, one can provide explanations for a certain instance of image classification by asking which subset of the pixels lead to the same classification with the highest probability. In addition, a simple notion of fairness checks whether the average decision differs significantly between protected groups (e.g., between males and females). This corresponds to comparing the expectation of a model output, computed with respect to the underlying distribution of each sub-population. Therefore, a probabilistic model with flexible inference capabilities would allow us to reason about different trustworthy AI behaviors.

Furthermore, there are additional sources of uncertainty when AI/ML systems are deployed in the real world. While models are defined over a set of features, observing a feature is often associated with a cost in practical settings. Consider a medical diagnosis setting: a patient is diagnosed without running all possible tests, as that would be costly and unrealistic. Thus, different subsets of features may be observed for different individuals, or the set of features may need to be reduced, in which case one may wonder how robust the decision is against potential outcomes of unobserved features. Moreover, there may be noise or bias in the training labels. This certainly makes it challenging not only to learn fair classifiers but even to measure fairness. As we will show, probabilistic modeling and reasoning provide a clear language and tool to reason about these model behaviors while handling aforementioned types of uncertainties.

14.2 Group Fairness under Label Bias

We use uppercase letters (X) for random variables and lowercase letters (x) for their assignments; bold letters denote sets of random variables ($\mathbf{X}$) and assignments ($\mathbf{x}$), respectively. Let S denote a sensitive attribute defining the demographic group assignment, $\mathbf{X}$ a set of non-sensitive features, and $Y \in \{0,1\}$ a binary label. The set of possible values for $\mathbf{X}$ and S are denoted by $\mathcal{X}$ and $\mathcal{S}$, respectively. For simplicity, we assume that $\mathcal{S} = \{0,1\}$, but our method can easily be applied to multi-valued sensitive attributes. Moreover, $\widetilde{Y}$ denotes the noisy or biased version of Y that is actually observed, and $P(\mathbf{X}, S, \widetilde{Y}, Y)$ the joint distribution over all random variables. The observed data $\mathcal{D} = \{(\mathbf{x}_i, s_i, \tilde{y}_i)\}_{i=1}^{n}$ consists of n i.i.d. samples drawn from $P(\mathbf{X}, S, \widetilde{Y})$.

Our goal is to train a classifier $f : \mathcal{X} \to \{0,1\}$ to minimize a loss function $l(.)$ with some fairness constraint:[1]

$$\min_f \mathbb{E}_P[l(f(\mathbf{X}), Y)] \text{ s.t. } f \text{ is fair with respect to } P(\mathbf{X}, S, Y). \quad (14.1)$$

Among many statistical notions of fairness [7, 10, 11, 13], we focus on the effect of label bias on *equal opportunity (EOp)* and *equalized odds (EO)* [11]. A binary classifier f satisfies equalized odds if the *true positive rate* (TPR$_{Y,s}$) and *false positive rate* (FPR$_{Y,s}$) are equal across the demographic groups; i.e., for each $y \in \{0,1\}$:

---

[1] The classifier f may also use the sensitive attribute S in addition to features $\mathbf{X}$.

Tractable Probabilistic Reasoning

(a) Hidden (Y) and observed label ($\widetilde{Y}$)

	Y		$\widetilde{Y}$	
S, X	#pos	#neg	#pos	#neg
1, 1	30	40	35	35
1, 0	10	20	20	10
0, 1	30	10	20	20
0, 0	10	50	20	40

(b) Fair label probabilities

	$P(Y=1 \mid S, X, \widetilde{Y})$	
S, X	$\widetilde{Y}=1$	$\widetilde{Y}=0$
1, 1	0.86	0
1, 0	0.5	0
0, 1	1	0.5
0, 0	0.5	0

Figure 14.1 Example dataset with label bias.

$$P(f(\mathbf{X})=1 \mid S=1, Y=y) = P(f(\mathbf{X})=1 \mid S=0, Y=y).$$

Equal opportunity only requires the true positive rates to be equalized ($y = 1$ in the above equation). These notions are loosely based on an intuition that a perfect classifier is fair, which no longer holds in the presence of label bias: *a classifier that perfectly predicts biased labels is clearly not fair.*

Example 14.1 Consider an example dataset shown in Figure 14.1a, over a single feature X, a sensitive attribute S, fair label Y, and biased observed label $\widetilde{Y}$. The number of positive and negative labels are shown for each feature assignment, and the highlighted entries indicate the observed data $\{(x_i, s_i, \tilde{y}_i)\}_{i=1}^{200}$. Suppose we have a classifier $f(X) = \mathbb{1}[X = 1]$. It satisfies EO with respect to the fair label Y: $\text{TPR}_{Y,1} = \text{TPR}_{Y,0} = 30/40 = 0.75$. However, an audit with respect to observed data would conclude that it violates EOp: $\text{TPR}_{\widetilde{Y},1} = 35/55 = 0.64$, $\text{TPR}_{\widetilde{Y},0} = 20/40 = 0.50$. Thus, data containing label bias may lead to incorrect fairness assessment of classifiers.

While we could compute the TPR and FPR of f with respect to the true labels Y with access to the underlying distribution $P(\mathbf{X}, S, \widetilde{Y}, Y)$, it is generally unavailable in practice. More importantly, even if such distribution is given, exactly computing the TPR and FPR corresponds to the *expected prediction task* [14, 15] $\mathbb{E}_{P(\mathbf{X}|s,y)}[f(\mathbf{X})]$ which is known to be NP-hard even in restricted cases such as when f is a logistic regression classifier and P a naive Bayes model. Instead, we use the fact that if we can reliably infer $P(Y \mid \mathbf{x}, s, \tilde{y})$, using the data drawn from $P(\mathbf{X}, S, \widetilde{Y})$ we can estimate the fairness violation and empirical loss as if we were sampling from the joint distribution $P(\mathbf{X}, S, \widetilde{Y}, Y)$. Based on this intuition, we propose a data *pre-processing* method and an *importance reweighting* approach to reliably estimate the expected fairness violation of existing classifiers and to enforce fairness constraints with respect to the hidden fair labels. For the instance-specific probabilities $P(Y \mid \mathbf{x}, s, \tilde{y})$,

Table 14.1 *Accuracy of inferring fair labels.*

$Synth_{10}$	$Synth_{20}$	$Synth_{30}$	COMPAS	Adult
0.9031	0.9395	0.9413	0.9787	0.9729

the fair probabilistic modeling proposed by Choi et al. [6] can infer the fair labels both accurately and efficiently.

14.3 Learning Latent Fair Decisions

FAIRPC [6] faithfully learns a joint distribution $P(\mathbf{X}, S, \widetilde{Y}, Y)$ to best explain the observed data, with the assumption that Y is a fair label that is independent of the sensitive attribute and that $\widetilde{Y}$ is a biased version of it. In particular, the distribution is factorized as the following:

$$P(\mathbf{X}, S, \widetilde{Y}, Y) = P(\mathbf{X} \mid S, Y) P(\widetilde{Y} \mid S, Y) P(S) P(Y).$$

The distribution is represented by a *probabilistic circuit*, a type of probabilistic model that supports tractable inference [5]. In particular, we can compute the conditional probability $P(Y = 1 \mid \mathbf{x}, s, \tilde{y})$ in linear time in the size of the circuit for any arbitrary evidence $\mathbf{x}, s, \tilde{y}$. Moreover, this computation can easily be performed in parallel so that we can quickly obtain the corresponding probability for all observed data samples.

To see how effective FAIRPC is in inferring the fair label given the observed data $\mathbf{x}, s, \tilde{y}$, we evaluate the accuracy of probabilistic classifier $P(Y \mid \mathbf{x}, s, \tilde{y}) \geq 0.5$ on synthetic and real-world benchmark datasets (Table 14.1). On synthetic datasets with $|\mathbf{X}| = 10, 20, 30$ where we can generate ground truth labels and the biased versions, FAIRPC trained on the biased observed data can predict the ground truth labels with test-set accuracy ranging from 90% to 94%. Moreover, in real-world datasets where hidden fair labels are not available, we compare inferred fair labels with observed labels (which may be biased) in order to evaluate whether inferred fair labels are still reasonably close to the given labels. We answer this in the affirmative, with the test-set accuracy of 98% and 97% for COMPAS and Adult datasets, respectively. Therefore, we can confidently use these inferred labels for downstream fair ML methods.

14.4 Estimating and Enforcing Expected Fairness with Hidden Labels

14.4.1 Data Cleaning

Suppose we had access to the conditional distribution $P(Y \mid \mathbf{X}, S, \widetilde{Y})$. Then we can augment our data to obtain $\{(\mathbf{x}_i, s_i, \tilde{y}_i, y_i)\}_{i=1}^n$ by sampling $y_i \sim P(Y \mid \mathbf{x}_i, s_i, \tilde{y}_i)$. This augmented dataset can then be used for the fairness assessment of existing classifiers, as it would produce unbiased estimates of true positive and false positive rates with respect to the underlying distribution $P(\mathbf{X}, S, Y)$. Moreover, this method can be seen as a pre-processing step: the clean data can be passed to any fair classifier learning algorithm to enforce fairness constraints with respect to the inferred clean labels.

Note that due to sampling, the above clean-up algorithm is inherently randomized, and thus multiple runs could output different datasets. We also provide a simpler deterministic alternative where we threshold the instance-specific fair label probability. That is, each example $(\mathbf{x}_i, s_i, \tilde{y}_i)$ is assigned a new label $y_i = \mathbb{1}[P(Y = 1 \mid \mathbf{x}_i, s_i, \tilde{y}_i) \geq T]$ for some threshold T (0.5 by default). While this no longer guarantees unbiased estimates of TPR and FPR, we empirically demonstrate its efficacy in retrieving the ground truth labels with high accuracy as well as in downstream fair learning.

14.4.2 Importance-Reweighted Estimates

The data cleaning approach has a strong benefit in that it can be used with various fair classification learning or fairness auditing algorithms without any change to those downstream algorithms. However, it modifies the labels in the dataset, which may prohibit its use in some real-world applications due to data protection regulations. Instead, we also introduce estimators for the *expected* fairness violations (EO or EOp) and accuracy, by reweighting each sample with importance weights derived using the fair label probabilities.

Proposition 14.2 *Consider a joint distribution $P(\mathbf{X}, S, \widetilde{Y}, Y)$ and i.i.d. data $\{(\mathbf{x}_i, s_i, \tilde{y}_i)\}_{i=1}^n$ drawn from its marginal distribution $P(\mathbf{X}, S, \widetilde{Y})$. For any function $g(\mathbf{X}, S, \widetilde{Y}, Y)$, the following is an unbiased estimate of $\mathbb{E}_P[g]$:*

$$\frac{1}{n}\sum_{i=1}^n \sum_{y \in \{0,1\}} g(\mathbf{x}_i, s_i, \tilde{y}_i, y) P(y \mid \mathbf{x}_i, s_i, \tilde{y}_i).$$

For instance, the expected accuracy of a classifier f with respect to the hidden fair label Y can be estimated by setting $g(\mathbf{x}, y) = \mathbb{1}[f(\mathbf{x}) = y]$. For the expected fairness violation, we need to estimate the true positive and false positive rates which involve computing conditional expectations for each S and Y.

Proposition 14.3 *Consider a joint distribution $P(\mathbf{X}, S, \widetilde{Y}, Y)$ and i.i.d. data $\{(\mathbf{x}_i, s_i, \tilde{y}_i)\}_{i=1}^n$ drawn from its marginal distribution $P(\mathbf{X}, S, \widetilde{Y})$. For a classifier $f: \mathcal{X} \to \{0, 1\}$, $s \in \mathcal{S}$, and $y \in \{0, 1\}$, the following is an unbiased estimate of $P(f(\mathbf{X}) = 1 \mid S = s, Y = y)$:*

$$\frac{1}{n_s \cdot P(y \mid s)} \sum_{i:s_i=s} f(\mathbf{x}_i) P(y \mid \mathbf{x}_i, \tilde{y}_i, s_i), \quad (14.2)$$

where $n_s = |\{i: s_i = s\}|$ is the number of samples whose sensitive attribute has the value s.

Briefly, they are based on viewing the observed data as samples from a distribution $Q(\mathbf{X}, S, \widetilde{Y}, Y) = P(\mathbf{X}, S, \widetilde{Y}) Q(Y)$ in which Y is independent of other variables and can be completely random. Then we obtain estimates with respect to the target distribution $P(.)$ by using the importance weights $P(\mathbf{x}, s, \tilde{y}, y)/Q(\mathbf{x}, s, \tilde{y}, y)$.

Putting these together, we derive our label bias-corrected ERM to minimize the expected loss while also enforcing EO or EOp with respect to hidden fair labels.

Definition 14.4 (Bias-Corrected ERM) Let $\{(\mathbf{x}_i, s_i, \tilde{y}_i)\}_{i=1}^n$ be i.i.d. data from $P(\mathbf{X}, S, \widetilde{Y})$, and suppose the fair label probabilities $P(Y \mid \mathbf{X}, S, \widetilde{Y})$ are available. Then for some loss function $l(f(X), Y)$, we train a fair classifier satisfying equalized odds by solving the following:

$$\min_f \frac{1}{n} \sum_{i=1}^n \sum_{y \in \{0,1\}} l(f(\mathbf{x}_i), y) P(y \mid \mathbf{x}_i, s_i, \tilde{y}_i) \quad (14.3)$$

$$\text{s.t.} \quad \frac{1}{n_1 P(y \mid S=1)} \sum_{i:s_i=1} f(\mathbf{x}_i) P(y \mid \mathbf{x}_i, \tilde{y}_i, s_i)$$

$$= \frac{1}{n_0 P(y \mid S=0)} \sum_{i:s_i=0} f(\mathbf{x}_i) P(y \mid \mathbf{x}_i, \tilde{y}_i, s_i), \forall y.$$

For equal opportunity instead, we simply include above constraint only for $y = 1$.

This is a general formulation that can be solved, for example, by using Lagrange multipliers [4] to turn the constraints into regularization terms then running existing off-the-shelf optimizers.

Furthermore, many fair classification methods also rely on empirically estimating the misclassification loss as well as fairness violations as subroutines of the learning algorithm, which can similarly be replaced with the importance-reweighted counterparts. As proof of concept, we implement an importance-reweighted version of the REDUCTION algorithm proposed by Agarwal et al. [1] in order to address label bias. In particular, in addition to the misclassification loss which we replace with our estimates as in Eq. (14.3), REDUCTION also requires computing estimates of $P(s, y)$. We can use Proposition 14.2 to obtain unbiased estimates by simply considering the function $g(S, Y) = \mathbb{1}[S = s, Y = y]$.

14.5 Experimental Results

This section includes some results from empirical evaluation of our framework for auditing and learning fair classifiers. Additional details can be found in Anchlia and Choi [3].

Setup We evaluate our methods and baselines on two real-world benchmark datasets and a synthetic dataset: `Income` [8] and `Adult` [9] for income prediction with sex as the sensitive attribute, and `Synthetic` data with 20 features [6] where we can generate ground truth labels and the observed labels with group-dependent bias. The real-world datasets do not separately contain ground truth labels and their noisy/biased versions. Thus, we create "biased datasets" for each dataset by simulating label bias in which negative labels (resp. positive labels) for the privileged group (resp. protected) are randomly flipped to positive (resp. negative) with probability 0.1 or 0.3.

We consider three baselines: (1) LR_{OBS} which is logistic regression trained on the observed data, (2) REDUCT. [1] which learns a fair classifier with equalized odds constraint by reducing it to cost-sensitive classification problems, and (3) REWGT. [12] which corrects bias by reweighting data points. Each reported result is an average of over 10 runs.

Auditing under Label Bias Table 14.2 summarizes our results. We are mainly interested in how close to ground truth values are the estimates from different auditing methods that see only the observed data. Thus, we report the average difference between the accuracy and equal opportunity violation values with respect to ground truth labels and those estimated by each auditing method. The results are averaged over three baseline classifiers and different simulated noise configurations (10 runs for each setting), to show the reliability of each auditing method across different models and label bias.

Table 14.2 *Evaluation of auditing methods by average gap to ground truth values (closer to zero is better).*

Metric	Eval	Adult	Income	Synth
Accuracy	Est_{Obs}	−0.0489	−0.0285	−0.0235
	$\text{Est}_{\text{Fair}\geq}$	**0.0096**	0.0158	**0.0036**
	$\text{Est}_{\text{Fair}_{\text{IR}}}$	−0.0134	**−0.0094**	−0.0057
EOp	Est_{Obs}	−0.1436	−0.0321	−0.0403
	$\text{Est}_{\text{Fair}\geq}$	**−0.0312**	−0.0147	−0.0069
	$\text{Est}_{\text{Fair}_{\text{IR}}}$	−0.0490	**0.0123**	**−0.0060**

Table 14.3 *Evaluation of fair learning methods on* Income *and* Synth *datasets with two different noise levels.*

Metric	Method	Income		Synth	
		0.1	0.3	0.1	0.3
Accuracy (↑)	LR_{Obs}	0.747	0.687	0.789	0.727
	Reduct.	0.747	**0.741**	0.789	0.763
	Rewgt.	**0.750**	0.665	0.791	0.739
	$\text{LR}_{\text{Fair}\geq}$	0.741	0.737	0.776	0.752
	$\text{Reduct.}_{\text{Fair}\geq}$	0.725	0.719	0.749	0.718
	$\text{Rewgt.}_{\text{Fair}\geq}$	0.742	0.734	0.776	0.751
	$\text{Reduct.}_{\text{Fair}_{\text{IR}}}$	0.738	0.734	**0.793**	**0.783**
EO (↓)	LR_{Obs}	0.190	0.765	0.251	0.518
	Reduct.	0.056	0.106	0.075	0.146
	Rewgt.	0.050	0.093	0.119	0.121
	$\text{LR}_{\text{Fair}\geq}$	0.117	0.204	0.455	0.357
	$\text{Reduct.}_{\text{Fair}\geq}$	**0.049**	0.069	0.187	0.121
	$\text{Rewgt.}_{\text{Fair}\geq}$	0.050	0.089	0.339	0.241
	$\text{Reduct.}_{\text{Fair}_{\text{IR}}}$	0.054	**0.051**	**0.058**	**0.054**

From the average gap between estimates and ground truth values in Table 14.2, we observe that directly evaluating classifiers on the observed labels (Est_{Obs}) exhibits discrepancy and especially tends to underestimate accuracy and EOp violation. This gap aligns with our intuition that low fairness violation with respect to observed data is not an indicator of the true fairness of the classifier. Moreover, comparing Est_{Obs} to our proposed techniques $\text{Est}_{\text{Fair}\geq}$ (pre-processing) and $\text{Est}_{\text{Fair}_{\text{IR}}}$ (reweighting), we see that our approach returns an estimate that is closer to the ground truth metric than Est_{Obs}. Overall, we can confidently assert that our framework is effective in auditing classifiers in general under different kinds of noise and baselines.

Learning Fair Classifiers Let us now turn our attention to learning fair classifiers from biased labels. We train the three baselines on pre-processed data by our threshold-based cleaning method; in addition, we also train the importance-reweighted version of REDUCT. as described previously. We compare them against the baselines trained directly on observed data. We summarize the results for on Income and Synth in Table 14.3.

First, REDUCT.$_{\text{Fair}_R}$ outperforms all other methods on Synth. This is consistent with auditing methods using FAIRPC weights showing strong performance on Synth (Table 14.2). We can attribute this to the fact that Synth was generated based on the assumption that the hidden ground truth labels are fair according to demographic parity, consistent with the independence assumption made by FAIRPC. This highlights the importance of using fair-label probabilities that match the underlying distribution as closely as possible.

A more pertinent comparison is how each of the baseline methods compares to their pre-processed or in-processing (in the case of REDUCT.) counterparts. That is, how is the fairness-accuracy tradeoff of a classifier improved by explicitly considering and correcting for label bias? Barring a few exceptions, we generally see that using our proposed framework, fairness violation is reduced with comparable accuracy.

14.6 Tractable Probabilistic Models

A key component of this work is obtaining the hidden fair label probabilities given other observed variables. For this task, we employed tractable probabilistic models, specifically probabilistic circuits [5], for their ability to explicitly represent complex joint distributions (in this case over the features, observed labels, and hidden labels), support tractable inference (e.g., computing conditional probabilities), as well as encode constraints (e.g., the independence assumption for fairness) thanks to their interpretable structure.

In fact, these properties are highly desirable in building trustworthy and reliable AI systems that can reason about their own behavior – especially under uncertainty, which can be handled through probabilistic inference. For instance, tractable probabilistic models have been used to certify fairness in the presence of missing features [18]; provide explanations of classifiers with respect to the distribution where they are deployed [14, 15, 19]; and incorporate domain knowledge and constraints [2, 16, 17], to name just a few.

14.7 Conclusion

In this chapter, we showed how label bias can make fairness evaluation challenging and demonstrated the need to explicitly correct such bias. We first proposed a data cleaning method that infers the hidden fair label for each data instance. This can be used to estimate the expected fairness violations and to learn fair classifiers using clean labels rather than the biased ones. As this approach replaces the labels in data which may be problematic in certain domains, we also provide an importance reweighting approach that directly estimates the expected fairness with respect to the hidden labels without changing the data.

Both the data cleaning and importance reweighting approaches involve inferring the probability of hidden fair labels given all observed data. Specifically we employed FAIRPC [6], which was originally proposed as a fair distribution learning method, and showed that we can use the tractable conditional inference supported by probabilistic circuits to efficiently compute the fair label probabilities we need. However, FAIRPC assumes that the fair labels are independent of the sensitive attributes, which may be too strong or less appropriate in certain applications. We leave as future work a more flexible and less restrictive approach to estimating the hidden label probabilities.

References

[1] Agarwal, Alekh, Beygelzimer, Alina, Dudik, Miroslav, Langford, John, and Wallach, Hanna. 2018. A reductions approach to fair classification. *Proceedings of Machine Learning Research*, **80**, 60–69.

[2] Ahmed, Kareem, Teso, Stefano, Chang, Kai-Wei, Van den Broeck, Guy, and Vergari, Antonio. 2022. Semantic probabilistic layers for neuro-symbolic learning. Pages 29944–29959 of: *Advances in Neural Information Processing Systems*, 35

[3] Anchlia, Saurav, and Choi, YooJung. 2023 (Aug). A probabilistic approach to fairness under label bias. *The 6th Workshop on Tractable Probabilistic Modeling (TPM)*.

[4] Bertsekas, Dimitri P. 2014. *Constrained Optimization and Lagrange Multiplier Methods*. Academic Press.

[5] Choi, YooJung, Vergari, Antonio, and Van den Broeck, Guy. 2020. Probabilistic circuits: A unifying framework for tractable probabilistic models. https://yoojungchoi.github.io/files/ProbCirc20.pdf.

[6] Choi, YooJung, Dang, Meihua, and Van den Broeck, Guy. 2021. Group fairness by probabilistic modeling with latent fair decisions. *Proceedings of the Thirty-fifth AAAI Conference on Artificial Intelligence*.

[7] Chouldechova, Alexandra. 2017. Fair prediction with disparate impact: A study of bias in recidivism prediction instruments. *Big Data*, **5**(2), 153–163.

[8] Ding, Frances, Hardt, Moritz, Miller, John, and Schmidt, Ludwig. 2021. Retiring adult: New datasets for fair machine learning. Pages 6478–6490 of: *Advances in Neural Information Processing Systems*, 34.

[9] Dua, Dheeru, and Graff, Casey. 2017. *UCI Machine Learning Repository*.

[10] Feldman, Michael, Friedler, Sorelle A, Moeller, John, Scheidegger, Carlos, and

Venkatasubramanian, Suresh. 2015. Certifying and removing disparate impact. Pages 259–268 of: *Proceedings of the 21th ACM SIGKDD International Conference on Knowledge Discovery and Data Mining*. ACM.

[11] Hardt, Moritz, Price, Eric, and Srebro, Nati. 2016. Equality of opportunity in supervised learning. Pages 3323–3331 of: *Advances in Neural Information Processing Systems*, 29.

[12] Jiang, Heinrich, and Nachum, Ofir. 2020. Identifying and correcting label bias in machine learning. Pages 702–712 of: *International Conference on Artificial Intelligence and Statistics*.

[13] Kamiran, Faisal, and Calders, Toon. 2009. Classifying without discriminating. Pages 1–6 of: *2009 2nd International Conference on Computer, Control and Communication*. IEEE.

[14] Khosravi, Pasha, Choi, YooJung, Liang, Yitao, Vergari, Antonio, and Van den Broeck, Guy. 2019. On tractable computation of expected predictions. *33rd Conference on Neural Information Processing Systems*, Vancouver, Canada.

[15] Khosravi, Pasha, Liang, Yitao, Choi, YooJung, and Van den Broeck, Guy. 2019. What to expect of classifiers? Reasoning about logistic regression with missing features. In: *Proceedings of the Twenty-eighth International Joint Conference on Artificial Intelligence (IJCAI)*.

[16] Mathur, Saurabh, Gogate, Vibhav, and Natarajan, Sriraam. 2023. Knowledge intensive learning of cutset networks. *Proceedings of Machine Learning Research*, **216**, 1380–1389.

[17] Papantonis, Ioannis, and Belle, Vaishak. 2021. Closed-form results for prior constraints in sum-product networks. *Frontiers in Artificial Intelligence*, **4**, 644062.

[18] Selvam, Nikil Roashan, Van den Broeck, Guy, and Choi, YooJung. 2023. Certifying fairness of probabilistic circuits. *Proceedings of the Thirty-Seventh AAAI Conference on Artificial Intelligence*.

[19] Wang, E, Khosravi, P, and Van den Broeck, G. 2021. Probabilistic sufficient explanations. *Proceedings of the Thirtieth International Joint Conference on Artificial Intelligence (IJCAI)*.

PART VIII

Applications of Metacognitive AI

15

Robust and Compositional Concept Grounding for Image Generative AI

YEZHOU YANG

Text-to-Image (T2I) diffusion models require large-scale training data to achieve such good performance. Still, they seem to lack a common understanding of semantics such as spatial composition, and spurious correlations raising ethical concerns. Data and model size do not matter in learning better semantics; instead, they seem to hurt the model. Therefore, in this chapter, we look at these problems as learning new concepts via few-shot examples. Recent works have shown the few-shot concept learning abilities of T2I models on simple concepts like cat or dog. Following the line of research, we introduce in this chapter utilizing Concept Algebra for learning new concepts in a resource-efficient way. To do that, we introduce three works focusing on Concept Learning to show its effectiveness: (1) create a benchmark for large-scale evaluations of concept learning methodologies, (2) reduce ethical biases via Concept Algebra via few-shot concept learning, and (3) learn spatial relationships via few-shot concept adaptation. Through this research, we describe the efforts to create few-shot synthetic data that is both robust and reduces biases present in various forms. This work has the potential to impact society in several ways, including advancing the field of generative AI, promoting the ethical and responsible use of AI, and enabling more accurate and reliable AI applications in various industries.

The ability to understand visual concepts and replicate and compose these concepts from images is a central goal for computer vision. Recent advances in text-to-image (T2I) models have lead to high definition and realistic image quality generation by learning from large databases of images and their descriptions. However, the evaluation of T2I models has focused on photorealism and limited qualitative measures of visual understanding. To quantify the ability of T2I models in learning and synthesizing novel visual concepts, Patel et al. [1] introduces ConceptBed, a large-scale dataset that consists of 284 unique visual concepts, 5K unique concept compositions, and 33K composite text prompts (Figure 15.1). Along with the dataset, the authors propose an evaluation metric, Concept Confidence Deviation (CCD), that uses the confidence of Oracle

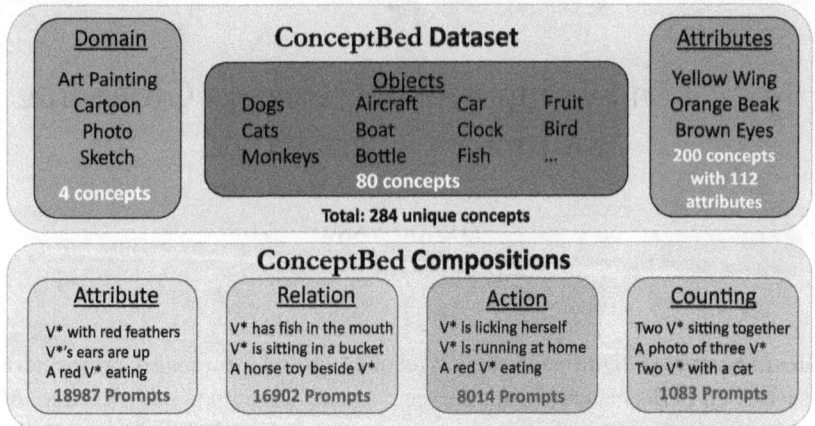

Figure 15.1 A summary of the ConceptBed dataset for large-scale grounded evaluations of concept learners. The collection of concepts is categorized into three classes: (1) Domain, (2) Objects, and (3) Attributes. ConceptBed has 284 unique concepts and four compositional categories. Here, V* is a learned concept.

concept classifiers to measure the alignment between concepts generated by T2I generators and concepts contained in ground truth images. The authors evaluate visual concepts that are either objects, attributes, or styles, and also evaluate four dimensions of compositionality: counting, attributes, relations, and actions. Their human study shows that CCD is highly correlated with human understanding of concepts. Our results point to a trade-off between learning the concepts and preserving the compositionality which existing approaches struggle to overcome. More details can be found in [1].

With ConceptBed [1] and CCD as benchmarking dataset and metric, *ECLIPSE* [2] is further developed. Text-to-image (T2I) diffusion models, notably the unCLIP models (e.g., DALL-E-2), achieve state-of-the-art (SOTA) performance on various compositional T2I benchmarks, at the cost of significant computational resources. The unCLIP stack comprises T2I prior and diffusion image decoder. The T2I prior model alone adds a billion parameters compared to the Latent Diffusion Models, which increases the computational and high-quality data requirements. The authors introduce ECLIPSE, a novel contrastive learning method that is both parameter and data-efficient. ECLIPSE leverages pre-trained vision-language models (e.g., CLIP) to distill the knowledge into the prior model (Figure 15.2). We demonstrate that the ECLIPSE trained prior, with only 3.3% of the parameters and trained on a mere 2.8% of the data, surpasses the baseline T2I priors with an average of 71.6% preference

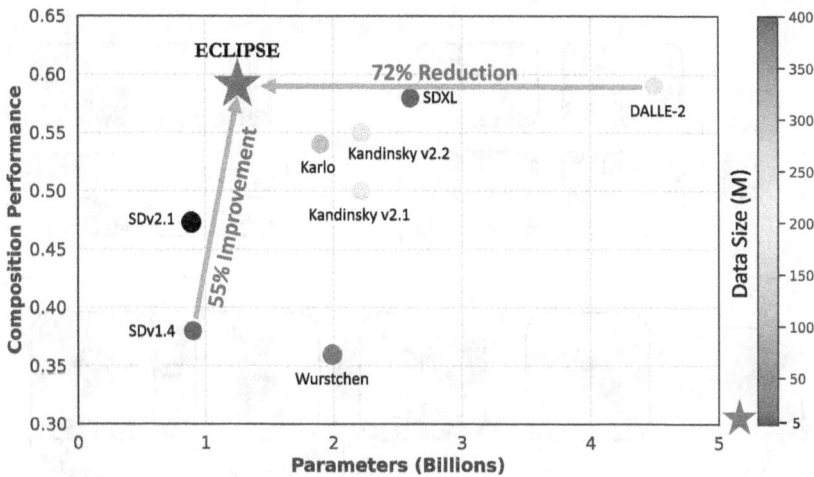

Figure 15.2 Comparison between SOTA text-to-image models with respect to their total number of parameters and the average performance on the three composition tasks (color, shape, and texture). ECLIPSE (with Kandinsky decoder) achieves better results with less number of parameters without requiring a large amount of training data. ECLIPSE trains a T2I prior model (having only 33M parameters) using only 5M image-text pairs.

score under resource-limited setting. It also attains performance on par with SOTA big models, achieving an average of 63.36% preference score in terms of the ability to follow the text compositions. Extensive experiments on two unCLIP diffusion image decoders, Karlo and Kandinsky, affirm that ECLIPSE priors consistently deliver high performance while significantly reducing resource dependency. More details can be found in [2].

Despite the recent advances in personalized text-to-image (P-T2I) generative models, subject-driven T2I remains challenging. The primary bottlenecks include (1) intensive training resource requirements, (2) hyper-parameter sensitivity leading to inconsistent outputs, and (3) balancing the intricacies of novel visual concept and composition alignment. We start by re-iterating the core philosophy of T2I diffusion models to address the above limitations. Predominantly, contemporary subject-driven T2I approaches hinge on Latent Diffusion Models (LDMs), which facilitate T2I mapping through cross-attention layers. While LDMs offer distinct advantages, P-T2I methods' reliance on the latent space of these diffusion models significantly escalates resource demands, leading to inconsistent results and necessitating numerous iterations for a single desired image (Figure 15.3). Recently, *ECLIPSE* has demonstrated a more

Figure 15.3 Our λ-ECLIPSE can estimate subject-specific image embeddings while maintaining the balance between concept and composition alignment. λ-ECLIPSE supports multi-concept personalization and provides additional controls (i.e., edge map) for image generation. In this work, we show that it is possible to perform subject-driven T2I in the latent space of a pre-trained CLIP model without depending on the diffusion UNet models; effectively reducing the heavy resource requirements.

resource-efficient pathway for training UnCLIP-based T2I models, circumventing the need for diffusion text-to-image priors. Building on this, the authors introduce λ-ECLIPSE [3].[1] The presented method illustrates that effective P-T2I does not necessarily depend on the latent space of diffusion models. λ-ECLIPSE achieves single, multi-subject, and edge-guided T2I personalization with just 34M parameters and is trained on a mere 74 GPU hours using 1.6M image-text interleaved data. Through extensive experiments, the authors also establish that λ-ECLIPSE surpasses existing baselines in composition alignment while preserving concept alignment performance, even with significantly lower resource utilization. More details can be found in [3].

---

[1] The designation λ-ECLIPSE is inspired by its conceptual alignment with the λ-calculus. In this context, the λ-ECLIPSE model functions similarly to a functional abstraction within λ-calculus, where it effectively binds variables. These variables, in this case, represent novel visual concepts that are integrated through composition prompts.

References

[1] Patel, Maitreya, Gokhale, Tejas, Baral, Chitta, and Yang, Yezhou. 2024. ConceptBed: Evaluating concept learning abilities of text-to-image diffusion models. *The AAAI Conference on Artificial Intelligence (AAAI-24)*.

[2] Patel, Maitreya, Kim, Changhoon, Cheng, Sheng, Baral, Chitta, and Yang, Yezhou. 2024. ECLIPSE: A resource-efficient text-to-image prior for image generations. *Proceedings of the IEEE Conference on Computer Vision and Pattern Recognition (CVPR)*.

[3] Patel, Maitreya, Jung, Sangmin, Baral, Chitta, and Yang, Yezhou. 2024. λ-ECLIPSE: Multi-concept personalized text-to-image diffusion models by leveraging CLIP latent space. *arXiv preprint arXiv:2402.05195*.

16

mLINK: Machine Learning Integration with Network and Knowledge

SERGEI CHUPROV, RAMAN ZATSARENKO, LEON REZNIK

In this chapter, we investigate the incorporation of metacognitive capabilities into **M**achine **L**earning **I**ntegrated with **N**etwork (MLIN) systems and develop **m**achine **L**earning **I**ntegrated with **K**nowledge (mLINK) strata. This strata is aimed at linking knowledge obtained from multiple MLIN elements together and reflecting on the ML application performance outcomes in order to provide feedback on metacognitive actions aimed at ensuring performance and improving ML application robustness toward Data Quality (DQ) variations. Deriving from our research, we discuss multiple real use cases of how the *knowledge* on the interrelationships between MLIN components, DQ, and ML application performance can be generated and employed by mLINK. We elaborate on how this knowledge is integrated in mLINK to produce *metaknowledge*, deemed as recommendations on concrete adaptation actions or strategies needed to be employed in order to improve ML application robustness towards DQ variations. We define the process of employing these recommendations by mLINK as *metacognition* and, based on our research, describe multiple examples of employing these metacognitive strategies in practice, such as: optimizing the data collection; reflection on DQ; DQ assurance; enhanced Transfer Learning; and employing Federated Learning for enhancing security, privacy, collaboration, and communication in MLIN.

16.1 Introduction

The contemporary cyberinfrastructure for industrial machine learning (ML) systems integrates the facilities for data collection, network communication, and data processing with ML-based intelligent end applications into ML Integrated with Network (MLIN) systems. These MLIN systems are now widely used in various real-time ML applications, from traffic signs detection in Intelligent Transportation Systems [4] to voice recognition and transcription by intelligent virtual assistants [3]. However, a critical challenge emerges when relying on MLIN cyberinfrastructure for data collection, communication, and processing, as the quality of this data may vary significantly and impact the performance of ML applications [3]. To address this issue, we have developed and patented a novel MLIN system architecture [8, 13] that goes beyond existing approaches by considering the interrelationships between different MLIN components. By analyzing these interrelationships, the system is capable of providing feedback on restructuring actions aimed at ensuring ML application performance and improving its robustness against Data Quality (DQ) degradations.

In this chapter, we switch our focus from pure interrelationships analysis to integrating them into *knowledge*, which is generated, accumulated and utilized to progress to MLIN with knowledge – mLINK (**m**achine **L**earning **IN**tegration with **K**nowledge) strata. In this structure, the knowledge becomes the core element linking together other elements and used to improve novel functionalities, performance, reliability, security, and robustness to possible variations of the operational conditions. In some applications, the cognition ability similar to one in natural systems could be demonstrated. Similarly to its natural counterpart, mLINK supports the concept of *metacognition*, which improves the quality of cognitive processes such as learning and reflecting on knowledge acquired. Metacognition could be incorporated in mLINK by analyzing and adapting the learning process to achieve better outcomes provided by these intelligent applications and by accumulating knowledge about the relationship between the components and utilizing it for specified goals. In our work, on a specific example of our mLINK system, we demonstrate how the knowledge can be utilized for learning how to improve the robustness of ML application. Under robustness in this case, we understand the ability of ML application to keep the performance level according to the user and application requirements in the cases of input DQ variations. By utilizing the *knowledge* on the relationships between the various components' parameters and operational conditions that lead to performance variations, the strategies for learning how to attain better robustness can be derived.

In this chapter, we incorporate metacognitive capabilities into mLINK and create a system that (1) accumulates the *knowledge* on the interrelationships between various environmental conditions, MLIN components' parameters, input DQ, and intelligent end application performance, and (2) integrates this *knowledge* to learn how to adjust MLIN structure in various situations to achieve better ML robustness when DQ varies. As particular examples of *knowledge* derivation and integration to improve learning, we consider cases of dynamic data collection infrastructure restructuring, adjusting network facilities, and changing learning paradigms employed in the ML application. By making the *knowledge* a core of the mLINK structure, we realize the concept of *metacognition* and allow to learn better strategies on dynamic and efficient intelligent end application's robustness improvement. Considering mLINK as a generic framework, we demonstrate on multiple cases how leveraging of the following *metacognition* strategies and methods allows to improve the outcomes of various intelligent systems: data collection optimization; assurance of DQ over data communication; enhanced transfer learning (TL); and improving security and privacy of the data and its communication with federated learning (FL).

16.2 Knowledge Accumulation in mLINK: Use Cases

In our research, we investigated various ways to derive knowledge from the distinct MLIN components, which we employed in mLINK. As mLINK is composed of diverse infrastructural and intelligent components from various domains, we claim that the knowledge derived from those domains can help improve the robustness and performance of the ML application. We considered a broad spectrum of sources to derive knowledge from, ranging from knowledge on network communication channels [8] to knowledge on clients participating in an FL training process [7]. We employ the derived knowledge on the MLIN interrelationships to produce metaknowledge, which is utilized by mLINK to adapt the MLIN components to changing DQ conditions in order to improve the robustness and performance of the ML application. In the following subsections, we present the methods we used to derive the knowledge from various domains. Table 16.1 provides details about the investigated use cases, and Table 16.2 summarizes examples of various metacognitive techniques' applications in mLINK.

16.2.1 Knowledge Derived through Infrastructure Adjustment

We have explored various ways to measure and improve Data Quality (DQ) in our previous research [5, 11]. We devised a method to calculate DQ scores that consider factors such as sensor accuracy and platform security for various smartphones [11]. We then extended this method to other mobile devices with embedded sensors, incorporated different kinds of data, and leveraged our knowledge base to build tools that select data sources in real-time using Genetic Algorithms (GA) [5].

Our GA-based tools allow to find the optimal combination of data sources in much less time than conventional methods, while preserving high performance

Table 16.1 *Real-world use cases for knowledge accumulation in mLINK.*

Investigated application use cases	Employed data collection and its modality	Employed ML system(s)	Investigated DQ variation conditions
1: Traffic sign images classification (Section 16.2.2)	Subset of traffic and stop sign images from the Open Image V6 dataset [12]	VGG16, InceptionV3, EfficientNet, YOLO, Faster-RCNN	**Network QoS variations:** packet loss, buffer size; **other cyberinfrastructure problems:** noise, grayscale, contrast variation
2: Medical images classification (Section 16.2.2)	Chest X-ray images dataset [10]	ResNet50, InceptionV3, VGG16	**Network QoS variations:** packet loss
3: Sound classification (Section 16.2.2)	Firework [14] and gunshot [15] sound recordings datasets	VGG16, VGG19, ResNet50, InceptionV3	**Network QoS variations:** packet loss
4: Voice recognition and transcription (Section 16.2.2)	English language subset of Mozilla Common Voice dataset [1]	DeepSpeech [9]	**Network QoS variations:** packet loss
5: Object detection and classification in videos (Section 16.2.2)	Subset of The Berkeley Deep Drive (BDD110K) Dataset [16]	AWS Rekognition	**Network QoS variations:** packet loss, buffer size
6: Anomalous financial transactions detection (Section 16.2.4)	Data on financial transactions provided by SWIFT[a]	Deep Neural Network	**Adversarial attacks against the data:** label flipping

[a] www.drivendata.org/competitions/105/nist-federated-learning-2-financial-crime-federated/page/589/

across diverse data types and formats. Moreover, we investigated how the overall DQ of the application changes when data from multiple sources are aggregated. Our findings [5] indicate that the final DQ depends on the function

Table 16.2 *Meta-cognition incorporation into mLINK.*

Use cases	Utility function based on	Metacognitive Strategies to Improve Robustness	Details
1: Knowledge Derived through Infrastructure Adjustment	Data Quality of the source	Data Quality Informed Sensor Selection	[5, 11]
2: Knowledge Derived from Relationship between Data Quality, Infrastructure, and ML application performance	ML Application Performance and network QoS metrics	Network Protocol Selection Rules based on Network QoS	[2, 3, 4, 6, 8, 17]
3: Knowledge Derived Based on TL Methods	ML Application Performance	TL with Mixed and Distorted Dataset	[4]
4: Knowledge Derived in FL Setting	Trust and Reputation metrics measured based on each client's Data Quality	Client Exclusion based on Observered Values of Trust and Reputation	[7]

that integrates the DQ scores from different sources, and this function can be adjusted according to the user's preferences and the application's needs.

16.2.2 Knowledge Derived from Relationship between Data Quality, Infrastructure, and ML Application Performance

We studied how different data modalities, such as voice, image [3], and video [17], affect the ML performance under network Quality of Service (QoS) degradation. We used the ML performance to tune the network parameters to avoid DQ deterioration. We examined the connections between the network QoS degradation during data transmission and the ML application performance, and showed how these connections can be used to generate network adjustment suggestions to enhance ML robustness. As a practical example, we used the knowledge on these connections to generate feedback that provides MLIN network adjustment recommendations when the input DQ varies resulting in ML performance drop.

We implemented our feedback component as a network adjustment rule set, which recommends switching to a suitable transport protocol, such as UPD, TCP, or QUIC, based on the network QoS metrics and the ML application performance. In our empirical study [8], we found that the UDP protocol can be used safely with minimal ML performance loss ($\leq 5\%$) and low packet loss rate ($\leq 2.5\%$), while TCP protocol can be used with moderate ML performance loss (between 6% and 9%), medium packet loss rate ($\geq 5\%$ and $\leq 10\%$), and tolerable delay. However, if the ML performance drops significantly ($\geq 10\%$), packet loss rate is high ($\geq 10\%$), and delay is unacceptable, QUIC protocol can be used to improve ML end application robustness to DQ variations.

16.2.3 Knowledge Derived Based on Transfer Learning

TL is a method that enables reusing pre-trained ML systems when their input data values or application domain change. We used TL to improve the performance of the ML application that suffers from poor DQ at its input due to network QoS variations [4]. We used the VGG16 open source image recognition architecture as our base model, which we improved by re-training it using TL. This model was pre-trained on the Open Images V6 dataset and then re-trained using three different TL procedures: with original, distorted, and mixed (both original and distorted) data. We evaluated the performance of each re-trained model. The model re-trained on original data was tested on distorted data; the model re-trained on distorted data was tested on distorted and mixed data; and the model re-trained on mixed data was tested on distorted data only. Our evaluations helped us to accumulate various types of knowledge related to the re-training process in the TL setting. We assessed the classification error rate per epoch during both training and testing phases. This showed us how well the model learned from the new data and generalized to unseen examples. In addition, we investigated the relationship between the learning rate and the classification error. This helped us to find the appropriate hyperparameters for the re-training process, avoid overfitting or underfitting, and improve the ML robustness toward the input DQ variation. In [4], we provide more specific recommendations on the particular re-training strategies and hyperparameters that allow for improving ML robustness.

16.2.4 Knowledge Derived in Federated Learning Setting

FL is a popular distributed ML paradigm that tackles privacy issues arising when the model is trained on sensitive data. In our work, we devised a novel approach to accumulate knowledge that we employ to identify potentially compromised local clients in FL using our Reputation and Trust metrics [7]. We clustered the local model updates after each training round in their parameter space, and computed the major cluster that is considered to contain non-compromised models. We calculated the distance between the major cluster's center and the outliers to compute our Trust and Reputation indicators for each local client. Reputation is a measure of how consistent a client's update is with the previous rounds, and Trust is a function based on Reputation to regulate the sensitivity toward the changes in the local client's Reputation. Reputation and Trust reflect the knowledge we derive about the quality of each client's models trained on the local data. We updated these indicators in each aggregation round and kept track of each client's historical performance. If the Trust of a local client falls below a certain threshold, we regarded this client as compromised and excluded it from the aggregation and further communication. This threshold can be set based on the application's and user's requirements. This knowledge-based approach helped us to enhance the FL security and privacy, and ML robustness toward DQ variations.

16.3 Meta-cognition and Its Realization through Feedback

To improve the robustness and performance of ML application in the cases of input DQ variation, we incorporate metacognitive capabilities into MLIN by developing our mLINK framework. mLINK employs various metacognitive strategies to derive and integrate *knowledge* used to inform feedback on MLIN adaptation aimed at ensuring the best performance even in dynamic environments with varying input DQ. The feedback is composed of MLIN adaptation actions needed to be implemented by the MLIN components. Since this set of meta-rules on adaptation actions is determined based on the initerrelationships between the mLINK infrastructure characteristics, input DQ, and ML application performance, the recommendations on adaptation actions is referred to as *metaknowledge*. The process of employing this *metaknowledge* in order to learn the strategies that result in better ML application performance is referred to as *meta-cognition*. Our mLINK framework realizes the concept of *meta-cognition* in practice and demonstrates its ability to continuously improve the effectiveness of the integrated MLIN system.

In Figure 16.1, we schematically represent our framework and demonstrate how the feedback is generated and applied in our practical example. Based on the interrelationships between input DQ, MLIN cyberinfrastructure conditions, and ML application performance, which we studied in our use cases described in Section 16.2, the *knowledge* is derived and accumulated in the *knowledge* base. We integrate the accumulated *knowledge* and derive *metaknowledge* reflecting particular strategies to improve ML application performance in the cases of DQ variation in MLIN. These strategies are aimed at addressing the situations when the performance of ML application goes down; we describe them in more detail later in this section. Based on the *metaknowledge*, the current MLIN components' characteristics, and ML application performance, feedback on MLIN adjustment actions is generated. These adjustment actions reflect changes in the current MLIN cyberinfrastructure (e.g., change data sources or network protocols) needed to be implemented in order to address DQ variation and improve ML application robustness. All these three components: *knowledge* base, *metaknowledge*, and feedback actions, make up the *meta-cognition*, which reflects the process of learning better strategies to improve ML application performance in MLIN.

To implement mLINK framework in practice, we developed the MLIN adjustment feedback system that entails the *meta-cognition* functionality by aggregating the derived *knowledge* and converting them into *metaknowledge*, which is used to produce recommendations for the MLIN components aimed at improving the ML application robustness. To accommodate the feedback provided to the MLIN components, we employed and realized the following particular metacognitive strategies:

- Data collection optimization (Section 16.2.1): we dynamically adapted the data collection infrastructure based on the knowledge we derived from analyzing the interrelationships between the DQ provided by the data sources and the performance of the ML application. This involved dynamic selection of the data sources, adjustment of the DQ evaluation methodology, and restructuring data fusion framework in order to satisfy the requirements toward the level of DQ established by the ML application.
- Reflection on DQ (Sections 16.2.1 and 16.2.2): we continuously monitored the impact of DQ, affected on each of the MLIN data life-cycle steps, on the performance of the ML application. We investigated and derived knowledge when DQ is affected both in the stages of the data collection and network communication. The knowledge derived from these interrelationships allowed us to identify the most appropriate conditions and MLIN parameters for achieving the best ML application performance.

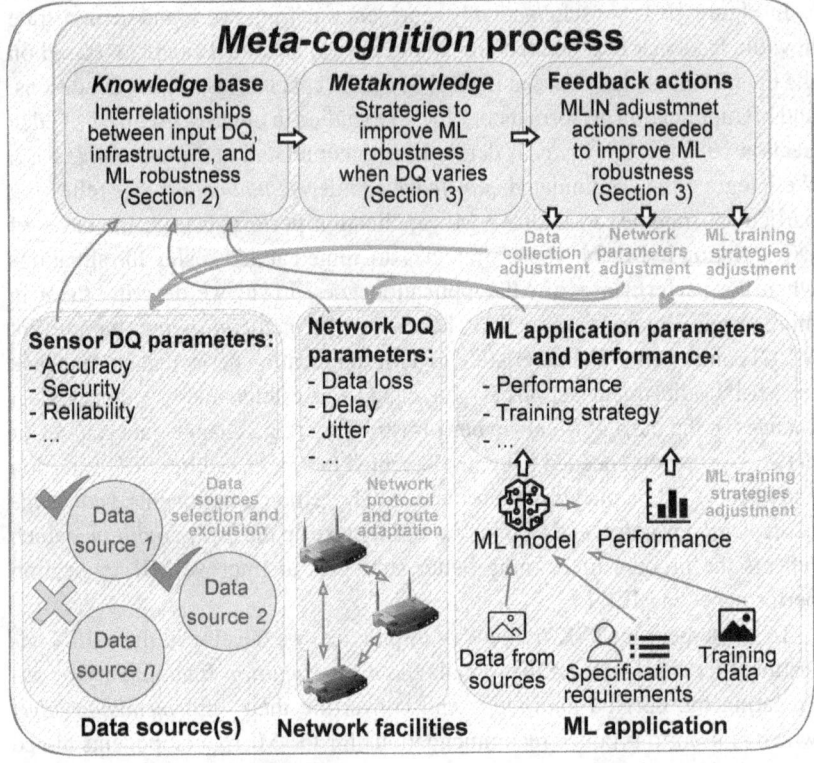

Figure 16.1 Meta-cognition process realization with mLINK framework to enhance ML application robustness.

- DQ assurance (Sections 16.2.1 and 16.2.2): we developed methods and tools to: evaluate DQ and ML performance in real-time; and provide the feedback based on these evaluations in order to adjust MLIN system parameters and improve ML application robustness and performance. In particular, based on the integrated knowledge, we developed a system that adjusts: the set of data sources used for collecting the data; and such network communication parameters as protocols and buffer resource allocation to improve the quality of data in the cases when DQ is affected by the changing network conditions. This ensured that the data reaching the ML application satisfied the quality required for the ML application.
- Enhanced TL (Section 16.2.3): we leveraged the TL strategy in order to improve performance on the input data of varied quality and reduce the time and effort required for data collection and model training. We employed

knowledge gained from re-training and testing the ML models on various DQ cohorts in order to compose appropriate re-training datasets that resulted in the higher performance demonstrated by ML applications when the quality of input data varied.
- FL for security, privacy, collaboration, and communication improvement (Section 16.2.4): for training the ML model, we employed FL paradigm, where training occurs on devices storing the local data, minimizing the need for data sharing. This allowed to enhance data security and data holders' privacy while enabling collaborative learning across multiple participants. Moreover, we developed an approach to enhance the security and privacy of the conventional FL by detecting and excluding the compromised local units from the learning procedure.

16.4 Conclusion

We utilized the concept of *metacognition*, which has been renowned in scientific research for enhancing the learning outcomes, for improving the robustness of intelligent applications in Machine Learning Integrated with Network (MLIN) systems in real-world conditions that might include possible cases of input Data Quality (DQ) variation. We proposed and developed the mLINK (**m**achine **L**earning **IN**tegration with **K**nowledge) framework, which (1) integrates the *knowledge* on the interrelationships between various factors affecting the input DQ and the intelligent end application's performance in MLIN, (2) produces *metaknowledge* based on this integration, and (3) generates MLIN feedback adjustment actions needed to address the ML application performance drop. We verified our generic mLINK framework with various metacognitive strategies on multiple real-world use cases. We employed them in mLINK to dynamically adjust the MLIN structure in order to optimize the data collection, network communication, and learning strategies for various DQ variation scenarios and diverse ML applications. Our results demonstrate the successful incorporation of *meta-cognition* capabilities into contemporary AI systems, enabling them to learn and adapt to varying DQ, thus mitigating performance degradation.

16.5 Acknowledgment

This work was partially supported by NSF award #2321652.

References

[1] Ardila, Rosana, Branson, Megan, Davis, Kelly, et al. 2019. Common voice: A massively-multilingual speech corpus. *arXiv preprint arXiv:1912.06670*.

[2] Chuprov, Sergei, Satam, Akshaya Nandkishor, and Reznik, Leon. 2022. Are ML Pages 1–4 of: *2022 IEEE Western New York Image and Signal Processing Workshop (WNYISPW)*.

[3] Chuprov, Sergei, Reznik, Leon, Obied, Antoun, and Shetty, Srujan. 2022. How degrading network conditions influence machine learning end systems performance? Pages 1–6 of: *IEEE INFOCOM 2022-IEEE Conference on Computer Communications Workshops (INFOCOM WKSHPS)*.

[4] Chuprov, Sergei, Khokhlov, Igor, Reznik, Leon, and Shetty, Srujan. 2022. Influence of transfer learning on machine learning systems robustness to data quality degradation. Pages 1–8 of: *2022 International Joint Conference on Neural Networks (IJCNN)*.

[5] Chuprov, Sergei, Reznik, Leon, Khokhlov, Igor, and Manghi, Karan. 2022. Multimodal sensor selection with genetic algorithms. Pages 1–4 of: *2022 IEEE Sensors*.

[6] Chuprov, Sergei, Mahajan, Shivam, Zatsarenko, Raman, Reznik, Leon, and Ruchkan, Aleksandr. 2023. Are industrial ml image classifiers robust to withstand adversarial attacks on videos? Pages 1–4 of: *2023 IEEE Western New York Image and Signal Processing Workshop (WNYISPW)*.

[7] Chuprov, Sergei, Memon, Moinuddin, and Reznik, Leon. 2023. Federated learning with trust evaluation for industrial applications. Pages 347–348 of: *2023 IEEE Conference on Artificial Intelligence (CAI)*.

[8] Chuprov, Sergei, Reznik, Leon, and Grigoryan, Garegin. 2023. Study on network importance for ML end application robustness. Pages 6627–6632 of: *ICC 2023-IEEE International Conference on Communications*.

[9] Hannun, Awni, Case, Carl, Casper, Jared, et al. 2014. Deep speech: Scaling up end-to-end speech recognition. *arXiv preprint arXiv:1412.5567*.

[10] Kermany, Daniel S, Goldbaum, Michael, Cai, Wenjia, et al. 2018. Identifying medical diagnoses and treatable diseases by image-based deep learning. *Cell*, **172**(5), 1122–1131.

[11] Khokhlov, Igor, Chuprov, Sergei, and Reznik, Leon. 2022. Integrating security with accuracy evaluation in sensors fusion. Pages 1–4 of: *2022 IEEE Sensors*.

[12] Krasin, Ivan, Duerig, Tom, Alldrin, Neil, et al. 2017. OpenImages: A public dataset for large-scale multi-label and multi-class image classification. Dataset available from https://storage.googleapis.com/openimages/web/index.html

[13] Reznik, Leon, and Chuprov, Sergei. 2022, U.S. Provisional application, pending. *Network adjustment based on machine learning end system performance monitoring feedback (63/406,514)*.

[14] Sound Effects. 2018. *Fireworks Sound Effect (1 Hour)*. Accessed on January 31, 2024.

[15] SoundEffectsFactory. 2014. *Ultimate military/weapon gun shot sound effect pack! [200+ sounds for 3 hours]*. Accessed on January 31, 2024.

[16] Yu, Fisher, Chen, Haofeng, Wang, Xin, et al. 2020. Bdd100k: A diverse driving dataset for heterogeneous multitask learning. Pages 2636–2645 of: *Proceedings of the IEEE/CVF Conference on Computer Vision and Pattern Recognition*.

[17] Zatsarenko, Raman, Marathe, Chirayu Anil, Chuprov, Sergei, Hyland, Matthew, and Reznik, Leon. 2023. Are industrial ML image classifiers robust to data affected by network QoS degradation? Pages 1–4 of: *2023 IEEE Western New York Image and Signal Processing Workshop (WNYISPW)*.

17
Military Applications of Artificial Intelligence Metacognition

BONNIE JOHNSON

17.1 Introduction

This chapter explores the military applications of metacognitive AI systems and the potential implications of this emerging technology on future warfare.

AI technologies are evolving rapidly and are poised to have far-reaching impacts in nearly every aspect of society around the globe, including the military domain. AI offers cognitive reasoning and learning about problem domains – processing large quantities of data to develop situational awareness, generate solution goals, recommend courses of action, and provide robotic systems with the means for sense-making, guidance, actions, and autonomy. A new and emerging technology is poised to revolutionize AI systems. This technology is metacognition which allows AI systems to become self-aware, or to think about their own cognition. Metacognition enables AI systems to monitor themselves, reason about themselves, and even control themselves. This chapter explores the potential for metacognition to improve AI systems and the potential implications of this for military applications.

This chapter is intended as a starting point for presenting some concepts and ways of thinking about this potentially revolutionary technology and how it may apply to future military operations. Four aspects of military operations were selected as having high potential for benefitting from future metacognitive AI systems. These are: (1) improving human interaction with AI systems, (2) providing safe and ethical AI behavior, (3) enabling autonomous systems, and (4) improving automated decision aids. As an introduction to metacognition and its potential applications for the military, the chapter begins by offering a brief overview of foundational concepts that include AI system applications and metacognition. This is followed by a discussion on the potential contribution of metacognition to improve the four aspects of military operations. The chapter concludes with speculations concerning the more distant future of metacognition and its implications on AI systems and warfare.

17.2 Foundational Concepts

17.2.1 AI System Applications

AI systems for the purposes of this chapter, are human-made computer systems that provide automated "intelligence" in terms of sense-making (perceiving a situation or environs through sensor data analysis), decision-making, and issuing commands. This chapter considers military AI systems to include any system that: (1) already supports or will support a military application, and (2) is intelligent enough to accommodate metacognitive capabilities. This last bit may not be fully defined yet, but this chapter makes some assumptions about how much "intelligence" is needed.

Three types of applications for AI systems (shown in Figure 17.1) are: (1) data products, (2) cyber-physical, and (3) decision sciences. AI systems that provide data products take in large quantities of data and process this data for pattern/image/speech recognition, information searches, anomaly detection, and matching/pairing information and users. The output or product of this type of AI application is often in the form of data or information provided to a user or another system. AI systems that enable and support cyber-physical systems are based on the sense-perceive-decide-act cycle. For this type of application, AI systems receive sensor data and construct a scene, or internal model, of the real-world environment and contextual situation related to their physical system of interest (i.e., robot, drone, etc.). Cyber-physical AI systems set goals for their physical systems of interest (or use predetermined, or user-provided goals), and develop actions for their physical systems to meet those goals. These AI systems then issue commands to their physical systems. This type of AI application often follows a dynamic cycle – with the AI adjusting the sense-making, goals, decisions, and actions as the situation unfolds. For the third type of application, decision sciences, AI systems process data and information concerning a situational problem domain to develop and assess solutions or courses of action. For this type of application, the AI produces decision recommendations (i.e., medical diagnoses, optimized shipping routes and courses of action). This type of AI system is often based on a sense-perceive-decide-recommend cycle.

There are similarities and differences in the cognitive processes across the three types of applications. These similarities and differences affect what types of metacognitive processes are best suited for each application. There is an inherent connection between AI system cognition and associated metacognition. Therefore, the metacognitive needs for AI systems differ depending on

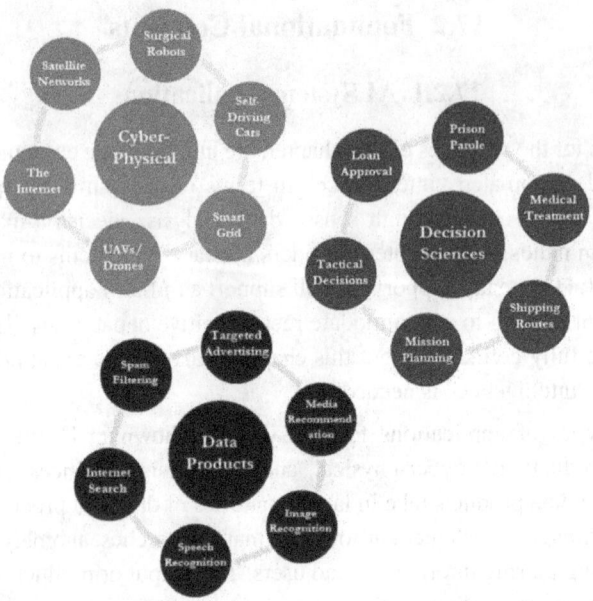

Figure 17.1 AI system application domains: cyber-physical, decision sciences, and data products [16].

the intended application. Metacognitive research is exploring how tailored or individualized these capabilities need to be for each AI system.

Military applications fall into all three of the AI system application domains. AI systems that produce data products can perform speech and image recognition, detect patterns of life, and identify anomalous behavior. Data products support intelligence and surveillance operations, physical security operations, and the detection of cyber-attacks. AI systems that support cyber-physical systems are critical for supporting the operation of autonomous and robotic systems, satellites, and smart networks. AI systems that provide decision science applications can make sense of highly complex problem domains and recommend courses of action to support logistics, mission planning, route planning, search and rescue missions, area and fleet defense, and real-time tactical decisions. This chapter explores how metacognitive capabilities can support the three types of AI systems to enhance a variety of military operations.

17.2.2 Metacognition

Metacognition has been described as "thinking about one's own thinking" [21, 24]. Metacognition or "knowledge of cognition" refers to what a system knows

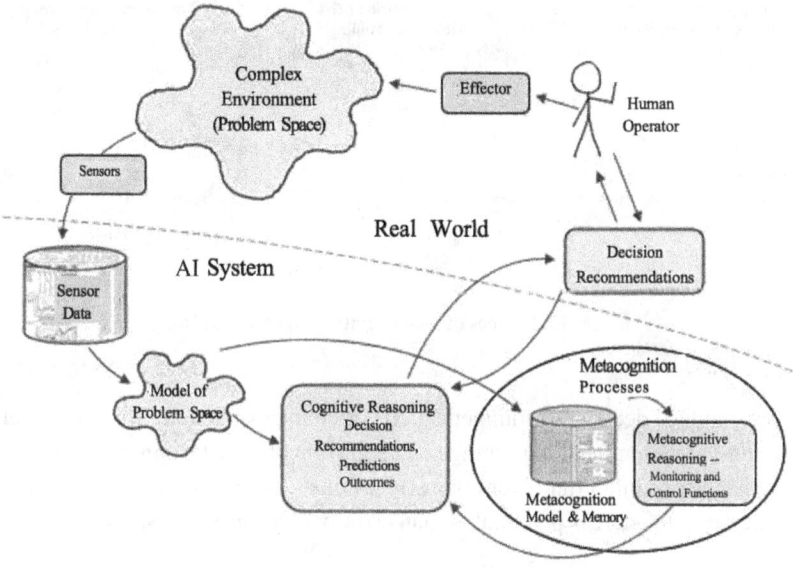

Figure 17.2 Use case example of metacognition within an AI system [18].

about itself and about its own cognition. Metacognition leverages the inherent intelligent features of AI systems to provide introspection and self-awareness [10]. Metacognition allows AI systems to monitor and reason about their internal components and processes. AI systems can use metacognition to identify reasoning failures [5] and enable self-healing and self-management for safe and desired behavior [16].

A high-level depiction of metacognition within an AI system is illustrated in Figure 17.2. The use case in this figure shows an AI system making decision recommendations to solve a problem space in a complex environment. The AI system (below the dotted line) receives sensor data, develops an internal model of the real-world problem domain and performs cognitive reasoning to develop a course of action recommendations. The metacognition processes monitor the cognitive processes within the AI. In doing so, the metacognitive subsystem monitors the AI processes, gathers data, and develops a metacognition model and memory. It enables the AI system to monitor and adjust itself and its outputs (in this case, decision recommendations) by controlling its own processes.

The human operator shown in Figure 17.2 represents the individual or group recipient of the AI system's output or decision recommendations. Other types of AI systems may involve human operators in teaming arrangements

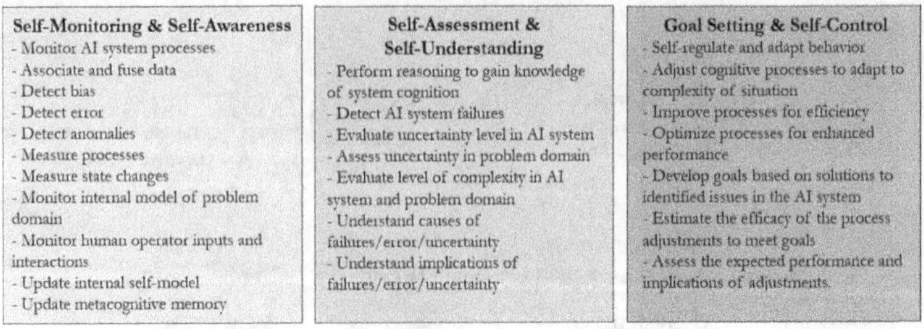

Self-Monitoring & Self-Awareness	Self-Assessment & Self-Understanding	Goal Setting & Self-Control
- Monitor AI system processes - Associate and fuse data - Detect bias - Detect error - Detect anomalies - Measure processes - Measure state changes - Monitor internal model of problem domain - Monitor human operator inputs and interactions - Update internal self-model - Update metacognitive memory	- Perform reasoning to gain knowledge of system cognition - Detect AI system failures - Evaluate uncertainty level in AI system - Assess uncertainty in problem domain - Evaluate level of complexity in AI system and problem domain - Understand causes of failures/error/uncertainty - Understand implications of failures/error/uncertainty	- Self-regulate and adapt behavior - Adjust cognitive processes to adapt to complexity of situation - Improve processes for efficiency - Optimize processes for enhanced performance - Develop goals based on solutions to identified issues in the AI system - Estimate the efficacy of the process adjustments to meet goals - Assess the expected performance and implications of adjustments.

Figure 17.3 Types of metacognitive capabilities [18].

to co-produce decision recommendations or courses of action. These types of human–AI system teaming arrangements may involve metacognitive capabilities that also monitor (and even control) the human contributions to the decision processes – to address potential human error or cognitive issues.

17.2.3 Metacognitive Capabilities

Metacognitive functions can be grouped into three broad capability categories: (1) self-monitoring and self-awareness, (2) self-assessment and self-understanding, and (3) goal setting and self-control [18]. This framework of metacognitive capabilities, shown in Figure 17.3, provides a means to understand the different types of lower-level functions that can contribute to AI system metacognition. Functions from each of the three categories will need to interact to provide the dynamic metacognition cycle of self-awareness, self-understanding, and ability to control and adapt for improvement and to avoid failures. Research is currently underway to develop these capabilities and advance the technology toward operationalization.

The full range of envisioned metacognitive capabilities is based on a continuous sense-perceive-decide-act cycle. Conceptually, as the real-world problem domain changes, future cognitive and metacognitive processes will be able to adapt to those changes accordingly and create solutions in the real-world problem domain through decision recommendations or commands to cyber-physical systems. Experts in the AI metacognition research and development community propose a set of constructs (or components) that will be needed to achieve the metacognitive capabilities. Table 17.1 identifies and describes metacognitive constructs that are under development. This chapter introduces these concepts at a high level to provide context for how future metacognition

Metacognition constructs	Descriptions
Metacognitive architecture	A metacognitive architecture is envisioned as the structures within the AI system that provide divisions and relations among modules [19, 26]; an internal framework for self-monitoring/assessment, sharing and storing self-knowledge, and self-reasoning/goal setting; and the meta-level control mechanisms needed for self-control [5].
Meta-semantic ontology	A high-level internal representation of states, events, and processes in the AI system, expressed in non-transient, re-usable information structures [25].
Meta-semantic competences and reasoning	The ability to use information about information, or information about things that acquire, derive, use, contain or express information, including the ability to represent things that represent, and what they represent [25].
Internal metacognitive model (or self-model)	An internal representation of the AI system's reasoning processes that store the data generated by the reasoning processes along with the rules that were used; it describes the internal state of the AI system [5].
Metacognitive memory	The meta-abilities support the AI system's memory capabilities, the AI system's self-awareness of memory. These abilities may include memory representation, retention, mining, retrieval, and self-monitoring [10].
Meta-learning	Meta-learning is the ability to "learn to learn" [27]. Meta-learning allows the AI system to monitor and self-regulate its own learning processes [6] – to select among different learning algorithms or improve its learning abilities over time and with more experience.
Metacognitive reasoning	The processes that oversee, manage, and coordinate cognition activities [21]. Reasoning that enables the processes of introspective monitoring and meta-level control [5]. Metacognition requires hierarchical causal reasoning that includes statistical inferences, relational reasoning, and counterfactual reasoning [23].

Table 17.1 *Metacognitive constructs.*

may affect military applications as they evolve and become part of future AI systems.

17.3 Military Applications of Metacognitive AI Systems

Military operations often present highly complex problem domains. AI systems are well-suited in many regards to support the development of solutions to these problems. Depending on the nature of the problem or mission, AI systems may produce data products, control autonomous systems, or develop decision recommendations. In all three applications, the solution space products or outputs can be "good" (accurate, effective, relevant) or "bad" (incomplete, laden with errors, or just plain wrong). The consequences of "bad" data products, commands, or decisions can lead to catastrophic consequences in the military domain. Costly warfare resources can be squandered, lives can be lost, and military operations gone awry can escalate into regional conflicts or larger geopolitical issues.

17.3.1 Human–AI Interaction

Metacognition is expected to significantly enhance human–AI interactions. Many AI systems for military applications will require human interaction. This "interaction" will range from simple AI-generated knowledge sent from the AI system to the user, to highly interdependent complex human–AI teaming with dynamic two-way sharing for cooperative knowledge generation and development of products, decisions, and commands. Metacognition can provide human operators with greater understanding and transparency of the AI system's inner workings and can provide a "confidence assessment" that indicates the level of error in AI system outputs. Metacognitive capabilities can also be developed to monitor the human's interactions with the AI system to detect operator error and anomalies.

Calibrated Confidence

Military operations often present highly complex problem domains [2]. AI systems are well-suited in many regards to support the solution space to address these problems [15]. AI systems may be producing data products, controlling autonomous systems, or developing decision recommendations. In all three types of applications, the solution space products or outputs can be "good" or "bad." They can be very accurate and effective, or they can be incomplete, full of errors, biased, or even just plain wrong. Confidence scores from standard ML models often provide an inaccurate representation of uncertainty [1]. The consequences of "bad" data products, commands, and decisions can be especially catastrophic in the military domain [17]. Resources can be wasted, lives can be lost, and in the worst case, military operations gone awry can escalate into regional conflicts and larger geopolitical issues.

Metacognition can calibrate the appropriate level of confidence associated with AI system output. Through self-assessment and assessment of the situation, the AI system's metacognition can assign a level of confidence to its products along with an explanation of how it arrived at the assessment. Table 17.2 identifies factors inherent in the real-world problem space and in the AI system's

Source of factors	Factors that affect confidence level of AI outputs
Real-world problem space	Complexity – dynamics, tempo, heterogeneity
	Sensors – error, limitations, uncertainty, bias
	Operators – induced error, cognitive limitations, slow pace, group think
Internal model	Error from input data – error, bias, incompleteness, insufficiency, latency
	Observed complexity in real-world problem space
	Uncertainty and unknowns
	Discrepancies between internal model and real world
AI system processing	Processing-induced errors
	Limitations in processing

Table 17.2 *Factors that affect the calibrated level of confidence in AI outputs [16].*

internal model and processing that can affect the confidence level of the AI system's outputs. As an example, metacognition can monitor and assess the complexity in the real world and limitations of the real-world sensors and operators and can factor this into the accuracy (or "goodness") of the AI system products.

Calibrated confidence is a metacognitive feature that is useful when human operators and users are recipients of AI system outputs. Therefore, this metacognitive capability is relevant in two of the AI system application domains: data products and decision sciences. In the military domain, a warfighter may wish to know an associated level of confidence in an AI system's recommended course of action. Perhaps the AI system calculated a large amount of error and uncertainty associated with a recommendation and provided a low confidence level associated with the recommendation. The warfighter can factor this into the decision of whether to act on the recommendation or wait until the AI system produces a recommendation with a higher confidence level (presumably more error-free).

Transparency

AI systems are notoriously opaque [23]. This characteristic of opacity is a result of the complex inner workings of AI systems. Opacity can wreak havoc on establishing and maintaining human trust in AI systems. For military applications, warfighters need trusted AI systems, often for time-critical, high-stakes decisions. DARPA recognized the need for transparency and managed a program called Explainable AI (XAI) from 2015 to 2021 [14]. The XAI program did not explicitly study metacognition, but it advanced the community's understanding of the inherent opacity of AI systems and the need for transparency for humans interacting with AI systems. One of the conclusions of XAI was that in general, the most effective AI methods (like deep learning) are also the most opaque. Metacognition can provide some transparency for humans interacting with AI systems.

The better an AI system understands itself and how its processes function to develop outputs and how errors and failures occur, the better it can provide explainability to human users. Through self-assessment, metacognition can provide humans with insight and access to information concerning the AI system that otherwise would only be available through a postmortem examination after events have occurred. Pearl [23] asserts that causal reasoning needs to be applied to address the opacity of AI systems. He writes that AI systems need to develop a "mental representation of their environment, interrogate that representation, distort it by mental acts of imagination, and finally answer the 'What if?' kind of questions" [23, 17]. Pearl is describing a metacognitive model that the AI system creates of itself and the metacognitive capability of reasoning about itself. Both are necessary for achieving transparency.

Military examples where AI system transparency is essential, include decision science applications. When warfighters use AI systems to help make critical decisions with potential life-or-death consequences, warfighters may desire AI system transparency. During certain time-critical operations, such as tactical conflicts involving the defense of assets, there may not be time available for humans to review the inner workings of the AI system. For other kinds of military operations in which time is available, such as mission planning, human planners may benefit from AI system transparency as complex courses of actions are planned. The AI system may develop multiple courses of action alternatives to fulfill the commander's intent. The human planners may benefit by having access to how the AI system arrived at the recommendation alternatives; to review the sources of error and uncertainty and to see how the "confidence level" was arrived at. This transparency may support human planners as they decide whether to act on the AI system's recommendations. As an example, if a course of action is accompanied by a low confidence level, the human may investigate the causal factors of the low confidence and decide to take the action if it is determined that the confidence level may increase as more data becomes available.

Understanding

Transparency gives humans more insight into the inner workings of the AI system. But transparency, alone, is not enough. Metacognition can go a step further by reasoning about how the AI, itself, is generating its output. This includes reasoning about how failures may occur, or how they have occurred (as a post-mortem). It includes reasoning to understand sources of error and how they propagate throughout the AI system's processes. It can also reason about

Military Applications of AI Metacognition

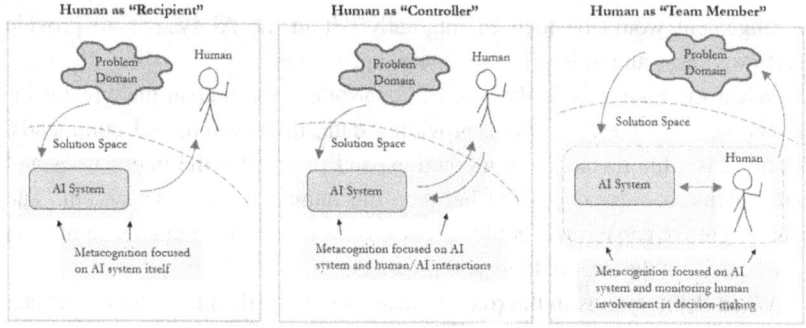

Figure 17.4 Metacognition for the different types of human–AI interaction.

the human's role in AI system processes (during human–AI teaming) including reasoning about human contributions to AI system errors and failures. Future metacognition can provide AI causal reasoning and human–AI reasoning to support AI self-understanding and human understanding of AI systems.

The ability for metacognition to allow AI systems to reason about themselves and gain self-understanding is particularly essential as AI systems tackle more complex problems. As an example, warfighters may come to rely on AI systems to generate combat action alternatives and weigh the pros and cons of each. Under time-restrictive conditions, human warfighters may need to understand which types of reasoning processes led the AI system to recommend a particular course of action. The "transparency" ability of the AI system will produce an associated confidence level with the recommendation, while the added ability of "understanding" will provide more details about the AI system's reasoning process.

Human Monitoring and Assessment

The role of the human in human–AI interactions can be characterized as: (1) a "recipient" of AI system outputs, (2) a "controller" of the AI system, or (3) a team member with the AI system. Figure 17.4 illustrates these three forms of human–AI interaction and highlights how metacognition will shift its focus depending on the role of the human. The purpose of these illustrations is to frame the thinking about the role of metacognition in enabling and enhancing all three forms of human–AI interactions.

When the human is in the role of "recipient," the AI system produces its outputs – likely data products or decision recommendations – and the human is primarily a receiver of this output information. Metacognition in this arrangement would be focused internally within the AI system, to provide self-awareness and to improve its internal processes.

When the human is in the role of "controller," the human has greater interactions with the AI system, providing data, information, and commands. In this case, the metacognition could expand to monitor the interactions and information/commands shared between the human and the AI system. The metacognitive capabilities could assess the impacts of the human's inputs and commands on the AI's outputs and products.

When the human is in the role of "team member," the interactions between the human and AI system are expected to be more dynamic with the human truly teaming with the AI system. Metacognition for this arrangement could expands its purview to include both the AI system and the human within its domain of self-awareness, assessment, understanding, and controlling. The metacognitive capability could take on the additional feature of monitoring the human as part of the decision process. As the human becomes more involved in the cognitive processes of developing solutions to address the problem space, this presents a greater opportunity for metacognitive processes to not only monitor and regulate the AI system, but also monitor and even regulate the human's role in the interaction.

17.3.2 AI Safety and Ethics

Metacognition is expected to provide solution methods that support AI safety and ethical behavior. Safe and ethical AI systems are particularly important for military applications where the stakes are high. Military operations involving weapon systems or other safety-critical warfare systems can lead to unintentional harm or conflict escalation. The use of weapons in the military can be described as the "kill chain" – which is a series of actions and decisions that involve observing a threat, orienting weapon systems, deciding it is appropriate to use a weapon, and acting – or firing the weapon. Research is underway to determine how AI systems, can be used by warfighters to support different steps in the kill chain. AI is very useful for the first step of observing a battlespace and identifying threats. An example is airspace deconfliction in which an AI system can assist in distinguishing between threat missiles and aircraft and friendly military and commercial aircraft. However, an AI failure in this task can lead to deadly friendly fire consequences. AI can support the kill chain "decide" function by applying rules of engagement (ROE) to determine whether its appropriate to use a weapon in a given situation. This is an example of AI assisting with an ethical decision. An AI failure in this task can also lead to serious consequences if the wrong decision is made.

Military Applications of AI Metacognition

Failure Category	Failure Mode Examples
AI system produces faulty/poor decision recommendations	Biased outcomes/predictions
	Skewed outcomes/predictions
	Uncertain outcomes/predictions
Human–AI interaction issues	Operators have lack of trust in the AI system
	Operators are overly trusting (over reliant) in the system
	Operators ignore the system
	Operators misunderstand the system recommendations/predictions
	Operators introduce errors into the system
	Operator misuse, inattentiveness, fatigue
AI system under cyber attack	System infiltrated by adversary
	System taken control by adversary
	System and its outputs corrupted by cyber attack
	Adversary jams or shuts down system
	Adversary gains access to system; AI system output is compromised

Table 17.3 *AI system failure modes [16]*.

Metacognition provides methods and means for preventing AI system failures and unethical behavior. Metacognition can provide indicators before failures or undesired behavior occur. The indicators can be used to alert human operators of a possible failure or to alert the AI system itself to then be used for self-control or self-regulation. Research shows that humans with more highly developed metacognitive skills are better at moral reasoning [21]; this indicates that metacognition may offer capabilities in ethics for AI systems.

As AI systems are introduced into new domains, such as the military domain, they introduce new possible safety issues and failure modes. The use of AI systems introduces three broad categories of failures: (1) faulty or poor AI system outputs (decision recommendations, data products, cyber-physical commands, (2) issues with human–AI interactions, and (3) new cyber-attack vulnerabilities. Table 17.3 lists examples of failure modes in each of the three categories.

The root causes of AI system failures can occur at different times in an AI system's lifecycle. Table 17.4 identifies examples of root causes, categorized by issues. Problems can creep into AI systems before they are deployed (or become operational), such as machine learning algorithms being trained by faulty or biased training datasets. Once AI systems are operational, problems can occur with errors in the operational datasets or in the algorithms. In some cases, if the real-world operational situation is too complex, this may overwhelm the AI systems and act as a root cause of failures. Failures also may be induced unintentionally by operators or intentionally by adversarial cyber-attacks.

Metacognition can enable AI systems to prevent failures through self-diagnosis that identifies indicators or red flags that a failure may occur [16]. Metacognitive anomaly detection can identify unusual data input into the AI

Type of Root Cause	Root Cause Examples
Issues within the training datasets	Biased training datasets
	Incomplete training datasets
	Corruption in the training datasets
	Mis-labeled data
	Mis-associated data
	Lack of rare examples – training data doesn't include all possible scenarios that may be encountered
	Unrepresentative datasets
Issues with the process of data validation	Poor data collection methods
	Poor data validation methods
	Improper data validation criteria
	Insufficient data validation
Issues with machine learning algorithms	Underfitting in the model
	Overfitting in the model
	Algorithm trained to the wrong model
	Wrong algorithm mathematical model used
Issues with operational datasets	Uncertainty/error in the operational datasets
	Corruption in the operational data
	Datasets that the AI system is not designed to handle
Operational complexity	Pace of situation overwhelms AI system (or human–AI team)
	Decision space of situation overwhelms AI system (or human–AI team)
Operator trust issues	Lack of explainability
	Lack of confidence
	Overreliance
	Insufficient operator training or experience with system
Operator induced error	Inverse trust issues (AI system loses "trust" in the human operator)
	Operator misuses AI system (accidentally or intentionally)
	Operator fails their part in the decision process (overwhelmed, negligent, confused, etc.)
Adversarial attacks	Hacking
	Deception
	Inserting false or corrupt data
	Gaining control of AI system

Table 17.4 *Root causes of AI system failures [16].*

system or unusual behavior patterns within the AI system processing. Metacognition can identify when an AI system encounters new situations or entities in the real world that are not recognized, and therefore not part of the machine learning training process. This recognition of unknowns can serve as an alert to human operators and can be considered by the AI system as it assesses itself and its products.

Metacognition can enable an AI system to perceive its own complexities and predict its own emergent behavior. AI systems consisting of multiple machine-learning components can lead to unsafe and unintentional (and even unethical) behavior that results from the interactions of components contributing to errors [22]. The metacognitive capabilities could be designed to perform a detailed self-characterization of the AI system functioning and performance of each internal component. This metacognitive self-assessment could support self-understanding of the internal dynamics of failures and undesired behaviors and provide options for correcting or avoiding these situations.

Metacognition can support AI safety and ethics by establishing behavioral "guardrails" within the AI system. The guardrails, which can rely on metacognitive self-control and self-regulation to limit AI system functioning to the desired behaviors (safe and ethical commands to robots, decision recommendations that ensure safe and ethical outcomes, etc.). The safe and ethical behavior guardrails could be established as the system is designed, or it could be designed as an evolutionary set of guardrails that change over time as the system learns and as problem domains evolve.

17.3.3 Autonomous Systems

Metacognition will be a game-changer for AI-enabled autonomous cyber-physical systems. Cyber-physical systems comprise a large category of systems that include a physical component or "effector," that exhibits behavior within its environment to accomplish some objective or set of objectives. These systems include sensor abilities to allow them to observe their environment. They also contain a "cyber" component that processes the sensor data to gain situational awareness and determine actions or effects to accomplish objectives. Autonomous systems may rely on varying levels of autonomy, but as systems become more autonomous, they shift away from human involvement and rely on their own intelligence to sense, decide, and act. Through situational awareness, self-awareness, goal-setting, and self-control, metacognition can improve the ability for fully autonomous systems to adjust their behavior and goals to better accomplish missions in complex environments.

Metacognitive constructionist learning (CL) is a concept for enabling an AI system to build an "internal illustration" or model of its learned knowledge, based on its own experiences, decision recommendations, and predictive outcomes [11]. The CL process is envisioned to enable an AI system to shift its "locus of control" from having to be externally controlled, to being self-sufficient and in control of its own actions and decisions. This "internal illustration" is effectively an internal metacognition model that the AI system could develop and maintain. Metacognitive CL is a central concept for enabling fully autonomous systems (with embedded AI systems) that do not require a human in the loop as a remote controller.

Metacognitive "reflection on affordances" is a concept for enabling an AI-enabled autonomous system to reason about the physical world and observe and discover possible actions and constraints for itself in its environment [25]. AI systems receive sensor data about the physical world and use this information to sense and perceive. Cyber-physical autonomous systems need to reason about this situation awareness information and determine what actions are available to them in different situations and which actions will produce desired results [13]. "Proto-affordances" are possible processes/actions of physical systems and constraints on those processes/actions. Metacognition can enable AI systems to reason about the proto-affordances of objects and structures in the environment of an autonomous system. Based on learned observations and discoveries as AI-enabled autonomous systems sense and interact with their environment, they can reflect on (or reason about) proto-affordances and then determine their own actions accordingly. As an example, an AI-enabled autonomous system may encounter an unfamiliar object that obstructs its path. By reflecting on observations and discoveries, the autonomous system may learn that the object can be manipulated or moved. "Action-affordances" are the possibilities for and constraints on possible actions that the AI-enabled autonomous system can perform [25]. In the example of the unfamiliar obstructive object, the autonomous system may identify possible actions it can take and may decide to retreat from the object or bypass it.

A metacognitive "integrated dual-cycle" architecture (MIDCA) is a concept that uses metacognition alongside cognition to enable autonomous systems to self-regulate for robust behavior in dynamic environments and to manage unexpected events [8]. This architecture concept is based on a dualcycle with a problem-solving and comprehension loop at the cognitive level and a control and monitoring loop at the metacognitive level. Each cycle or loop is based on a "note-assess-guide" process in which "note" detects discrepancies, "assess" hypothesizes causes for discrepancies, and "guide" performs a suitable response [8]. This architecture is intended to enable AI-enabled autonomous systems to rely on the monitoring metacognitive loop to support recovery when unfamiliar events occur and to encourage more effective learning through purposeful knowledge gathering when information gaps arise [8]. The MIDA architecture presents an opportunity for the AI-enabled autonomous system to reason about its goals at the cognitive level, and reason about its reasoning at the metacognitive level.

Autonomous systems are becoming a mainstay in modern military operations. They are used for many missions and uses, including surveillance, communications, deterrence, strikes, bombing, deliveries, logistics, and search and rescue. Their use by nation states and terrorists in conflicts is on the rise. As AI processes continue to advance, especially enabled with metacognition, autonomous systems will become more intelligent – better able to adapt to complex situations, better able to handle uncertain and unfamiliar environments and better at solving different problems without humans in the loop. Metacognition will be a critical enabler of achieving greater forms of autonomy and intelligence. Metacognition will enable AI-enabled autonomous systems

Elements of decision recommendations	Descriptions
Real-world situation	Awareness and assessment of situation, environment, and context
Threats	Threat identification, characterization, tracking, prioritization, attribution
Courses of Action	Recommended actions, commands, tasks for warfare assets (real-time or near-real-time)
Plans	Assigned locations, missions, objectives for warfare assets (usually ahead of time or proactive)
Replanning	Dynamic adjustments to plans in place to react and respond to changing circumstances
Defensive engagement strategies	Weapon assignments, weapon-target pairing, coordination of defensive courses of action
Resource optimization	Optimization of warfare resources often involving multiple resources optimized across multiple missions and operational domains

Table 17.5 *Elements of decision recommendations.*

to self-optimize and ultimately learn and evolve over time and even dynamically during operations. As a result, autonomous systems will be smarter, more capable, more adaptive, and more relied on for future military operations.

17.3.4 Decision Aids

Metacognition can enhance the decision recommendations produced by AI systems that are intended for decision science applications. For military operations, AI systems that are being developed as decision aids can support a wide range of military decisions including tactical, planning, and even strategic missions [20]. Military decision-making is often fraught with uncertainty, stress, complexity, time-criticality, and high-stake consequences [3]. In many military applications, decision-making will be a collaborative endeavor between humans and AI systems. However, the AI system, as a participant in this interaction, is intended to develop decision recommendations, in the form of tactical courses of action, mission plans, or strategies. These decision recommendations incorporate some or all of the following elements contained in Table 17.5.

Human–AI decision-making was discussed in an earlier section of this chapter. This section delves into the inner decision-making processes within decision science AI systems and explores the role of metacognition to enhance these processes.

The primary purpose of decision science AI systems is to produce the best possible decision recommendations for the situation at hand. Metacognitive capabilities within the AI system can aid in this endeavor. There are limitations to decision-making processes that can lead to poor recommendations: natural limitations (inherent complexity and uncertainty in the situation), limitations in sense-making/data gathering (sensor limitations, errors, incompleteness), and limitations in the decision-making process. Metacognition cannot overcome all the limitations, but it can enhance the decision-making processes within the AI system.

AI systems can use metacognitive capabilities to enhance their decision-recommendation outputs through self-awareness, self-assessment, adaptive goal setting, and by controlling their own processes to develop innovative decision alternatives and to self-select optimum alternatives through causal reasoning. Metacognition can provide the means for self-assessment by developing and assigning confidence levels to decision recommendations as discussed earlier in this chapter. The confidence levels are not only useful for human operators, but also for the AI system itself to gain better self-understanding. Metacognition can detect states of high complexity – either in the real-world situation or within the AI system and can use this knowledge for better self-understanding and for establishing reasonable goals for self-adaptation. Metacognition can provide causal reasoning to assess likely effects of different decision recommendation alternatives and then use these assessments to select the most effective recommendation. Finally, metacognition can support a dynamic and adaptive cognitive approach to the decision-making process within the AI system. Metacognition can support on-going process improvements to task sensors to gain more and better data and to regulate internal AI algorithmic methods to improve the processes of generating decision options and assessing them.

17.4 Speculation on Metacognition and Future Warfare

This chapter presents a speculative assessment of the implications of future metacognitive AI systems on military applications. Four aspects of military AI that are expected to be nearer-term applications for metacognition were explored: (1) human–AI interaction, (2) AI safety and ethics, (3) autonomous systems, and (3) decision aids. This list of focus areas is far from complete. Other military applications for metacognition are likely to include capabilities for reliability and maintenance that can leverage self-diagnosis abilities and cyber warfare that can leverage cyber-attack detection, mitigation, and protection. Metacognition is beginning to be studied for specific applications, like the cognitive radio [12], which indicates promising advances in this technology. Metacognition is expected to provide enhancements and new capabilities in many areas as AI systems become metacognitive and advance to the point of operationalization.

Future concepts and capabilities	Description of metacognition as an enabler
Meta-learning	Meta-learning is the oversight and improvement of AI system learning processes [27]. Future meta-learning capabilities are envisioned as a form of metacognition that may control the learning processes, selecting among algorithmic methods. Meta-learning may also enable adaptive learning, optimizing the learning processes based on the situation and needed courses of action.
Continuous life-long AI learning/learning "in situ"	Metacognition is likely to be a critical enabler of AI learning systems – to enable them to learn "in situ" (during operations) and to adapt continuously and continually during their lives as they gain experience and expand their memories [9].
Ethical human attributes	The addition of a metacognitive dimension for AI systems is predicted to provide "emotional maturity" types of human attributes, such as moral reasoning, common sense, emotional self-healing, ethical learning, and moral reasoning [21]. These attributes can support AI ethical behavior.
Self-organizing altricial AI systems	Metacognitive capabilities may enable AI systems to associate different perceived, learned, and remembered "information chunks" and knowledge to enable AI reasoning about unfamiliar situations (events and entities) to investigate and discover new knowledge and to self-organize in order to learn more and gain more experience [7]. Altricial systems begin as immature or "helpless" (requiring external care and support) – perhaps as less autonomous and needing human remotecontrol. This concept provides altricial systems with the innate abilities to learn and mature autonomously.
AI empathy	AI systems of the future may behave with empathy with the aid of metacognitive capabilities. Metacognitive skills such as self-knowledge and metacognitive experiences can combine with meta-affective knowledge and skills to enable AI system empathy [4]. Meta-affective knowledge is knowledge about how decisions or actions affect external situations and how they affect the AI system itself. This knowledge and the skills to be self-aware of this and its effects, can provide AI empathy capabilities.
Metacognition requiring its own metacognition	As metacognition offers meta-level control over AI internal cognitive processes, this begs the question of who (or what) will control the metacognition [25].

Table 17.6 *Future metacognition-enabled concepts and capabilities.*

Farther-into-the-future speculation leads to some highly interesting and revolutionary possible concepts and capabilities. Table 17.6 provides some descriptions of possible AI capabilities that may exist in the future with the aid of metacognition.

In conclusion, AI technologies are evolving rapidly and are poised to have far-reaching impacts on nearly every aspect of society, including future military operations. The co-evolving discipline of metacognition is developing abilities for AI to reason about its own reasoning. Metacognition, in essence, offers guardrails to address challenges inherent to AI systems. And the beauty of metacognition is that it takes advantage of the very essence of AI systems – the "intelligence" – to address the challenges through self-knowledge.

References

[1] Ahmed, Kareem, Teso, Stefano, Chang, Kai-Wei, Van den Broeck, Guy, and Vergari, Antonio. 2022. Semantic probabilistic layers for neuro-symbolic learning. *Proceedings of the Thirty-sixth Conference on Neural Information Processing Systems*.

[2] Bar-Yam, Yaneer. 2004. *Making Things Work: Solving Complex Problems in a Complex World*. NECSI, Knowledge Press.

[3] Bjurstrom, Erik, and Bakken, Bjorn T. 2022. Decision-making under uncertainty: A Brehmerian approach. *Journal of Behavioural Economics and Social Systems*, **4**.

[4] Burleson, Winslow. 2006. Affective learning companions: Strategies for empathetic agents with real-time multimodal affective sensing to foster meta-cognitive and meta-affective approaches to learning, motivation, and perseverance. Ph.D. thesis, Massachusetts Institute of Technology.

[5] Caro, Manuel Fernando, Josyula, Darsana P, Madera, Dalia Patricia, Kennedy, Catriona M, and Gómez, Adan A. 2019. The Carina metacognitive architecture. *International Journal of Cognitive Informatics and Natural Intelligence (IJCINI)*, **13**(4), 71–90.

[6] Caro, Manuel Fernando, Josyula, Darsana P, Cox, Michael T, and Jimenez, Jovani A. 2014. Design and validation of a metamodel for metacognition support in artificial intelligent systems. *Biologically Inspired Cognitive Architectures*, **9**, 82–104.

[7] Chappell, Jackie, and Sloman, Aaron. 2005. Natural and artificial meta-configured altricial information-processing systems. *International Journal of Unconventional Computing*, **3**(3), 211–239.

[8] Cox, Michael, Alavi, Zohreh, Dannenhauer, Dustin, Eyorokon, Vahid, Munoz-Avila, Hector, and Perlis, Don. 2015. MIDCA: A metacognitive, integrated dual-cycle architecture for self-regulated autonomy. *Proceedings of the Thirtieth AAAI Conference on Artificial Intelligence (AAAI-16)*.

[9] Crowder, James, and Carbone, John. 2020. Methodologies for continuous life-long machine learning for AI systems. *Artificial Psychology: Psychological Modeling and Testing of AI Systems*, 129–138.

[10] Crowder, James, and Friess, Shelli. 2011. Metacognition and metamemory concepts for AI systems. Pages 1–6 of: *Athens: The Steering Committee of the World Congress in Computer Science, Computer Engineering and Applied Computing (WorldComp)*.

[11] Crowder, James, and Friess, Shelli. 2012. Extended metacognition for artificially intelligent systems (AIS): Artificial locus of control and cognitive economy. Pages 1–6 of: *Athens: steering committee of the world congress in computer science, computer engineering and applied computing (WorldComp)*.
[12] Gadhiok, Manik, Amanna, Ashwin, Price, Matthew J, and Reed, Jeffrey H. 2011. Metacognition: Enhancing the performance of a cognitive radio. Pages 198–203 of: *2011 IEEE International multi-disciplinary conference on cognitive methods in situation awareness and decision support (CogSIMA)*.
[13] Gibson, James. 1979. *The Ecological Approach to Visual Perception*. Houghton Mifflin.
[14] Gunning, Dave, Vorm, Eric S, Wang, Yunyan, and Turek, Matt. 2021. DARPA's explainable AI (XAI) program: A retrospective. *Applied AI Letters*, **2**(4).
[15] Johnson, Bonnie. 2019. *A framework for engineered complex adaptive systems of systems*. Ph.D. thesis, Naval Postgraduate School. https://hdl.handle.net/10945/63463.
[16] Johnson, Bonnie. 2022. Metacognition for artificial intelligence system safety – an approach to safe and desired behavior. *Safety Science*, **151**.
[17] Johnson, Bonnie. 2023. Challenges in implementing artificial intelligence for the Naval Warfighter. *Naval Engineers Journal*.
[18] Johnson, Bonnie. 2024. *Metacognitive capabilities for artificial intelligence systems*. White Paper, Naval Postgraduate School.
[19] Langley, Pat, Laird, John E, and Rogers, Seth. 2009. Cognitive architectures: research issues and challenges. *Cognitive Systems Research*, **10**, 141–160.
[20] Lee, Bryan. 2022. *An automated and dynamic decision aid for fleet commander and combatant command strategic and operational planning*. Naval Postgraduate School, Monterey, CA.
[21] Negi, Sunder Kala, Rajkumari, Yaisna, and Rana, Minakshi. 2022. A deep dive into metacognition: Insightful tool for moral reasoning and emotional maturity. *Neuroscience Informatics*, **2**, 100096.
[22] Nushi, Besmira, Kamar, Ece, and Horvitz, Eric. 2018. Towards accountable AI: Hybrid human-machine analysis for characterizing system failure. Pages 126–135 of: *Proceedings of the AAAI Conference on Human Computation and Crowdsourcing*, vol. 6.
[23] Pearl. 2019. *Possible Minds: 25 Ways of Looking at AI*. Penguin Press.
[24] Prather, James, Becker, Brett, Craig, Michelle, Denny, Paul, and Loksa, Dastyni. 2020. What do we think we think we are doing?: Metacognition and self-regulation in programming. *Learning Sciences Faculty Publications*, 36. https://scholarworks.gsu.edu/cgi/viewcontent.cgi?article=1033&context=ltd_facpub
[25] Sloman, Aaron. 2011. Varieties of meta-cognition in natural and artificial systems. Pages 307–323 of: *AAAI Workshop on Metareasoning: Thinking about Thinking*.
[26] Sun, Ron, Zhang, Xi, and Matthews, Robert. 2006. Modeling meta-cognition in a cognitive architecture. *Cognitive Systems Research*, **7**, 327–338.
[27] Wang, Jane. 2021. Meta-learning in natural and artificial intelligence. *Current Opinion in Behavioral Sciences*, **38**, 90–95.

Printed by Integrated Books International,
United States of America